Biodiversity in India
Volume 7

The Editors

Prof. T. Pullaiah obtained his M.Sc. (1973) and Ph.D. (1976) degrees in Botany from Andhra University. He was a Post Doctoral Fellow at Moscow State University, Russia during 1976-78. He traveled widely in Europe and USA and visited Universities and Botanical Gardens in about 17 countries. Professor Pullaiah joined Sri Krishnadevaraya University as Lecturer in 1979 and became Professor in 1993. He held several positions in the University, which include Dean, Faculty of Life Sciences, Head of the Department of Botany, Chairman, BOS in Botany, Head of the Department of Sericulture, Co-ordinator and Chairman, BOS in Biotechnology, Vice Principal and Principal, S.K.University College. He retired from active service on 31st May 2011. He was selected by UGC as UGC-BSR Faculty Fellow and is working in Sri Krishnadevaraya University. Prof. Pullaiah has published 53 books, 295 research papers and 35 popular articles. His books include *Flora of Andhra Pradesh* (5 volumes), *Flora of Eastern Ghats* (vol. 1-4), *Biodiversity in India* (vol. 1-6), *Encyclopaedia of World Medicinal plants* (5 volumes), *Indian Medicinal Plants* (2 volumes), *Herbal Antioxidants* (3 volumes) *Taxonomy of Angiosperms* (presently in 3rd edition), *Plant Development, Plant Reproduction, Plant Tissue Culture, Flora of Guntur district, Flora of Kurnool, Flora of Anantapur district, Flora of Nizamabad, Flora of Ranga Reddi district* etc. He is Principal Investigator of 20 major Research Projects totaling more than a Crore of Rupees funded by DBT, DST, CSIR, UGC, BSI, WWF, GCC etc. Under his guidance 52 students obtained their Ph.D. degrees and 35 students their M.Phil. degrees. He is recipient of Best Teacher Award from Government of Andhra Pradesh, Prof. P. Maheswari Gold Medal and Prof. G.Panigrahi Memorial Lecture of Indian Botanical Society and Prof. Y.D.Tyagi Gold medal of Indian Association for Angiosperm Taxonomy. He is President of Indian Botanical Society and President of Indian Association for Angiosperm Taxonomy. He was Member of Species Survival Commission of International Union for Conservation of Nature and Natural Resources (IUCN).

Dr. S. Karuppusamy obtained his M.Sc. (1996) degree from Madurai Kamaraj University and Ph.D (2002) from Gandhigram University, Tamil Nadu. He was a Post Doctoral Fellow of Sri Krishndevaraya University (2004 – 2008), Anantapur. At present he is working as Assistant Professor, Department of Botany, The Madura College, Madurai since 2008. He has published 95 scientific papers in nationally and internationally reputed journals. He is the co-author of Flora of Andhra Pradesh (Vol. 5) and Flora of Eastern Ghats (Vol. 3 & 4). He is the recipient of Young Botanist Award 2006 from Indian Botanical Society. He has conferred many awards from Scientific bodies in India such as Fellow of Indian Association for Angiosperm Taxonomy, Fellow of Indian Botanical Society, Yong Scientist award from SERB, New Delhi, etc. He has completed two major research projects funded by UGC, New Delhi and NIF, Ahmedabad. He is the Principal Investigator of three major research projects from SERB, New Delhi, NIF, Ahmedabad and Forest Department of Tamil Nadu (TBGP- Tamilnadu Biodiversity Greening Project). He is life member of many scientific societies in India.

Dr. S. Sandhya Rani did her master degree (1991) with first rank in Botany from Sri Venkatswara University, Tirupati and thereafter obtained her Ph.D (1998) from Sri Krishnadevaraya University, Anantapur, Andhra Pradesh, India. At present she is working as Assistant Professor in the Department of Botany, Sri Krishnadevaraya University, Anantapur. She has 15 years of research experience in Plant Taxonomy and 4 years of teaching experience. Her publications include 46 research papers and 3 books viz., Trees of Andhra Pradesh (1999), Flora of Guntur district (2001) and Flora of Eastern Ghats Volume 4 (2011). She is the Principal Investigator of a Project funded by BRNS, Mumbai and life member of IAAT, IBS and NESA.

Biodiversity in India
Volume 7

— *Editors* —

T. Pullaiah
Department of Botany
Sri Krishnadevaraya University
Anantapur – 515 003
Andhra Pradesh

S. Karuppusamy
Department of Botany
The Madura College
Madurai
Tamil Nadu

S. Sandhya Rani
Department of Botany
Sri Krishnadevaraya University
Anantapur – 515 003
Andhra Pradesh

2014

Regency Publications
A Division of
Astral International Pvt. Ltd.
New Delhi – 110 002

ISBN 9789351302506

Published by : **Regency Publications**
A Division of
Astral International Pvt. Ltd.
– ISO 9001:2008 Certified Company –
4760-61/23, Ansari Road, Darya Ganj
New Delhi-110 002
Ph. 011-43549197, 23278134
E-mail: info@astralint.com
Website: www.astralint.com

Laser Typesetting : **Classic Computer Services**, Delhi - 110 035

Printed at : **Replika Press Pvt. Ltd.**

PRINTED IN INDIA

Preface

Since the passage of Biodiversity Act by Indian Parliament interest on Biodiversity has increased not only among scientists but also among administrators, politicians and lay man. A broad sprectrum of Biodiversity is covered in this book covering from lower Macrofungi to complex orchids, parasitic Loranthaceae to snakes. Their *ex situ* conservation is also dealt with.

Readers are requested to give their suggestions for improvement in the future volumes. Authors are invited to contribute their articles for the coming volumes.

T. Pullaiah

S. Karuppusamy

S. Sandhya Rani

Contents

Chapter 1

Systematic Studies on the Family Balsaminaceae of Tamil Nadu

R. Ramasubbu* and K. Saranya

Department of Biology,
The Gandhigram Rural Institute: Deemed University,
Gandhigram, Dindigul – 624 302, Tamil Nadu

ABSTRACT

Balsams are mostly herbaceous annuals rarely becoming shrubs and epiphytes completing their life cycle within the monsoon period. They are growing in mountain niche specific areas like wet dripping rocks, ravines and clefts of rocks, soils of rocky crevices, cushions of moss on wet rocks, evergreen shola forests and as epiphytes on mosscloud tree trunks. Balsams comprises two wide spread genera, *Hydrocera* Blume, a monotypic genus and *Impatiens* L. comprising more than 1,000 species worldwide. Among these, 230 species are reported from India and 109 species occur in South India. For the survey, exploration and documentation of wild balsam wealth of Tamil Nadu, a systematic study has been framed through frequent field collection and analysis of herbarium specimens of K (Kew Herbarium) and TBGT (Tropical Botanic Garden Travancore Herbarium). From the above research work, 73 species of *Impatiens* were documented from different forest regions of Tamil Nadu. Among them, 90 per cent of the species are restricted to hills of Western Ghats and 15 scapigerous, 6 epiphytes, 44 terrestrial herbs and 8 shrubs were included. Interestingly, *Impatiens munnarensis*, a possibly extinct species has been rediscovered from High Wavies mountains (Megamalai hills) of Tamil Nadu for the first time from its type locality

* Corresponding Author: E-mail: racprabha@yahoo.com

and it is a new distributional record to Tamil Nadu also. *Impatiens hookeriana*, a shrub described by Arnott in 1835 had been merged with *I. grandis* due to its few similar characters. The present study revalidated it as *I. hookeriana* due to its several distinct characters with *I. grandis*.

Keywords: *Balsaminaceae, Impatiens, Systematic study, Tamil Nadu.*

Introduction

The balsams are most attractive group of plants bearing curious and variously coloured and sized flowers with peculiar floral structure. The members are more or less succulent annuals or perennial herbs rarely becoming shrubs or epiphytes. Majority of the wild *Impatiens* have great horticultural potential due to their diversified forms, shapes and spectacular colours of their flowers. It belongs to the family Balsaminaceae and has about 1,000 species distributed in two genera namely, *Hydrocera* Blume and *Impatiens* L. (Fischer, 2004; Stevans, 2004). The genus *Hydrocera* is monotypic and characterized by its free petals and the berry like capsular fruits. The *Impatiens* has four lateral petals connate in pairs and five valves with loculicidal capsule. The sister group relationship between *Hydrocera* and *Impatiens* is also confirmed by recent molecular analysis (Yuan *et al.*, 2004). The balsams are mainly distributed in the tropics and subtropics of the old world, but a few species are also occurring in temperate Eurasia and North America. In India, the genus is represented by 203 species and mainly distributed in three major centers of diversity *i.e.*, Western Himalayas, Hills of North Eastern States and Western Ghats (Pandurangan, 1996; Bhaskar, 2012). The Western Ghats is a chain of mountains of 1600 kilometers of length running parallel to west coast of Peninsular India from the river Tapti to Kanyakumari, the southern tips of the Peninsular India. There are 92 species of balsams so far been recorded from peninsular India, of which more than 80 are endemic and confined to Western Ghats. Thirty species of *Impatiens* have already in threatened category including 19 critically endangered ones (Pandurangan, 1996; Pandurangan and Sreekala, 2008; Vajravelu, 1983). Though, the ideal climatic conditions prevailing in the Western Ghats region provide suitable habitat for balsams, the populations are rapidly declining due to various factors like habitat degradation, fragmentation of populations, lack of specific pollinating agents, poor fruit set, improper seed dispersal and other reproductive problems.

The genus *Impatiens* L., the taxonomically significant genus has been studied by many eminent botanists. Van Rheede (1689-90), a Dutch Governor of Malabar made an outstanding contribution towards flora of Peninsular India through his 12 volume *Hortus Malabaricus Indicus* consisting of text and plates from his own collections. This include *I. oppositifolia, I. latifolia, I. rheedei, I. kleinii*, and *I. fasciculata*. Linnaeus (1753) published four species of *Impatiens* of South India *viz. I. chinensis, I. oppositifolia, I. balsamina* and *I. latifolia* in his Species *Plantarum*. Roxburgh (1824) described *I. umbellata* in his Flora India and De Candolle (1824) described *I. kleinii* and *I. fruticosa* in his monumental *Prodromus*. Wallich Nathaniel (1831) prepared a Catalogue of more than 20,000 specimens of plants including 18 species of *Impatiens*. Wight, made a landmark contribution as for as Peninsula Indian *Impatiens* is concerned and described

15 new species of *Impatiens* (1838-1853). Wight together with George A Walker-Arnott (1834) gave abridged description of some of the *Impatiens* species found in the Peninsular India in their *Prodromus Florae Peninsulae Indiae Orientalis*. R. H. Beddome (1869-74) made impressive illustrations of *Impatiens* species of the Peninsula and described and illustrated 10 new species based on his own collections. J. D. Hooker described largest number of *Impatiens* species in the world and at least 21 species from Western Ghats, South India. Hooker and Thomson (1859-60) for the first time divided the genus *Impatiens* into two broad groups, Scapigerae and Caulescents.

Hooker was the pioneer researcher who made significant attempts to classify the species of *Impatiens* and discovered maximum number of *Impatiens* in Western Ghats (21spp.) followed by Wight (15 spp.), Beddome (10 spp.), Fischer (8 spp.), Heyne (7 spp.), Barnes (3 spp.) and others with one or two species. Fyson (1915, 1932), Gamble (1915), Fischer (1930–1938) and others have collected and classified *Impatiens* through their local floras. Barnes (1938, 1939) enumerated 42 species of Geraniaceae which comprised 37 species of *Impatiens* including a new species *I. johnii* occurring on the Travancore High Range. In continuation of research, an extensive work on the genus *Impatiens* was carried out by Bhaskar (1975) who studied in addition to taxonomy, geographic distribution, cytology, palynology, anatomy and reproductive biology including two new species *I. raziana* and *I. agumbeana.* Vivekananthan *et al.* (1997) compiled the taxonomic treatments made on *Impatiens* of India as a whole for the revised Flora of India. Desai and Janarthanam (2011) studied the genus *Impatiens* of northern and parts of Central Western Ghats and reported a new species. In addition several new species have also been reported by Chandrabose (1978-79), Chandrabose *et al.* (1984), Pandurangan and Nair (1995), Kumar and Sequiera (1996, 1999, 2001), Viswanathan and Manikandan (2003), Bhaskar (2006), Desai *et al.* (2009), Anil Kumar *et al.* (2011) and Narayanan *et al.* (2011).

Song *et al.* (2003) studied chromosomal evolution in Balsaminaceae, with cytological observations on 45 species from Southeast Asia. Grey -Wilson (1980) and Rao *et al.* (1986) confined the synonyms of *Impatiens.* Balsaminaceae is diversified family with two genera *Hydrocera* and *Impatiens* and other generic names are confined such as *Petalonema, Semeiocardium* and *Impatientella.* Nair (1991) highlighted the endemism in Western Ghats with special reference to *Impatiens.* According to him, in India there are more than 6 well defined regions of distribution as far as *Impatiens* concerned. The distribution pattern, individuality of northern and southern species, narrow endemism, neo-endemics etc. have contributed to make *Impatiens* a key genus for phytogeographical studies. Rajalal *et al.* (1996) reported that in India Balsaminaceae is represented by 203 species and distributed in three major centers of diversity. The balsams as they are commonly called and have been one of the most perpleximg groups of the Indian flora. Turril (1953) opined that their taxonomy has not progressed much beyond its 'alpha' stage being based on dried herbarium specimens. It is observed that the specimens were very often only to be distinguished by very minute differences in the shape of the flowers and the relations of their parts, which were too often masked or crushed in the dried specimens. A large number of species have been described by several workers by the traditional methods; but an objective taxonomic revision of the genus has, seriously lagged behind. Much interest has developed in recent years to revise the Flora of India as large numbers of new plants have been described since Hooker's Flora of British India was published. The project for a complete revision of the Flora of India was undertaken in different taxonomic laboratories in India.

Mathew (1991) analysed the floristic wealth of selected regions in Tamil Nadu and recorded five species of *Impatiens* from the plains. Gamble (1915 – 1919) studied the entire floristic wealth of Madras Presidency and documented nearly 83 species of *Impatiens*. The Anaimudi is a paradise for the diversity and distribution of plant wealth in which more than 60 species of *Impatiens* were reported (Ramasubbu *et al.*, 2011). Bhaskar (2012), studied the systematic analysis of genus *Impatiens* of Western Ghats, South India and documented around 109 species of *Impatiens* and 69 species exclusively from Tamil Nadu. Mohanan and Henry (1994) studied the entire floristic wealth of Thiruvananthapuram, Kerala and documented 14 species of *Impatiens*. Mathew (1981) also studied the entire floristic wealth of Tamil Nadu Carnatic and documented 5 species of *Impatiens*. Nayar *et al.* (2006) documented the entire floristic wealth of Flowering plants of Kerala and reported 80 species of *Impatiens* and one species of *Hydrocera*. Rathakrishnan *et al.* (2005) revised the flora of Kerala, partially analysed the flowering plants diversity and documented 64 species of *Impatiens*. Desai and Janarthanam (2011) studied the genus *Impatiens* in the northern and parts of central Western Ghats and documented around 38 species of *Impatiens*. Mathew (1981) also illustrated few *Impatiens* like *I. acaulis* and *I. fruticosa* in The Illustrations on the Flora of the Tamil Nadu Carnatic. Vajravelu (1983) documented the entire floristic wealth of Tamil Nadu, India and reported 62 species of *Impatiens* and one *Hydrocera* species. Mathew (1999) documented the entire floristic wealth of Palni hills and documented 11 species of *Impatiens*. Shimizu (2005) studied several types of seed coat on the basis of comparison with each other characters such as chromosome numbers, floral characters and seedling morphology, systematic problems concerning *Impatiens*. Nair (1991) stated that half of the balsam diversity occurred in South India and more than 86 species are endemic to the Western Ghats. Due to these reasons, nearly 80 species of *Impatiens* are already threatened with uncertain future. Narayanan *et al.* (2011) discovered *Impatiens johnsiana* from the Wayanad District, Kerala, India, a new endemic epiphytic scapigerous balsam related to *I. dendricola* and *I. scapiflora*. Vivekanandan *et al.* (1997) recently studied the *Impatiens* for the flora of India revealed that India harbours 203 species. Of these, about 137 species are endemic to India and 86 species are distributed in Western Ghats and adjoining areas.

Geographical Distribution of the Genus *Impatiens* L.

Impatiens, the sub-cosmopolitan genus belonging to the family Balsaminaceae is distributed in the highlands and mountains of the old world of Asia and Africa and tropics and subtropics (Grey Wilson, 1980). It is also reported from Myanmar, Java, Madagascar, Philippines, Sri Lanka and Sumatra (Bhaskar, 1981). Hooker and Thomson (1859) pointed out that *Impatiens* attains its maximum development in India and the Western Ghats is the most prolific area. Barnes (1939) made extensive collection on the high ranges of Kerala and published an account of Geraniaceae occurring in the area. He recorded 12 species of *Impatiens* endemic to high ranges and also found more than 30 species concentrated within a radius of 16 km of Munnar and most of them within a much smaller area. Idukki district, a major part of the Anamalai high ranges forms a center of endemism as far as balsams are concerned. It is the type locality for more than 25 species of *Impatiens*. The high range is one of the richest areas in the Western Ghats with respect to the species of balsams.

The highly undulating mountainous configuration and the formation of a well marked upland shola forest with very high rainfall provide suitable habitat for *Impatiens* (Barnes, 1939). Many species of *Impatiens* are found restricted to specific altitudinal zones and most of the exclusive endemics are found restricted in isolated pockets in the high altitudes (2000m asl) in Western Ghats. According to Bhaskar (1981), the high rate of endemism of this genus is due to its orophytic nature, extreme adaptation to some particular microclimate. This perplexing genus has two main centers of distribution in India; the Himalayas in the north and the Western Ghats in the south and each area has a natural home for 100 species each. It generally occurs at high altitudes (1500 m asl), where there is active monsoon (rainfall), optimum temperature (20°C) and sunlight. The hilly ranges of Western Ghats are phytogeographically an important topographical unit in southern India and consist of areas with intermittent rain, innumerable valleys, grassy slopes, perennial streams and water falls form an appropriate habitat for the balsams. Most of the Indian *Impatiens,* particularly in the Western Ghats are narrow endemics and confined to specialized ecological habitats. Their colonizing abilities and regenerative capacities are very poor due to human interference. Hence, they become extinction prone. When the habitat is disturbed or destroyed, these endemics may become rare, vulnerable, endangered or even extinct.

Endemism and Rarity in *Impatiens* L.

Most of the species of flowering plants endemic to Peninsular India are confined to the Western Ghats (Nayar, 1980). Subramanyam and Nayar (1974) considered that, the summits of the Western Ghats are comparable to islands regarding endemic species. Endemic species are of great interest to taxonomists as some of them may be very ancient in origin while others may be comparatively new. They make phytogeographical zones distinct and unique. It is well known that the endemic plants determine whether a region has a typical or characteristic flora or not. The high percentage of endemic elements shows that the Western Ghats have a typical flora of their own.

The *Impatiens,* a non-endemic genus is one of the largest genera among the flowering plants and has more than 1000 species in the world. It is interesting to note that such a non-endemic genus having considerable number of endemic species make the Western Ghats phytogeographically a significant zone in India. Majority of the *Impatiens* species are endemic to India. As mentioned elsewhere, out of the 92 species available in Peninsular India, 82 have restricted distribution in Western Ghats. Based on the concentration of these species three endemic areas *viz.*, Anamalai (Anaimudi), Coorg, Nilgiris were identified (Pandurangan, 1996; Nayar, 1997). According to Vajravelu and Daniel (1983), 30 species of balsams are in threatened category including 19 critically endangered ones. Endemics being closer to rare groups are more vulnerable for extinction. The main causes of rarity appear to be fragmentation of population, habitat degradation, other biological problems etc. Several factors may be responsible for disturbing and destroying habitats. Some may be natural such as land upheavals, land slides, drought, forest fire etc., while others are man-made which include deforestation, developmental activities, excessive exploitation and indiscriminate collection (Jain and Sastry, 1980). According to them (l.c.) "if endemic species are eliminated from our country it mean they will be eliminated from the whole world, will be lost to science, will be struck of the roll of biological resources of this earth". Recent studies on *Impatiens* for the flora of India revealed that, India harbours 203 species (Vivekanandan *et al.*, 1997). Of these, about 137 species are endemic to India of which 86 species are distributed in Western Ghats and adjoining areas. Nair (1991) highlighted the endemism in Western Ghats with special reference to *Impatiens.* According to him, in India there are more than 6 well defined regions of distribution as far as *Impatiens* concerned. The distribution pattern, individuality of northern and southern species, narrow endemism, neo-endemics etc. have contributed to make *Impatiens* a key genus for phytogeographical studies.

According to Vajravelu and Daniel (1983) and Nair (1991), atleast 30 species of *Impatiens* have already entered into the endangered or threatened category. Of the 106 species and 13 varieties of *Impatiens* in South India, nearly 90 species of *Impatiens* are threatened today. About 37 species fall under the IUCN (2001) Category of Critically Endangered and 17 species under vulnerable status. About 8 species have not been collected since their type collections and their data are deficient as far as their taxonomic status is concerned. There are only 15 species under the Category of Least Concern and 8 species are near threatened. The extensive cleaning of forests areas over a period of time and protected natural habitat areas into tea gardens, coffee estates, monoculture cultivation of trees, removal of roadside bunds for widening urbanisation and tourism are the main causes for notable changes in the habitats. Most of the populations of *Impatiens* species are reduced to small pockets or completely wiped out from the nature. At least a significant size of populations of few rare and endemic epiphytic balsams which dwelled on lofty tree trunks and branches have vanished due to coupe felling for commercial timber during the las few years.

Unfortunately, the climate change in Western Ghats, decreasing the number of rainy days and increasing the hot and dry summer, have caused further havoc to the balsam populations. Shallow perennial streams are drying up early and remain dry during most part of the year. The cascading streams which were spraying water to the balsams could hardly be seen after the close of monsoon. Therefore, balsams dry up due to desiccation even before the life cycle is completed, especially the seed production, germination and establishment have severely affected. The family Balsaminaceae is an isolated group of plants and it is a neglected group among flowering plants. Due to their diverse shape, colour, morphology of flower and habit, high altitude dwelling nature, forest habit, very few taxonomical studies were conducted on the family Balsaminaceae. Tamil Nadu is the state having more number of species of *Impatiens* and diversity among South Indian states. Therefore, an attempt has been made for the systematic studies on the family Balsaminaceae in Tamil Nadu.

Study Area

Location and Physical Features

Tamil Nadu is situated at the south-eastern extremity of the Indian Peninsula. The state lies between 8° 5′ and 13° 35′ N latitudes and 76° 15′ and 80° 20′ E longitudes and covers an area of 1, 30,058 sq km, occupying 4.08 per cent of the total area of India. It has a coastline of about 990 km and a land boundary of 1200 km and bounded on the north by Karnataka and Andhra Pradesh, south by the Indian ocean, east by the Bay of Bengal and west by the Arabian sea along with the states of Kerala and Karnataka. The geographical area of the state has a roughly rhomboidal appearance with the longer diagonal of it stretching from Pulicat lake in the north, Kanyakumari in the south and the shorter diagonal from Gudalur in the west to Point Calimere in the east. A total number of 30 districts has been demarcated and topographically the landmass of the state can be divided broadly into two divisions namely, the eastern coastal plains and the north and the west hilly region. The eastern coastal plains run almost parallel to the coast and gradually rise from it. This can further be subdivided into coromandal plains of lowland in the north, the alluvial fertile plains of the Cauvery delta, dry southern plains and the inner plains extend almost up to the feet of the ghats. The eastern and western parts of Tamil Nadu are flanked by Eastern Ghats and Western Ghats respectively. In between these hills, an elaborate plateau with variable topography and vegetation was also recorded (Map 1.1).

Geology and Soil

About 73.5 per cent of the total area of the state is underlined with crystalline rocks of Precambrian period. The sedimentary belt occupies about 26.5 per cent of the total area all along the coast flanking the main crystalline masses on the west. The northern and western hilly regions of old crystalline rocks of the plateau have important mineral deposits. A wealth of minerals like lime stone, lignite, granite, clay, gypsum, fledsparband graphite are found abundantly in Tamil Nadu. Besides, small quantities of gold, copper, kaolin, bauxite, asbestos etc. are also found. Granite gneisses are reported in many places of hills. Varieties of soils are found in the state along with barren rock out-crops and devoid of soil cover.

Black soils are cracking clayey soils found in the Deccan trap covering southern parts of the state. Alluvial soils cover riverbanks and Cauvery delta and red soils cover most parts of the state. Laterite soils occur in Nilgiris and Anamalais and red sandy soils are observed in few districts. A long stretch of coastal sandy soil forming dunes are met with chiefly in Ramanathapuram, Thanjavur, Nagapattinam, Cuddalore, Kanchipuram and Kanyakumari districts. The costal districts of the state and the Western Ghats have a rich content of humus soil.

Climate and Rainfall

Tamil Nadu has an equatorial maritime climate in its coastal regions. In inland, the temperature may go to extreme in some places while it tends to moderate in the coastal areas. Usually, the climate of Tamil Nadu is very warm and dry except on the hill tops. December and January are the coolest months and April-June are the hottest. The average temperature for most parts of the state range between 28°C and 40°C in

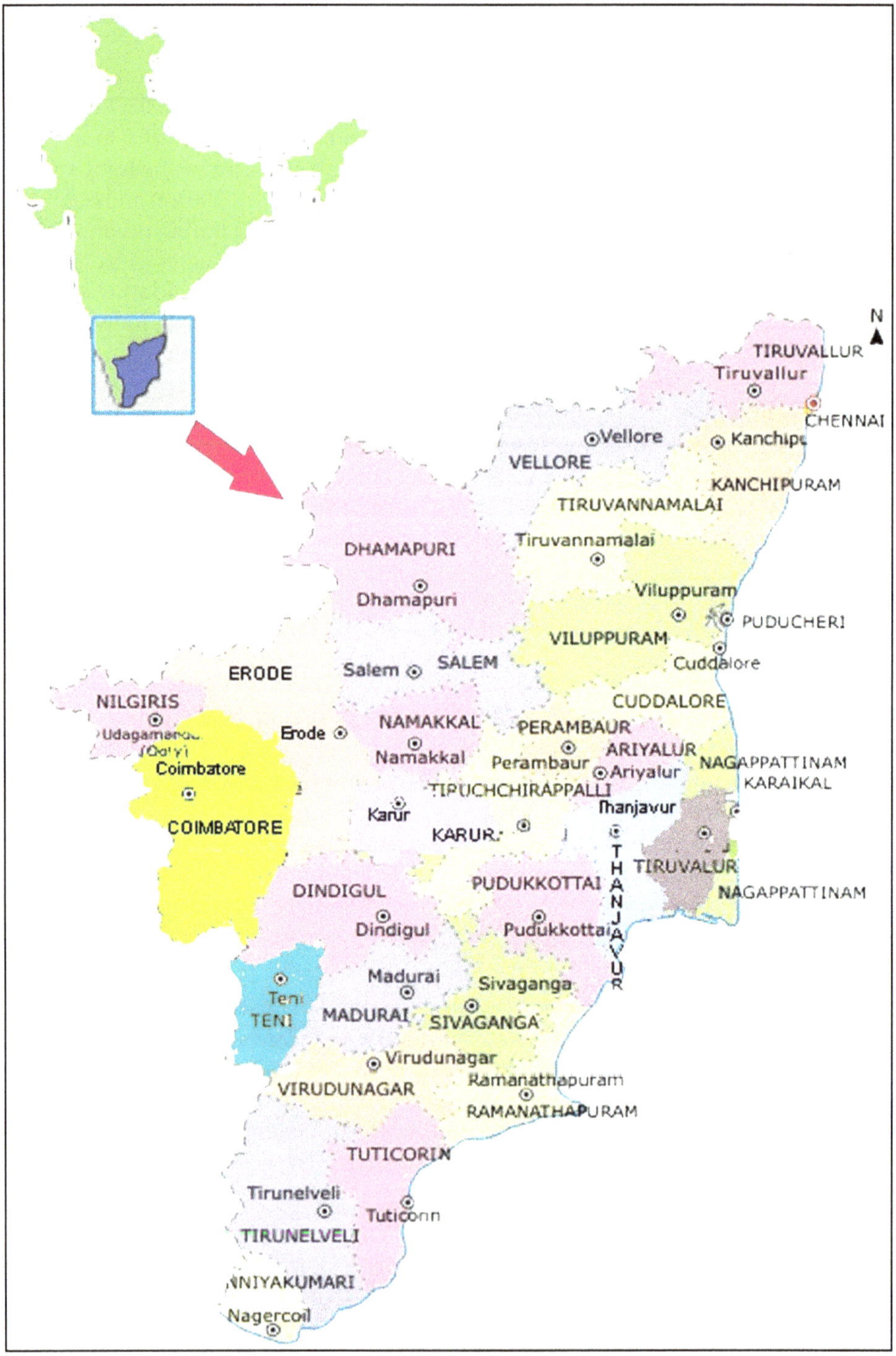

Map 1.1: Study Area: Tamil Nadu.

the summer season and between 18°C and 26°C in the winter season. In the more hilly terrains, the maximum temperature may be as low as 26°C in the minimum temperature may go down to 3° C. The normal annual rainfall received in the state varies from 640 to 1910 mm and the average rainfall is 945 mm rains are received from both monsoons. North-east monsoon gives the major portion of rainfall to the state during October to December and the rest from south-west monsoon during June to September. Nilgiris receives annual rainfall of 152 to 178 cm from both the monsoons. Many districts are drought-prone and large tracts with semi-arid conditions. Depending on the rainfall and crop distribution, the state has been grouped under 7 agroclimatic zones, namely northeast, northwest, west, southern, high rainfall, high altitude hilly and Cauvery delta.

Vegetation

The vegetation of Tamil Nadu is very varied depending upon the climate, altitude and edaphic factors. It can be broadly divided into four categories such as coastal vegetation, island vegetation, vegetation of the interior plains and vegetation of hills and mountains.

Coastal Vegetation

The entire coast of Tamil Nadu stretching from the Pulicat lake in the north to the southern most point of India, Kanniyakumari and on the west coast from Kanniyakumari to Poovar is chiefly sandy with outcrops of rocky headlands at Mahabalipuram, Mandapam and Kanniyakumari. There are also some tidal creeps and flats at the riverine mouths, of which the Vellar and Cauvery complex are significant.

Island Vegetation

The islands of Rameswaram, Kurusadai, Shingle, Hare and Church exhibit strand flora which is almost similar to that of Sri Lanka. Pure strand scrub of *Pemphis acidula* is characteristic to all islands except Rameswaram. There are also *Cordia subcordata, Halopyrum mucronatum, Ipomoea pes-caprae, Polycarpaea spicata* and *Scaevola sericea. Ipomoea violacea* is possibly a new entrant to Rameswaram islands.

Vegetation of the Interior Plains

The vegetations of the plains at the foot or sometimes on the gently undulating slopes of hills and hillocks in the rocky area boardering the coastal planes come under Southern tropical thorn forest. They are common in Chengalpattu, North and South Arcot, Pudukkottai, Tiruchchirappalli, Thirunelveli and less so in Salem and Coimbatore districts and these are often called as scrub jungles.

Vegetation of Hill and Mountains

Dry deciduous forests generally occur above 400 m on shallow soils of well drained hill sides. The canopy is closed though uneven and not dense. Most of the species are deciduous.

South Indian moist deciduous forest lie above the zone of dry deciduous forests. The trees attain a height of 30-36 m or more and the species are predominantly deciduous. The number of species constituting the upper canopy is small. There may be some evergreen elements dominating the lowest storey giving the false impression of evergreen nature to these forests. Bamboos are usually common.

Sholas are the mountain forests mostly confined to moist and sheltered valleys, glens and hollows as in the Anamalais, Nilgiri and Pulney. The trees found are evergreen, usually short-boled and of less than 12 m height with dense round crowns, coriacious leaves and branches covered with epiphytes, mosses and lichens. Woody climbers are also frequent.

Eastern Ghats

The Eastern Ghats beginning in Orissa state and run through Andhra Pradesh, Karnataka including Tamil Nadu with an average elevation of 700 m asl. The Eastern Ghats enter the state from Andhra Pradesh in the north and cut across the state and merge with Nilgiri hills on the Western Ghats. In Tamil Nadu, Eastern Ghats are in the form of detached hill groups, *viz.* Javadu hills, Shevaroy hills, Kalrayans, Pachamalai hills, Kollimalai hills, Kanjamalais and Sirumalai hills.

Shevaroy Hill

These hills form an interesting strip of discontinuous chain of Western Ghats. They receive both south west and north east monsoon rainfall, but the latter is more vigorous starting from mid September which is more optimal period for the growth of balsam like *I. shevaroyensis*, *I. yercaudensis* and *I. aquatica.*

Western Ghats

Along the whole length of the western part of Tamil Nadu, Western Ghats run southward and terminate at Kanyakumari, the southernmost extreme tip of Peninsular India with enhanced mountain ranges like Nilgiris, Anamalais, Palni hills, Cardamom hills, Varusanadu and Andipatti hills ranges and Agasthiyamalai at a distance of 80–160 km from the coastline of Arabian sea. The altitudinal range of Western Ghats from Kanyakumari towards the north, becomes elevated higher, averaging about 1220 m asl and rising up to 2554 m at Mukurthi and 2637 m at Doddabetta in Nilgiris, the highest peak of Tamil Nadu. They are joined by the Eastern Ghats and at the point of junction form the high plateau of Nilgiris. The Western Ghats have an almost continuous chain of mountain ranges interrupted by Palghat gap of about 25 km width near Palghat of Kerala and Coimbatore district of Tamil Nadu and also with smaller passage nearby Shenkottai, Aryankavu and Aramboly. The hills lying south of the gap are known by the names Anamalais and Cardamom hills. On the east, Palnis hills are found on the offshoots of Anamalais. Western Ghats have tropical evergreen, semi-evergreen and deciduous forests as well as savannahs intermixed with cultivated lands and settlements. Slopes of Western Ghats are the source of many rivers which flow eastwards towards the Bay of Bengal.

Nilgiris

The Nilgiris, commonly known as blue mountains is a part of the Western Ghats extended towards eastern side. The area has been occupaid with extensive grassy downs and lands mixed with densely wooded sholas. The grassy slopes possess rich balsam flora like *I. herbicola, I. tenella, I. scabriuscula, I. chinensis, I. leschenaultii* etc. The rocky outgrowths and slopes support *I. modesta, I. clavicornu* and *I. pseudo-acaulis,* but the richest assemblage of balsams is found mostly in the sholas. The Kundah shares good number of endemic species like *I. clavicornu, I. nilgirica, I. orchioides, I. levingei, I. laticornis, I. neo-barnesii, I. denisonii* and interestingly, all these species belong to taxonomic Section: 'Scapigerae'. The main reason for the occurrence of large number of species in Nilgiris is due to the rainfall received (average of 187.96 cm) during both south-west and north-east monsoons with major amount received through north east monsoon (October–December).

Anamalai

Anamalai Range is closer to Pulni hills and is an ecologically rich place as it covers hills and valleys ranging in altitude from 200 m to 2513 m asl with many streams. Akkamalai is in the high ranges touching 2513 m altitude and rich in balsams. More than 27 species of *Impatiens* like *I. crenata, I. parasitica, I. leptura, I. elegans, I. ligulata, I. henslowiana, I. macrocarpa, I. wightiana, I. verecunda, I. chandrasekharanii* and *I. concinna* have been reported from Anamalais.

Pulni Hills

It is special botanically rich place with specific temperate flora rare in Peninsular India. The plateau consists of undulating hills over three quarters of which are covered with grasslands. The annual rainfall is about 170 cm and its distribution almost throughout the year keeps the vegetation always fresh, feature so striking in Peninsular India. Kodaikanal has a rich balsam species (11 species) and especially shrubby species like *I. phoenicea, I. viscida, I. henslowiana,* and other herbs like *I. campanulata, I. cordata,* and *I. elegans.*

Shivagiri, Mahendragiri and Courtallum

Shivagiri, Mahendragiri and Courtallum area comprises an innumerable number of water falls in which a number of endemic balsams have also been reported from this region which include *I. auriculata, I. viridiflora, I. modesta, I. rufescens, I. umbellata,* etc. The Kakachi forest areas has typical monsoon forest and comprises species like *I. cordata, I. uncinata, I. maculata, I. grandis,* etc. Geographically, Mahendragiri hill occupies almost the extreme southern end of Western Ghats and a part of the chain of Eastern Ghats with species like *I. travancorica,* and *I. disotis.* Patchakumatchi hills situated to the south west of Madurai Town are highly humid (1600 m). Cumbum Valley is horse-shoe shaped, surrounded by hills on three sides, Western Ghats in the west and Varushanad and Andipatti hills on the east and these two ranges are jointed at the basin of Periyar lake near Kumali and Thekkadi of Kerala.

Agasthyamalai

The Agastyamalai situated at the southern most tips of the Western Ghats and its western slopes forms a compact block of hilly range. It is notable place for its very rich and diverse vegetation dotted with high concentration of endemic species. The highest peak is the Agasthyakudam, reported with an altitude of 1868 m above msl. Major portion of the area belongs to Neyyar and Peppara wild life santuary and the

Agastyamalai has been declared as Biosphere Reserve since 2001. IUCN has identified Agasthyamalai and its environ as one among the three centres of plant diversity within India (IUCN, 1987). The dense forests of these hills provide home for many endemic balsams. An interesting *Impatiens rufescens* var. *agasthyamalayensis* is also reported from this range (Bhaskar, 1975).

Material and Methods

For the Systematic studies on the family Balsaminaceae in Tamil Nadu, intensive exploration trips were conducted twice a week in the beginning and once in a week later for the thorough collection of ephemerals in the different forest areas of Tamil Nadu.

Collection of Plants

The plants were collected both in flowering and fruiting condition. In the case of no flowering and fruiting conditions during collection, the plant twig with few leaves were collected for proper identification. The collected plant twigs were tagged properly with accurate accession number.

Field Observation and Records

The collected plants were tagged properly and the data regarding the details of the plants and the collected informations were recorded properly. After the proper collection of specimens from the different forest areas especially in the hilly areas of Tamil Nadu, the specimens were properly identified with its specific floral and important key characters. The plants which were identified in the field were also confirmed by dissection microscope and line drawing if necessary.

Doubtful specimens collected from the study area were compared with herbarium specimens of Tropical Botanic Garden and Research Institute, Thiruvananthapuram (TBGRI) herbarium. Further, all the specimens collected from the field work were identified with standard floras published by different taxonomist (Gamble, 1915–1919; Mathew, 1981; Nayar *et al.*, 2006; Vajravelu, 1983; Grey Wilson, 1980; Augustine, 1999; Desai and Janarthanam, 2011; Bhaskar, 2012; Ganeshaiah, 2012).

Herbarium Preparation and Documentation

The identified specimens were processed and pressed properly as per the method suggested by Jain (1963). The processed specimens were pasted on the herbarium sheets (size: 28 × 42 cm). The prepared herbarium sheets were deposited in Gandhigram Rural Institute – Deemed University Herbarium (Department of Biology).

Systematic Enumeration

For the systematic studies on the family Balsaminaceae in Tamil Nadu, the details regarding complete taxonomic description along with citation of individual species were given properly. The description details along with phenological data (Flowering and Fruiting) were also provided for proper identification. The taxonomic key was also prepared to the genus and species level for the identification and confirmation. 73 species from *Impatiens* and 1 species from *Hydrocera* were reported from different parts of Tamil Nadu.

Among them, 90 per cent of the species reported are restricted to hills of Western Ghats and 15 scapigerous, 6 epiphytes, 41 terrestrial herbs and 8 shrubs were reported. Interestingly, *Impatiens munnarensis*, a possibly extinct species has been rediscovered from High Wavies mountains (Megamalai hills) of Tamil Nadu for the first time after its type collection and new distributional record to Tamil Nadu also. *Impatiens hookeriana*, a shrub described by Arnott in 1834 had been merged due to its few similar characters with *I. grandis*. The present study revalidated it as *I. hookeriana* due to its several distinct characters with *I. grandis*.

Most of the Indian *Impatiens* particularly in the Southern Western Ghats are narrow endemics and confined to specialized ecological habitats. The high rate of endemism in this genus is due to its orophytic nature, extreme adaptation to some particular microclimate. Their colonizing abilities and regenerative capacities of wild *Impatiens* are very poor due to human interference. When the habitat is disturbed or destroyed, these endemics may become rare, vulnerable, endangered or even extinct. Several factors, therefore, may be responsible for the distribution and subsequent establishment in the Southern Western Ghats.

BALSAMINACEAE

Shrubs, annual or perennial herbs, sometimes succulent, inhabiting moist and shady places, rarely epiphytes and aquatic habit; stems erect or procumbent, often rooting at lower nodes, occasionally acaulous. Leaves alternate, opposite, verticillate or spiral, simple, petioled, sometimes sessile, crenate, dentate or serrate at margin with teeth or crenatures generally apiculate, the lower most sometimes with glands, glabrous or pubescent; petioles occasionally glandular. Flowers bisexual, zygomorphic, in axillary racemes or umbels or solitary, binate or fascicled. Sepals 3, rarely 5, free; lateral 2 small; lip large in lower one, petaloid, often tapering or abruptly constricted into a nectariferous spur. Petals 5; posterior one standard, flat or cucullate; lower 4 usually connate in lateral pairs; wings entire, 2–3 lobed, sometimes with an auricle descending into spur. Stamens 5; anthers cohering into a ring. Ovary superior, 5-loculed; ovules many; style short or absent; stigma sessile, 5-toothed. Fruit berry or capsule; seeds smooth or tubercled.

Artificial Key to the Genera

1 a. Plants aquatic, fruit fleshy pseudoberry, not turgid .... *Hydrocera*

1 b. Plants semi aquatic or not aquatic, fruit capsule, turgid .... *Impatiens*

HYDROCERA Bl.

Hydrocera triflora (L.) Wight & Arn., in Wight, Cat. Ind. Pl.: 28 (445). 1833 & Prodr. 140. 1834 emend. Venkatesw. & Dutt. in J. Bombay Nat. Hist. Soc. 58 (2): 544–546. 1961; Hook. f., Fl. Brit. India 1: 483.1875. *Impatiens triflora* L., Sp. P1. 938. 1753. *Tytonia triflora* (L.). C.E Wood in Arn. Arb. 56 (4): 413.1975.

Herbs, perennial, semi-aquatic up to 1m high; stems erect, 5–angular; lower nodes bearing long fibrous roots; internodes hollow, fistular, floating. Leaves spirally arranged or alternate, linear or linear-lanceolate to linear-elliptic, cuneate at base, acute to long -acuminate at apex, serrate at margins, 11–28 x 0.6–3.7 cm, glabrous, deep-green above, pale beneath, base and midnerve sometimes tinged red; lateral nerves 10–17 pairs, oblique; sessile or with petiole up to 1 cm long, base with a pair of sessile glands. Inflorenscence axillary, short, 5–10 flowered raceme or cyme. Flowers pink, tinged with carmine or purple, inside of throat yellowish, rarely greenish-yellow with reddish-purple nerves; bracts lanceolate, acute to acuminate, 6–9 mm long. Sepals 5, petaloid, imbricate; lateral sepals 4, in 2 pairs; lower pair elliptic to elliptic-oblong, obtuse: upper pair elliptic-oblanceolate, obtuse; lower sepal navicular, spurred; spur 6-8 mm long, curved, slightly swollen at tip. Petals 5, free; lower end narrowly oblong, unevenly belowed; upper lobe entire, lower lobe irregular, toothed, emarginated and apiculate. Stamens 5, filaments slightly shorter than gynoecium, broader and united towards connective; anthers connate. Ovary glabrous, 5- locular with 3 ovules in each locule on axile placenta. Stigmas 5, sessile, persisting in fruit. Fruit fleshy pseudo-berry, 8–10 x 7–8 mm, pentangonal in cross section, truncate at base, short- beaked, pale green becoming reddish or purplish-red with a hard endocarp; dehiscing septicidally leaving the seeds attached to column; seeds usually solitary in each cell, curved, corrugated; albumen absent; cotyledons thick, radical short.

Fl. and Fr.: Almost throughout the year.

Distribution: Thanjavur.

IMPATIENS L.

Terrestrial, semi-aquatic or occasionally epiphytic annuals, perennial herbs or shrubs in moist or shaded places. Leaves alternate, opposite or verticillate, petioled or not, simple, entire or serrate, pinnately or rarely palmately nerved. Flowers axillary, solitary, fascicled or racemes or umbels. Lateral sepals 2, much smaller than lip. Lip saccate, cymbiform funnel-shaped or rarely bowl-shaped with long, short or obsolete spur. Standard obovate or suborbicular, sometimes keeled on dorsal side. Lateral petals 4, united into 2 pairs. Fruit capsule, dehiscing loculicidally; valves 5, elastically springing away from placentariferous axis; seeds smooth or tubercled, glabrous or hairy, non endospermous.

Artificial Key to the Species of *Impatiens*

I. a. Plants scapigerous.

1. a. Plants with bulbous root stock, leaves radical, flowers with tuft of hairs .... *I. crenata*

   b. Plants with tuberous root stock, leaves radical 2

2. a. Pedicels glabrous, wings bilobed, standard petal humped at base, forming a hood above column .... *I. acaulis*

   b. Pedicels not glabrous, wings not bilobed 3

3. a. Root stock tuberous with fibrous root, leaves palmately lobed, wings 3-lobed .... *I. chandrasekharanii*

   b. Root stocks without fibrous roots 4

4. a. Leaves orbicular, lateral sepals obliquely ovate, spur incurved .... *I. clavicornu*
   b. Spur not incurved 5
5. a. Tuft hairs present above conjunction of lobes, lobes are spured .... *I. denisonii*
   b. Tuft hairs absent in conjunction lobes 6
6. a. Wings vaulted standard petal, hairs on seeds short and stout .... *I. pseudo-acaulis*
   b. Wings not vaulted 7
7. a. Bracts small, 2-3 mm long, glabrous .... *I. stockii*
   b. Bracts not small 8
8. a. Hairs on seeds long and filamentous .... *I. shevaroyensis*
   b. Filamentous hairs are absent 9
9. a. Spur cylindrical, slightly bent in the middle part, pink pollen .... *I. tirunelvelliensis*
   b. Spur not cylindrical, pink pollen absent 10
10. a.Bracts oblong, semi amlexicaulous, scapes solitary .... *I. laticornis*
    b. Bracts not oblong 11
11. a. Dorsal auricle of wings spiniferous .... *I. lawsonii*
    b. Dorsal auricle of wings not spiniferous 12
12. a. Lip not spurred, flowers reddish brown or pale brown .... *I. orchioides*
    b. Flowers not reddish brown or pale brown 13
13. a. Bracts ovate, semiamplexicaulous, sepals 1-2 .... *I. niligirica*
    b. Bracts not semiamplexicaulous .... *I. scapiflora*

I. b. Plants epiphytes

1. a. Leaves cuneate at base, wings 2-lobed, swollen fleshy stem with leaf scar .... *I. auriculata*
   b. Leaves not cuneate, wings 2-lobed, leaf scars absent 2
2. a. Filiform pedicels, seeds minute with spiral hairs, lateral sepals ovate, spur of lip straight .... *I. modesta*
   b. Filiform pedicels absent, absence of seeds hairs 3
3. a. Leaves crowded at ends of branches, lateral sepal linear green with foliaceous crest .... *I. parasitica*
   b. Leaves crowded at ends of branches, lateral sepal linear lanceolate, acuminate, yellowish green 4
4. a. Stem tumoid, lip rounded and ventricose spur .... *I. jerdoniae*
   b. Stem not tumoid, ventricose spur absent 5

5. a. Bracts papillose, capsules ellipsoid .... *I. neo-barnesii*
   b. Bracts not papillose .... *I. viridiflora*

I. c. Plants shrubs, leaves alternate

1. a. Leaves serrulate, reddish petiole, lip straight .... *I. hookeriana*
   b. Leaves not serrulate, lip not straight 2
2. a. Pedicels larger than leaves, standard petal orbicular, bracteate .... *I. fruticosa*
   b. Pedicel not larger than leaves, standard orbicular 3
3. a. Standard two-lobed, violet coloured .... *I. latifolia*
   b. Standard not two-lobed 4
4. a. Succulent stem with leaf scars, pedicel papillose .... *I. henslowiana*
   b. Stem not succulent, pedicel not papillose 5
5. a. Lip funnel-shaped, seeds papillose .... *I. wightiana*
   b. Lip not funnel-shaped 6

I. d. Plants shrubs, leaves opposite

1. a. Plants aquatic, stem quadrangular, cordate shaped at base of the leaves, seeds shining .... *I. diversifolia*
   b. Stem not quadrangular 2
2. a. Inflorescence subumbels, bracts persistant, spur of lip straight .... *I. goughii*
   b. Bracts not persistant, spur of lip not straight 3
3. a. Leaves mucronate, stipule setaceous, saccate lip, standard petal with pubescent line at centre .... *I. ligulata*
   b. Leaves not mucronate, pubescent line is absent on standard petal 4
4. a. Standard orbicular, keeled on dorsal side, wing hirsute .... *I. munronii*
   b. Wing not hirsute 5
5. a. Bracts triangular, ovate, seeds shiny, black to brown .... *I. minor*
   b. Triangular bracts are absent 6
6. a. Leaves sessile .... *I. oppositifolia*
   b. Leaves not sessile 7
7. a. Lateral sepal falcate, spur short blunt .... *I. parviflora*
   b. Lateral sepals not falcate 8
8. a. Villous pedicels present, saccate lip without spur .... *I. rufescens*
   b. Villous pedicels absent 9

9. a. Wings sub sessile, pedicels shorter than leaves, dorsal auricle decurved .... *I. tenella*
   b. Pedicels not shorter than leaves 10
10. a. Lip conical, spur cylindrical .... *I. tomentosa*
   b. Lip not conical 11
11. a. Capsules ovoid and turgid stem viscous .... *I. viscosa*
   b. Capsules not turgid, stem not viscous 12
12. a. Wings long, stipulate, pedicels longer than leaves, glabrous .... *I. lenta*
   b. Wings short, not stipitae 13
13. a. Stipules filiform, sparingly branched stem .... *I. munnarensis*
   b. Stipules not filiform 14
14. a. Stem succulent, spur absent, presence of bristles on leaves .... *I. herbicola*
   b. Stem not succulent, absence of bristles on leaves .... *I. flaccida*

I. e. Plants herbs, leaves opposite

1. a. Inflorescence fascicle, leaves serrated, petiole with stipule .... *I. floribunda*
   b. Inflorescence not fascicle 2
2. a. Seed glabrous, lip cymbiform .... *I. leschenaultii*
   b. Seeds not glabrous 3
3. a. Lateral sepals minute .... *I. cuspidata*
   b. Lateral sepals not minute 4
4. a. Spur equal to lip, puberulous capsule .... *I. debilis*
   b. Spur not equal to lip, puberulous capsules are absent 5
5. a. Spur tubular, glabrous, lobed wing petals present .... *I. fasciculata*
   b. Spur not tubular, lobed wing petals are absent 6
6. a. Plants aquatic, wing bilobed .... *I. aquatica*
   b. Plants not aquatic 7
7. a. Auricle small, away from spur, ascending side by side facing each other, seeds black .... *I. aliciae*
   b. Auricle not small and near to the spur 8
8. a. Pedicels deflexed in fruits, lateral sepals acute with a callous point .... *I. chinensis*
   b. Pedicels not deflexed in fruits, lateral sepals acute without a callous point 9
9. a. Stipules setaceous, spur short and curved, capsules erect, elliptic .... *I. concinna*
   b. Stipules not setaceous, spur not curved, capsules not erect 10

10. a. Stems woody, fruiting pedicel hairy, fruit glabrous .... *I. yercaudensis*

b. Stems not woody, fruiting pedicels not hairy 11

11. a. Lip boat-shaped, spurless .... *I. pendula*

b. Spur present 12

12. a. Lip spathulate, standard suborbicular .... *I. omissa*

b. Lip not spathulate .... *I. gardneriana*

I. f. Plants herbs, leaves alternate

1. a. Bracts oblong, spur shorter than lip .... *I. disotis*

b. Bracts not oblong 2

2. a. Inflorescence umbel, lip cymbidium or cymbiform .... *I. elegans*

b. Inflorescence not umbellate 3

3. a. Bracts well developed, flowers pink, standard not spurred .... *I. leptura*

b. Bracts not well developed 4

4. a. Stem purple- blotched, basal lobe of wings minute, seeds shortly haired .... *I. maculata*

b. Stem not blotched 5

5. a. Auricle curved outwardly .... *I. balsamina*

b. Auricle not curved 6

6. a. Bracts sub foliaceous, irregular petiole .... *I. campanulata*

b. Plants with regular petiole 7

7. a. Leaves cordate at base, nerves slender with scattered hairs on both surfaces .... *I. cordata*

b. Scattered hairs absent 8

8. a. Lateral sepals cordate shaped, capsule cylindrically ellipsoid .... *I. dasysperma*

b. Lateral sepals not cordate shaped 9

9. a. Lip tip swollen and incurved .... *I. phoenicea*

b. Lip tip not swollen 10

10. a. Leaves whorled, basal lobe broadly truncate at apex .... *I. tangachee*

b. Leaves not whorled 11

11. a. Flowers white, lip boat shaped, distal lobe large and sessile .... *I. travancorica*

b. Distal lobes not sessile 12

12. a. Lip cymbiform, spur robust, shorter than lip, dorsal auricle decurved .... *I. trichocarpa*

b. Spur not robust 13

13. a. Wings with subequal lobes, obtuse, seeds many, echinate .... *I. umbellata*
    b. Wings with equal lobes 14
14. a. Spur inflated .... *I. uncinata*
    b. Spur not inflated 15
15. a. Clawed wings, spur slender, cymbiform .... *I. verticillata*
    b. Clawed wings are absent 16
16. a. Leaves covered with stiff hairs .... *I.viscida*
    b. Stiff hairs absent 17
17. a. Petioles with numerous stalked glands, lip funnel shaped .... *I. wightiana*
    b. Funnel shaped lip absent 18
18. a. Leaves coriaceous .... *I. inconspicua*
    b. Leaves not coriaceous 19
19. a. Seeds with dwarf bulbous dome-like hairs, spiral reticulate bands .... *I. levingei*
    b. Dwarf bulbous dome-like hairs absent .... *I. scabriuscula*

Impatiens acaulis Arn., Companion Bot. Mag.1: 325. 1835; Hook.f. & Thomson, J. Proc. Linn. Soc. Bot. 4: 119.1859; Dalzell & Gibson, Bombay F1.: 42.1861; Hook.f., F1. Brit. India 1: 443.1874 & Rec. Bot. Surv. India 4: 44.1906; Cooke, F1. Bombay 1: 170.1901; Gamble, F1. Pres. Madras 1: 138. 1915; Blatter, J. Bombay Nat. Hist. Soc. 33: 309, t. 1.1933; Santapau, F1. Khandala: 30. 1967; Vajravelu in Nair & Henry, F1. Tamil Nadu 1: 52.1983; Sharma *et al.*, F1. Karnataka: 36.1984; Grey-Wilson in Dassanayake, Rev. Handb. F1. Ceylon 5: 82, F.2 A–E. 1985; Rao, F1. Goa 1: 56.1985; Kulkarni, F1. Sindhudurg District: 58. 1988; Almeida, F1. Savantwadi 1: 76.1990; Lakshmi-Narasimhan & Sharma, F1. Nasik District: 113. 1991; Deshpande *et al.*, F1. Mahabaleshwar 1: 113. 1993; Kothari & Moorthy, F1. Raigad District: 46. 1993; Almeida, F1. Maharashtra: 188.1996; Saldanha, F1. Karnataka 2: 257. 1996; Vivekananthan *et al.*, in Hajra *et al.*, F1. India 4: 111. 1997; Mudaliar & Prasad in Singh & Karthikeyan, F1. Maharashtra State Dicotyl. 1: 445. 2000; Yadav & Sardesai, F1. Kolhapur District: 93, t. 3, f. 19. 2002; Bhat, F1 Udupi District: 87, t.21. 2003; Rathakrishnan *et al.*, in Daniel, F1. Kerala 1: 530. 2005; Bhaskar Tax. Mon. *Impatiens* of Western Ghats 48. 2012. (Plate 1.1A).

Terrestrial or epiphytic, scapigerous, tuberous or rhizomatous herb, 8–27cm high; tubers oblong, 0.5–0.8 x 0.5–1.5 cm. Leaves radical, 3–6 per tuber, ovate, oblong to orbicular, 2.5–16 x 2–9 cm, obtuse, cordate to truncate at base, distantly apiculate–crenate at margins, acute to acute-apiculate at apex, hairy mainly on nerves above, glabrous below; petioles 2.5–9 cm long, glabrous. Inflorescence racemose scape, 7–27 cm long, 4–10 flowered. Flowers 1.5–5 cm across, pink to lilac with white throat; bracts ovate, 2–3.5 x 1–3 mm, green; pedicels 0.7–3 cm long, glabrous, deflexed in fruits. Lateral sepals asymmetrically ovate, 2–5 x 1.5–4 mm, acute at apex, glabrous, pale green. Standard petal broadly obovate, 0.5–1 x 0.6–1.6 cm, concave, humped at base, forming a hood above column, obcordate to emarginated, glabrous, pink outside, white inside; wing petals 1.6–3.2 x 1–2.5 cm, unequally bilobed; basal lobe oblong, 0.9–2.5 x 0.3–0.8 cm, rounded to obtuse at apex; distal lobe asymmetrically obovate, 1.1–2.6 x 0.6–1.5 cm; lip boat shaped, 6–12 x 4–6 mm, 0.2–6 mm deep, acute at apex, glabrous, pinkish white; spur basal, tubular, 3–5.8 cm long, rounded at apex, glabrous, pink to white. Column 3–6 x 1–3 mm, curved; anthers 1 x 1mm, filaments white, 3–5 x 0.5–1 mm, pink. Pistil 2.5 -5 x 1–3 mm; ovary ellipsoid to broadly lanceoloid, glabrous. Capsules asymmetrically ellipsoid to lanceoloid, 1–1.3 x 0.4–0.7 cm, glabrous; pedicels 1–3.5 cm long; seeds oblongoid to lanceoloid, 1 x 0.5 mm, brown, hairy; hairs coiled forming a cone like structure.

Fl. and Fr.: July–Dec.

Distribution: Nilgiri, Salem, Tirunelvelli.

Impatiens aliciae Fisch. var. ***pandavaramalayensis*** Bhaskar, Tax. Mon. *Impatiens* of Western Ghats. 120. 2012.

Erect annual herbs, often procumbent, 18–40 cm tall, branched. Leaves opposite, short petioled, petiole 0.3–0.5 cm long, lamina linear-lanceolate, 5.0–8.0 x 0.5–1.0 cm, margin distantly serrate, sparsely hairy above, glaucous beneath. Flowers axillary, solitary; pedicels 1–2 cm long, hairy along one side; lateral sepals linear, acute; lip cymbiform, spur up to 3 mm long, stout, thick, straight or slightly bent, black when dried, cylindrical; standard suborbicular, acute; wings bilobed, anticous lobe largest, posticous lobe small, falcate, dorsal auricle small, away from spur, ascending side by side facing each other. Capsules ellipsoid, stalk up to 2.5 cm long; seeds black shiny.

Fl. and Fr.: November.

Distribution: Endemic to Pandavaramalai in Anamalai Hills, Coimbatore Dist., Tamil Nadu above 1300 m.

Impatiens aquatica Bhaskar, Tax. Mon. *Impatiens* of Western Ghats 123. 2012.

Annual or perennial aquatic herbs, 160 cm or more tall, most of the plant submerged in water, stem under water flaccid, above water stiff, terete, erect, red, internodes long, 10 cm, rooted at nodes, glabrous. Leaves opposite-decussate, aggregated at apex, stipulate, reddish, sessile, leaf oblong or elliptic, 6 x 1.8 cm, acute, serrate, base rounded or sub-cordate, venation impressed above, scabrous on the upper surface, pale and glabrous beneath. Flowers axillary, solitary or 2 in each axil, 3–3.5 cm across, pink or pale pink, rarely white, with deep pinkish blotches at throat, honey guides inconspicuous; pedicels equal or longer than leaves, 6 cm long, deflexed in fruit, minutely puberulous, red like on the stem; sepals linear-acute; lip funnel-shaped, white, mucronate, without an yellow eye at the centre inside, minutely puberulous; spur slender, 3 cm long, straight or slightly incurved, white or pale pink, tip darkened, often bifid at tip, glabrouse or minutely puberulous; standard orbicular, acuminate or horned, 1 cm across, glabrous; wing bilobed, larger anterior lobe 1.5 x 1.3 cm, lobed again with one small oblong lobe and a bigger lobe, smaller basal lobe ovate-oblong, 5 mm long, 3 mm wide, tip round or obtuse, dorsal auricle like projection not prominent. Pollen pink. Capsules 1.2–2 cm long, ellipsoid, turgid in middle, fruiting pedicel and fruit glabrouse; seeds 8–13, smooth, black, shiny.

Fl.: Nov.–Jan. Fruits and seeds are not formed yet during January.

Distribution: Endemic to Yercaud.

Impatiens auriculata Wight in Madras J. Lit. Sci.Ser. 1. 5: 8. t. 3. 1837; Hook. f., Fl. Brit. India 1: 460.1874; Bhaskar, Tax. Mon. *Impatiens* of Western Ghats 99. 2012. (Plate 1.1B).

Herbs, perennial, succulent, glabrous, epiphytic on tree trunks; stem fleshy, swollen, with leaf- scars, rarely branched, rooting at lower ends of stems, alternate, broadly ovate- lanceolate or elliptic, obtuse or acute at apex, obscurely crenulated with few glandular setae at base, 2–17 x 1–6 cm, petiole 1–4 cm long. Flowers in axillary, few flowered peduncles, 2.5 cm across. Lateral sepals very large, pendulous, obliquely ovate or oblong, acute, scarlet. Lip bright green, saccate, wrinkled; spur short, hooked, thick. Standard small, dark – green, hooded. Wings bilobed, hooded, purple. Capsules ovate; seeds many.

Fl. and Fr.: Aug. – Oct.

Distribution: Endemic to North-East monsoon zone in Courtallum, in the Western Ghats of Tirunelveli, Aghastyamalai.

Impatiens balsamina L., Sp. P1.: 938.1758; Wight & Arn, Prodr. Fl. Ind. Orient.: 135.1834; Hook.f. & Thomson, J. Proc. Linn. Soc. Bot. 4: 130.1859; Hook.f., F1. Brit. India 1: 453.1874 and Rec. Bot. Surv. India 4: 47.1906; Cooke, F1. Bombay 1: 174.1901; Gamble, Fl. Pres. Madras 1: 142. 1915; Ramamoorthy in Saldanha & Nicolson, F1.Hasan District: 400.1978; Vajravelu in Nair & Henry, F1.Tamil Nadu 1: 52.983; Almeida, Fl. Savantwadi 1: 77.1990; Murthy & Yoganarasimhan, F1. Coorg: 86.1990; Almeida, F1. Maharashtra 1: 189.1996; Saldanha, F1. Karnataka 2: 249.1996; Vivekananthan *et al.* in Hajra *et al.*, Fl. India 4: 123.1997; Mudaliar & Prasad in Singh & Karthikeyan, Fl. Maharashtra State Dicotyl. 1: 445. 2000; Bhat, F1.Udupi District 87. 2003; Rathakrishnan *et al.*, in Daniel, F1. Kerala 1: 532.2005. Bhaskar, Tax. Mon. *Impatiens* of Western Ghats 188. 2012. (Plate 1.1C)

Herb, up to 1.5 m high, stem erect, much branched or not, terete, swollen at nodes, pubescent, pinkish red. Leaves alternate, linear-lanceolate, oblanceolate, linear–oblong or elliptic to linear-oblanceolate, 2.5–15 x 0.5–4 cm, cuneate to attenuate at base, crenate to serrate at margins, acuminate at apex, sparsely hairy above, hairy only on nerves below; nerves 6–12 pairs; petioles 1–1.5 cm long, hairy; glands 1–5 pairs. Flowers 1–4 per axis, 2–2.5 cm across, pink; bracts linear – lanceolate, 1–3 x 0.5–1 mm, acute to acuminate at apex, dorsally hairy, dark pink; pedicels 1–1.5 cm long, hairy, deflexed in fruits. Lateral sepals ovate or lanceolate, 2.5 x 1.5 mm, concave, acute to acuminate at apex, dorsally hairy. Standard petals broadly ovate to orbicular, 5 x 5 mm, concave, dorsally keeled; keel hairy, mucronate with 2 mm long mucro; wing petals 1.5–2 x 1–1.6 cm, 2–lobed, auricled near base; auricle curved outwards; basal lobe oblong ovate, 5–6 x 3–5 mm, obtuse to apiculate at apex; distal lobe asymmetrically obovate, 1.2–1.6 x 1–1.5 cm, deeply notched at apex; lip conical, 8–13

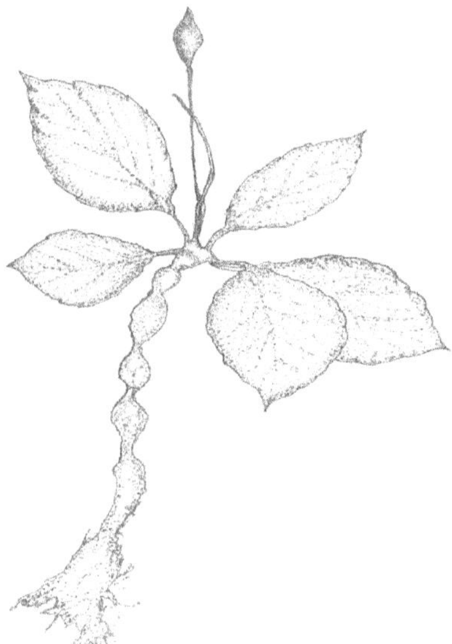

Figure 1.1: *Impatiens auriculata* Wight.

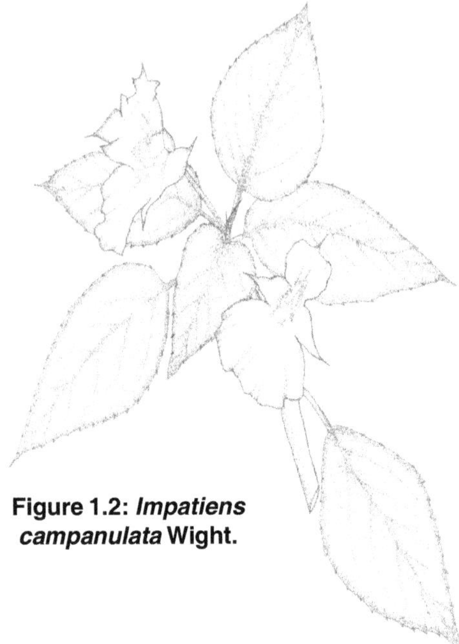

Figure 1.2: *Impatiens campanulata* Wight.

Figure 1.3: *Impatiens chandrasekharanii* Chandrabose.

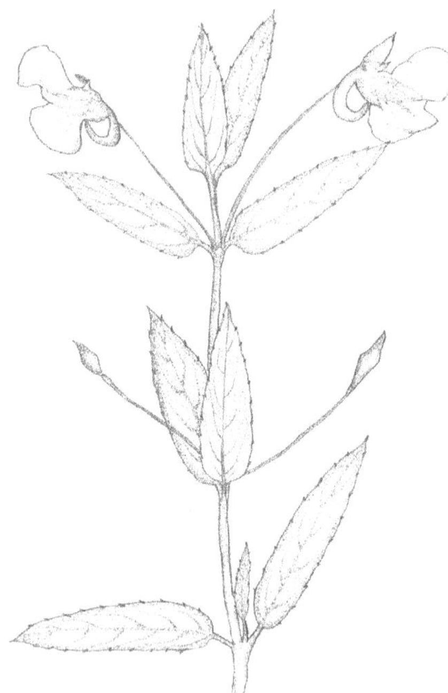

Figure 1.4: *Impatiens chinensis* L.

x 6 mm, 5–8 mm deep, acuminate to cuspidate at apex, hairy; spur basal, tubular, 3–3.5 cm long, rounded to bulged at apex, sparsely hairy. Column 4–6 x 1–3 mm, curved; filaments 3–5 x 1 mm; anthers 1 x 1mm. Pistil 3–5 x 1–2 mm, ovary ellipsoid to lanceoloid, villous. Capsules broadly ellipsoid, 1–1.5 x 0.6–0.9 cm, wooly; pedicels 2 cm long, seeds globose, 2.5 mm, granulate, dark brown.

Fl. and Fr.: Almost throughout the year.

Distribution: Yercaud, Poonachi, Anamali, Kodaikanal, Hassnur, Kollimalai, Naterikal, Nilgiris, Ponnaikadu, Aliyar, Pudur, Sengalteri, Gavigundumalai, Theni, Megamalai, Sirumalai.

Uses: Seed is edible. Alcoholic extract of the flower possesses marked antibiotic activity against pathogenic fungi and bacteria. Flower is reportedly used as a substitute for henna for dyeing finger nails.

Impatiens campanulata Wight in Madras J. Lit. Sci. Ser. 1.5: 11.t.7.1837 & Icon 3 (1): 2.t.744.1844; Hook. f., F1. Brit. India 1: 463.1874; Fyson, Fl. South Indian Hill St. 2: t. 74. 1932; Gamble, Fl. Pres. Madras 1: 102. 1957; Bhaskar, Tax. Mon. *Impatiens* of Western Ghats 236. 2012. (Plate 1.1E)

Herbs, glabrous, robust, erect, 1.5 m high, sparingly branched. Leaves alternate, elliptic or broadly ovate-lanceolate, shortly and abruptly acuminate, incurved, bristly serrate, 7 – 15 cm long, membranous, glaucous beneath, nerves prominent; petiole irregular 2.5 – 5 cm long. Peduncle axillary, erect, 3 – flowered; bracts very large, deciduous. Flowers creamy-white and yellow, speckled with purple. Lateral sepals broadly ovate, aristately acuminate, very large. Lip smooth, boat-shaped; spur very short, up to 2 mm long or absent. Standard orbicular, keeled on back. Wing–lobes small, incurved. Capsules ovoid to ellipsoid, attenuate at both ends; seeds globose, echinate.

Fl. and Fr.: April – Jan.

Distribution: Coimbatore, Madurai, Nilgiri, Tiger Shola, Kodaikanal, Pulneys, Anamalais, High wavies.

Impatiens chandrasekharanii Chandrab. in J. Bombay Nat. His. Soc. 75: 901. 1978 (1979); Bhaskar, Tax. Mon. *Impatiens* of Western Ghats 58. 2012. (Fig.3)

Scapigerous herbs; root stock tuberous with fibrous roots. Leaves radical; lamina suborbicular in outline, palmately 3- 9-lobed; lobes oblong- obovate or obovate, 1.5–5 x 0.3–1.5 cm, membranous, divided into lobules; lobules emarginate, palmately 3–5 nerved, petioles 1-6 cm long, scape 3- 20 cm long, erect. Flowers in racemes, 0.8- 1.2 cm across, white with pink tinged; pedicels 0.7- 2 cm long; bracts ovate-oblong. Lateral sepals ovate- oblong, subacute, 2 x1.2 mm. Lip broadly ovate, concave, 3 mm long; spur cylindric, 3 mm long. Standard sub-orbicular, 2 – 3 x 4 – 5 mm, membranous, keeled at centre. Wings 3-lobed; lobes oblong- obovate, subequal, 4 x 2 mm. Capsules ellipsoid.

Fl. and Fr.: July.

Distribution: Coimbatore, Akkamalai (Anamalais).

Impatiens chinensis L. var. ***brervicornis*** E. Barnes in J. Indian Bot. Soc. 18: 99. 1939; Vivek. *et al.*, in Hajra *et al.*, Fl. India 4: 131. 1997; Bhaskar, Tax. Mon. *Impatiens* of Western Ghats 126. 2012. *I. fasciculata* Lam., Encycl. 1: 363. 1785; Wight, Icon. Pl. Ind. Orient.: t. 748. 1844. *I. chinensis* auct. non L. 1753: Rama Rao, Fl. Pl. Travancore: 59. 1914; Gamble, Fl. Pres. Madras: 139. 1915; Manilal, Fl. Silent valley: 39. 1988; V. S. Ramach. & V. J. Nair, FI. Cannanore: 78. 1988; Nicolson *et al.*, in Regnum Veg.119: 70.1996; N.Mohanan & Sivad., Fl. Agasthyamala: 130. 2002; (Plate 1.1D).

Annual, 10–60 cm high; stems erect from a prostrate rooting base, angled, glabrescent or villous. Leaves opposite, subsessile, linear-oblong or lanceolate, acute or rounded at base, serrate at margin, 3–7 × 0.5–1.5cm, villous above, glaucous beneath, pinkish-nerved; stipules subulate, recurved, glandular. Pedicels 3–5 cm long, deflexed in fruit. Flowers axillary, solitary or in fascicles, 1.2–2.5cm across, rose or pink. Lateral sepals linear or falcate, acute with a callous point, 8 mm long. Lip cymbiform, 1.2 × 0.7cm, slender, incurved; spur to 3.5cm long. Standard orbicular, acuminate or horned. Wings orbicular or obovate, auricled at base. Capsules ellipsoid, tapering at both ends, 1.2 – 1.8cm long; seeds oblong, shining, black.

Fl. and Fr.: July – Feb.

Distribution: Madurai, Nilgiri and Salem.

Uses: The whole plant is made into a paste and applied to burns; it is taken internally with milk in gonorrhoea.

Impatiens clavicornu Turcz. in Bull. Soc. Imp. Naturalistes Moscou 32 (1): 271.1859; Bhaskar, Tax. Mon. *Impatiens* of Western Ghats 61. 2012. *I. beddomei* Hook. f., F1. Brit. India 1: 442.1874; Vivek. *et al.*, in Hajra *et al.*, Fl. India 4: 133.1997. *I. beddomei* Hook. f., Fl. Brit. India 1: 442. 1874 & in Rec. Bot. Surv. India 4: 45. 1906.

Herbs, 10- 30 cm high, scapigerous; root stock tuberous. Leaves radical, orbicular or reniform, entire or obscurely crenate- serrate, cordate at base, 2.5–10 cm across, fleshy, glabrous; nerves flabellate; petioles 2.5–8 cm long. Scape longer than leaves, up to 15 cm long. Flowers in racemes, 1.2 cm across, white with yellow patches on the wings or pink; bracts broadly ovate, obtuse subacute; pedicels slender. Lateral sepals broadly-obliquely ovate. Spur 1 cm long, incurved, tip inflated. Standard broader than long. Wings with 3 subequal, truncate, retuse lobes, dorsal auricle of wing obsure. Capsules ellipsoid, 0.8 cm long; seeds minute, granulate.

Fl. and Fr.: June – Oct.

Distribution: Endemic to Nilgiris.

Impatiens concinna Hook. f., Fl. Brit. India 1: 449. 1874; Rec. Bot. Surv. India 4 (3): 46. 1906; Gamble, Fl. Pres. Madras 141. 1915; Vivek. *et al.*, Fl. India 4: 134-135. 1997; Bhaskar, Tax. Mon. *Impatiens* of Western Ghats 130. 2012.

Small glabrous herbs, up to 30 cm tall, leaves opposite-decussate, 1.2–2 x 0.5–1 cm, subsessile, stipules setaceous or obsolete, lamina ovate or ovate cordate, typically triangular in shape, acute, sharply closely serrate, lower serratures almost setaceous, nerves strong beneath. Flowers axillary, solitary, about 8 mm across, purple; pedicels shorter than leaves, slender, sepals dimidiate, ovate-lanceolate, as long as lip; lip boat-shaped with a very short curved spur, Standard orbicular, spurred, wings bilobed, terminal clawed, dimidiate, ovate. Capsules erect, about 8 mm long, elliptic, turgid in middle, membranous; seeds few, 2-seeded, large compressed, testa black, shiny.

Fl. and Fr.: Feb. (North-East monsoon season).

Distribution: Atumalai, Ayyamalai hills in Coimbatore, Tamil Nadu at 1600 m altitude.

Impatiens cordata Wight in Madras J. Lit. Sci. Ser. 1. 5: 10. 1837; Hook. f., Fl. Brit. India 1: 462.1874 & in Rec. Bot. Surv. India 4: 48. 1906, excl. Syn. *I. leptura* Hook. f. 1875; Rama Rao, Fl. Pl. Travancore: 60. 1914; Gamble, Fl. Madras: 144. 1915; Vivek. *et al.*, in Hajra *et al.*, Fl. India 4: 135. 1997; N. Mohanan & Sivad., Fl. Agasthyamala: 131. 2002; Bhaskar, Tax. Mon. *Impatiens* of Western Ghats 238. 2012. (Plate 1.1F)

Herbs, flaccid, erect, 15 – 30 cm high; stems prostrate and rooting below, glabrous. Leaves alternate, ovate, cordate, acuminate, obscurely crenate, crenatures with minute cilia, 4–7.5 x 3–4.5 cm, membranous, nerves slender with scattered hairs on both surfaces; petiole 2.5 to 7.5 cm long. Peduncles 5–10 cm long, subterminal and axillary, 3–4 flowered; bracts broadly ovate, green. Flowers in umbels, lilac with purple centre or white with crimson near centre. Lateral sepals almost equalling standard, large, ovate, acuminate. Lip cymbiform; spur slender, incurved, 2 cm long. Wings with basal lobes small, incurved, fleshy; distal lobes almost orbicular, spreading. Capsules ellipsoid, beaked, glabrous; seeds almond shaped, 11–18 in capsule, 2.5 mm long, brown with a tuft of white hairs around the margin.

Fl. and Fr.: March – Nov.

Distribution: Endemic to Western Ghats of Kerala and Tamil Nadu; Coimbatore, Kanniyakumari, Madurai, Kakachi (Tirunelveli) Anamalais, High wavies, etc.

Impatiens crenata Beddome in Madras J. Lit. Sci. ser. 2. 20: 69.t. 7.f. 10. 1859 and Icon.t. 31. 1868 – 1874; Hook. f., Fl. Brit. India 1: 442.1876; Bhaskar, Tax. Mon. *Impatiens* of Western Ghats 62. 2012.

Herbs, scapigerous; root stock bulbous. Leaves radical, orbicular-reniform or cordate, rounded at base, crenate with a bristle on each crenature and shaggy hairs on nerves above, glabrous. Scape slender, 2 – 3 flowered at apex; pedicels short; bracts ovate- oblong. Flowers white with a tuft of blue hairs. Lateral sepals small, ovate. Lip with a short, obtuse spur. Wings deeply 2- lobed; inner surface with a dense tuft of hairs. Standard broader than long. Capsules tapering at both ends, glabrous.

Fl. and Fr.: July – Oct.

Distribution: Endemic to Western Ghats of Tamil Nadu; Coimbatore, Akka Mountains (Annamalais).

Impatiens cuspidata Wight & Arn. in Companion Bot. Mag.1: 321.1836; Wight, Icon. t. 741. 1844; Hook. f. in Rec. Bot. Surv. India 4: 47. 1906; Gamble, Fl. Madras 142.

1915; E. Barnes in J. Indian Bot. Soc. 18: 101. 1939; Vivek. *et al.*, in Hajra *et al.*, Fl. India 4: 137. 1997; Bhaskar, Tax. Mon. *Impatiens* of Western Ghats 191. 2012. *I. latifolia* sensu Hook. f., Fl. Brit. India 1: 450. 1874; (Plate 1.2A).

Shrubs, 50–100 cm high; stems and branches glaucous. Leaves opposite, alternate or whorled, elliptic- lanceolate, serrate, 5–11 x 2–4.5 cm, glandular at lower margins; nerves hairy beneath; petioles 1.5–5 cm long, glandular. Flowers axillary, solitary or in fascicles, pale pink or white; pedicels up to 8 cm long. Lateral sepals minute. Lip with 1.5–2 cm long spur. Standard 1.6 x 0.8 cm, larger than wings. Wings bilobed; basal lobe larger than distal ones. Capsules ellipsoid, cuspidate, 2.2 x 1.2 cm; seeds 7 -11.

Fl. and Fr.: Dec–June.

Distribution: Endemic to Western Ghats of Lambs Rock Nilgiri and Idukki.

Impatiens dasysperma Wight, Madras J. Lit. Sci. 5: 7, t.2.1837; Wight, Icon. P1. Ind. Orient. 3 (1): 1, t. 742. 1844; Hook.f. & Thomson, J. Proc. Linn. Soc. Bot. 4: 134. 1859; Hook. f., F1. Brit. India 1: 457.1874 & Rec. Bot. Surv. India 4: 47.1906; Gamble, F1. Pres. Madras 1: 148. 1915: Vajravelu in Nair & Henry. Fl. Tamil Nadu 1: 53. 1983; Sharma *et al.*, F1. Karnataka: 37. 1984; Saldanha, F1. Karnataka 2: 251. 1996; Vivekananthan *et al.*, in Hajra *et al.*, f1. India 4: 139. 1997; Rathakrishnan *et al.*, in Daniel, F1. Kerala 1: 538. 2005; Bhaskar, Tax. Mon. *Impatiens* of Western Ghats 193. 2012 (Plate 1.2B).

Herbs, 15–50 cm high. Stem quadrangular, glabrous to hairy at base, glabrous above, light green. Leaves simple, alternate, broadly lanceolate to ovate-lanceolate, 1.8–6.5 x 1.1–3.5 cm, attenuate at base, apiculate-crenate at margin, acute to apiculate at apex, dark green and hairy above, pale green and hairy on nerves below; petioles with two pairs of glands (rarely absent), 2–3.5 cm long, glabrous. Flowers 2 in each axils, 1.5–2 cm across, pink; bracts linear, 1.5mm long, concave, acute at apex; pedicels 2.5–4 cm long, glabrous, light green, deflexed in fruits. Lateral sepals ovate, 2.5 x 1.5 mm, acute at apex, glabrous, light green, 3-nerved (2 distinct, 1 obscure). Standard petal cordate, 8–10 x 6–8 mm, obcorcade at apex, dorsally hairy, pink; costa mucronate, 1 mm long; wing petals 1–1.5 x 0.8–1 cm, unequally 2-lobed; basal obovate, 6–8 x 4–6 mm, 2-lobulate; distal lobe broadly oblanceolate, 8–10 x 4–6 mm, rounded at apex; distal oracle absent; wing petals slightly protruding into a spur at base; lip conical, 7 x 5 mm, 4 mm deep, acuminate at apex, sparsely hairy light pink; spur tubular, 2–2.5 cm long, rounded at apex, glabrous to sparsely hairy. Column 4 x 1.5 mm; filaments 2.5 x 1 mm; anthers 1 x 1 mm. Ovary ellipsoid, glabrous. Capsules asymmetrically ellipsoid, 1–1.3 x 0.3–0.6 cm, glabrous; pedicels 2–4.3 cm long; seeds ovoid to globous, 1.5 mm across, compressed, papillate, hairy, dark brown.

Fl. and Fr.: June – Dec.

Distribution: Sholaiyar, Courtallum, High wavies, Kodaikanal, Bodi metu.

Impatiens debilis Turcz. in Bull. Soc. Imp. Nat. Mosc. 32. 1: 272. 1859; Hook. f. in Hook. Ic. Pl. t. 2911.1910; Bhaskar, Tax. Mon. *Impatiens* of Western Ghats 132. 2012.

Plate 1.1

A: Impatiens acaulis **Arn.;** ***B: Impatiens auriculata*** **Wight;** ***C: Impatiens balsamina*** **L.;** ***D: Impatiens chinensis*** **L.;** ***E: Impatiens campanulata*** **Wight;** ***F: Impatiens cordata*** **Wight.**

Plate 1.2

***A: Impatiens cuspidata* Wight & Arn.; *B: Impatiens dasysperma* Wight; *C: Impatiens diversifolia* B.Heyne ex Wight & Arn.; *D: Impatiens elegans* Beddome; *E: Impatiens floribunda* Wight; *F: Impatiens flaccida* Arn.**

Annual herbs, 20–30 cm tall, erect, glabrous, stem branched from the base, smaller branches slender, erect. Leaves 2–3 cm long, opposite, sessile, rigid, elliptic-lanceolate, acuminate, remotely serrulate or spinulate, rarely ciliate, base acute or obtuse, midrib underneath prominent, lateral veins obscure, intra-petiolar glands 0. Flowers small, 5–6 mm across, white, solitary or 2 in higher axils, pedicels in flower 4–5 mm long, 5–10 mm long in fruits; sepals 2, narrowly linear-acuminate, 3–4 mm long, 1-nerved; lip cymbiform, ovate, acuminate, 3–5 mm long, spur equal to lip or shorter, straight or incurved, mouth horizontal, long acuminate. Standard orbicular, cucullate, obtuse, cuspidate, 3 mm broad, multicostate, border puberulous, wing petals bilobed, subsessile, 5–7 mm long, nerves linear, dorsal auricle minute, opposite to basal lobe. Filaments small, filiform, anthers didynamous. Ovary falcate, cylindrical, obtuse. Capsules 8–10 mm long, scarcely stipitate, bulged in the middle, long acute, apex rostrate, puberulous, 1–2 seeded. Seeds orbicular, compressed, 3 mm diameter, smooth, black and shining.

Fl. and Fr.: Sep. – Nov.

Distribution: Endemic to Nilgiris.

Impatiens denisonii Beddome in Madras J. Lit. Sci. Ser. 3. 1: 41.1864 & Icon.t.151.1868–1874; Hook. F., F1. Brit. India 1: 444. 1874 & in Rec. Bot. Surv. India 4: 45. 1906; Rama Rao, Fl. Pl. Travancore: 59. 1914; Gamble, Fl. Pres. Madras: 139. 1915; Manilal, Fl. Silent Valley: 40. 1988; Vivek. *et al.*, in Hajra *et al.*, Fl. India 4: 140. 1997; Bhaskar, Tax. Mon. *Impatiens* of Western Ghats 65. 2012.

Herbs, scapigerous. Leaves radical, ovate-cordate, base deeply 2-lobed, sinus broad, obtuse or acute at apex, bristly crenate, 7–15 x 7.5–10 cm, hairy above, glabrous below; petioles shorter than leaves. Scapes twice to four times as long as leaves. Flowers numerous, 2.5 cm long, pink raceme, pedicels 2.5–3.5 cm long. Lateral sepals ovate. Lip broadly ovate with 3 broad spreading lobes with dense tuft of petalloid hairs above conjuction of lobes; spurred. Wings with filiforms auricle enclosed in long recurved spur. Capsules glabrous, seeds numerous with spiral hairs.

Fl. and Fr.: July – Nov.

Distribution: Endemic to Nilgiris, in Sisparah Ghat above 1000 m.

Impatiens disotis Hook. f. in Rec. Bot. Surv. India 4: 48. 1906; Gamble, F1. Pres. Madras 144.1915; Bhaskar, Tax. Mon. *Impatiens* of Western Ghats 239. 2012.

Succulent erect herbs, glabrous; stem thick, with number of leaf scars at base, furuncles (raphide sacs) prominent all over. Leaves alternate, spirally arranged, petioles 1.5–7 cm, with a pair of globular glands just below lamina, glabrouse, lamina 4–13 x 3–8 cm, ovate-acute or rounded, crenate, glabrous. Flowers sub-umbellate, 3 cm across, campanulate, white speckled with deep red inside; peduncles axillary, shorter than leaves; pedicels 2–5, up to 4.5 cm long; sepals ovate, aristately acuminate, 9 mm, greenish yellow; lips yellow, funnel-shaped, mucronate, spurs short, stiff, straight or incurved, 6–13 mm long, standard orbicular, emarginated, 10 mm dia. yellow, at midrib region without, red on the inner side with a white margin; wings with small basal lobe, mucronate at tip, terminal much larger and bilobed at apex, emarginate and mucronate, 2 cm long, red on the inner side, inner side of upper lobe white along the margin, the lower lobe white with an yellow patch near the throat. Capsules 2 cm long, tapering at both ends; seeds 7, 2.5–3mm long, brownish, surface appears smooth in the ocular observation as the hairs are sparse and short.

Fl. and Fr.: Dec. – Jan.

Distribution: Coimbatore, Tirunelveli, Mahendragiri in Kanyakumari.

Impatiens diversifolia B.Heyne ex Wight & Arn., Prodr. F1. Ind. Orient.: 139. 1834; Hook.f. & Thomson, J. Proc. Linn. Soc. Bot. 4: 121. 1859; Hook. f., F1. Brit. India 1: 446.1874 and Rec. Bot. Surv. India 4: 46.1906; Gamble, F1. Pres Madras 1: 140. 1915; Blatter, J. Bombay Nat. Hist. Soc. 33: 310.1933; Ramamoorthy in Saldanha & Nicolson, F1. Hassan District: 402. 1978; Vajravelu in Nair & Henry, F1.Tamil Nadu 1: 53. 1983; Sharma *et al.*, F1. Karnataka: 37. 1984; Alemida, F1. Maharashtra 1: 192.1996; Saldanha, F1. Karnataka 2: 252, t. 33. 1996; Vivekanathan *et al.*, in Hajra *et al.*, F1. India 4: 141. 1997; Mudaliar & Prasad in Singh & Karthikeyan, F1. Maharashtra State Dicotyl. 1: 463. 2000; Bhat, F1. Udupi District: 88, t.22. 2003; Rathakrishnan *et al.*, in Daniel, F1. Kerala 1: 539. 2005; Bhaskar Tax. Mon. *Impatiens* of Western Ghats 134. 2012. (Plate 1.2C)

Herb, 20–40 cm high, rooting at lower nodes. Stem quadrangular, glabrous red. Leaves opposite-decussate, linear-lanceolate, 3–4.5 x 1–1.5 cm, rounded to truncate at base, distantly crenate at margins, acute to acuminate at apex, pubescent above, glabrous below; midrib distinct; lateral nerves obscure; petioles 2 mm long, decurrent on stem and terminating in glands, glabrous. Flowers 3 in each axils, 2.1 cm across, pink; bracts 1.5 x 0.5 mm, acute at apex, glabrous; pedicels 2.5–3 cm long, with two rows of hairs, deflexed in fruits. Lateral sepals 2, linear to linear-oblanceolate, 6 x 1 mm. acuminate at apex, glabrous, 3-nerved. Standard sub-orbicular, 6.5 x 5 mm, deeply concave, mucronate at apex, glabrous; wing petals 1.2–1.5 x 0.7 -1 cm, bilobed; basal lobe linear to lanceolate, much smaller than distal lobe, 3 x 1mm, acute to mucronate at apex; distal lobe broadly obovate, 1.0 x 1 cm, elevated at base, shortly stipitate, obtuse at apex; dorsal auricle 1 mm, rounded; lip saccate, 9–13 x 3–4 mm, 3–4 mm deep, acuminate at apex, glabrous; spur tubular, 1.5–2.2 cm long, tapering from base to apex, coiled in to a semi-circular ring, straight or curved upwards, notched or rounded at apex. Column 4 x 2 mm, slightly curved; filaments 5, 3 x 1 mm, light pink; anthers 1 x 1mm. Pistil 3.5 mm long; ovary ellipsoid, 3 x 1 mm, glabrous. Capsules asymmetrically ellipsoid, 1–1.3 x 0.4–0.6 cm, glabrous; pedicels 3–3.4 cm long; seeds ovoid 1.5 x 1 mm, shining, glabrous, black to brown.

Fl. and Fr.: July–Oct.

Distribution: Naduvattam (Nilgiris), Kanyakumari Districts.

Impatiens elegans Beddome in Madras J. Lit, Sci. Ser. 2. 78: 67. 1859 & Icon. 29. t. 145.1868 – 1874; Hook. f., F1. Brit. India 1: 465. 1875. & in Rec. Bot. Surv. India 4: 49. 1906; Rama Rao, Fl. Pl. Travancore 60. 1914; Gamble, Fl. Pres. Madras: 145. 1915; Vivek. *et al.*, in Hajra *et al.*, Fl. India 4: 144. 1997; Bhaskar Tax. Mon. *Impatiens* of Western Ghats 240. 2012 (Plate 1.2D)

Herbs, erect, up to 40 cm high; stems glabrous, simple or branched, often rooting at lower nodes. Leaves alternate, ovate or ovate-cordate, rounded or cordate at base, acuminate, crenate with incurved bristles, 4–12 x 1–4 cm, hairy on nerves above, pale glabrous beneath; petioles up to 5 cm long. Peduncle slender, axillary and terminal. Flowers 3–5, umbelled or shortly racemed, 2.5 cm across, pale rosy-white with dark purple eye; bracts ovate-lanceolate, acuminate, recurved. Lateral sepals ovate with green ridge ending in an acumen. Lip cymbiform, very small, spur absent, standard broad, ovate, flat with a green ridge on its back. Capsules ellipsoid, beaked, turgid in middle, 1.2 cm long, glabrous; seeds subglobose, hairy.

Fl. and Fr.: July – March.

Distribution: Endemic to Anamalais, Kambam valley.

Impatiens fasciculata Lam. Encycl. Method. 1: 363. 1785; Wight. & Arn. Prodr. Fl. Ind. Orient 138. 1834; Wight, Ic. t. 748. 1844; Hooker, W. J. in Curtis Bot. Mag. 8: p. and t. 4631; Bhaskar. Tax. Mon. *Impatiens* of Western Ghats 136. 2012. *Balsamina fasciculata* DC. Prodr. 1: 686. 1824. *Impatiens vivekananthanii* J.Dessai & Janarth., Rheedea 21 (1): 58.2011. excl. *I. heterophylla* Wall. in Roxb. Fl. Ind. ed. Carey 2: 458. 1824; et Cat. 4748. 1831.

Herb, 60–70 cm high. Stem erect, quadrangular, succulent, subsucculent at base, slightly swollen at nodes, glabrous, pale green, internodes short at base, long at apex. Leaves opposite-decussate, sessile or subsessile, linear-oblanceolate to spathulate, 3.5–7 × 0.5–1.2 cm, truncate at base, distantly serrate at margins, acute to apiculate at apex, coriaceous, scabrid above, glabrous below, obscurely nerved; petioles decurrent on stem and terminating in glands. Flowers 2 or 3 per axil, 2–3 cm across, pink with maroon throat; bracts triangular, 1.5 × 1 mm; pedicels subterete, 3–4.5 cm long, glabrous to hairy with row of hairs. Lateral sepals linear to linear to linear-oblanceolate, 0.7–1.2 × 0.1–0.2 cm, acuminate at apex, pinkish white, 3-nerved; nerves minutely hairy, pinkish red. Standard petal reniform, 0.7–0.8 × 0.9–1.3 cm, concave, glabrous, pink to pinkish white, dorsally keeled; keel mucronate at apex; mucro 1mm long; wing petals 1.7–2.5 × 1–1.7 cm, 2-lobed; base lobe ovate, 4–7 × 2–3 mm, acute to rounded at apex; distal lobe much larger than basal lobe, 1.1–1.9 × 0.9–1.8 cm, obtuse at apex, notched towards inner side much below apex; lip conical, 9–14 ×5–6 mm, 5–7 mm deep, glabrous to sparsely minutely hairy, dark pink-nerved, pink; spur tubular, 2–4 cm long, notched at apex, glabrous to sparsely minutely hairy, thick, pinkish white, column 5–6 ×2–3 mm; filaments 4–5 × 1–15 mm, pink to white; anthers 1 × 0.75 mm, yellow, pistil 5 mm long; ovary asymmetrically obolong, 4.5 × 1 mm, glabrous.

Fl. and Fr: Sept – Oct.

Distribution: Nilgiris.

Impatiens flaccida Arn. in Companion Bot. Mag. 1: 322. 1836; Hook. in Bot. Mag. 87: t. 5276. 1861; Hook.f., Fl. Brit. India 1: 457. 1874, excl.syn. *I. bipartita* Arn and *I. floribunda* Wight. (Plate 1.2F)

Erect or decumbent herbs up to 50 cm high; stems slender, sparingly branched in lower half, green or flushed with pink, rooting at lower nodes. Leaves alternate, ovate-lanceolate or lanceolate-elliptic, crenate, 3–8 x 1.2–3.5 cm, sparsely pubescent; petioles 1.2–3.5 cm long with stipitate glands. Flowers solitary or in fascicles of 2–3, pale to deep pink; pedicles 2–3 cm long; bracts inconspicuous. Lateral sepals ovate,

Figure 1.5: *Impatiens cuspidata* Wight & Arn.

Figure 1.6: *Impatiens dasysperma* Wight.

Figure 1.7: *Impatiens flaccida* Arn.

Figure 1.8: *Impatiens goughii* Wight

Lip 8–10 mm long; spur 1.5–2.5 cm long, filiform or slender, incurved. Standard obcordate with a keel on dorsal side terminating into on acute point. Wings bilobed. Capsules 1.5 cm long, ellipsoid, glabrous; seeds globose, glabrous or papillose.

Fl. and Fr.: July – Dec.

Distribution: Sivagiri Hills of Tirunnevelly.

Impatiens floribunda Wight in Madras J. Lit. Sci. Ser. 1. 5: 7. 1837; Hook. f. in Rec. Bot. Surv. India 4: 47.1906; Gamble, Fl. Pres. Madras: 142. 1915, excl. syn.; Bhaskar, Tax. Mon. *Impatiens* of Western Ghats 197. 2012. *I. flaccida* sensu Hook. f., F1. Brit. India 1: 457.1874, p.p. non Arn. 1835. (Plate 1.2E)

Erect or decumbent herbs, up to 70 cm high; stems branched, green. Leaves alternate or opposite, rarely whorled, elliptic-lanceolate, caudate-acuminate at apex, 2–10 x 12–14 cm, serrate or crenate- serrate along margins, apiculate from apex of teeth, glaucous beneath; petioles 1.5–3.5 cm long with stipulate glands. Flowers solitary or in fascicles, purple pedicels up to 2.5 cm long. Lateral sepals small, ovate. Lip cymbiform; spur slender, straight, wings bilobed; seeds globose.

Fl. and Fr.: December.

Distribution: Shevagherry Hills (Tirunelveli), Nilgiris, Mahendra hill in Tamil Nadu.

Impatiens fruticosa Leschen. ex DC., Prodr. 1: 687. 1824; Wight, Icon. Pl. Ind. Orient.: t. 966. 1845; Hook. f., F1; Brit. India 1: 459, 1874 & in Rec. Bot. Surv. India 4: 48. 1906; Rama Rao, Fl. Pl. Travancore: 60. 1914; Gamble, Fl. Pres. Madras: 143. 1915; Vivek. *et al.*, in Hajra *et al.*, Fl. India 4: 149.1997; Bhaskar, Tax. Mon. *Impatiens* of Western Ghats 222. 2012. (Plate 1.3A)

Shrubs, perennial, erect, 2.5–3 m high, much branched; branches glabrous, glaucous, nodose at scars. Leaves alternate, ovate or elliptic, attenuate at base, acuminate at apex, crenate along margins, ciliate and glandular towards base, 7–13 x 3.5–5.5 cm, pubescent on both surfaces, sometimes silky beneath, many nerved, nerves stout; petiole 2–7 cm long, stout, often villous, naked or glandular. Flowers axillary, paired or solitary, 4 cm broad, pink; bracts linear-subulate; pedicles often much larger and exceeding leaves. Lateral sepals ovate, acuminate, 1.5 x 0.6 cm, tomentose. Lip trumpet-shaped, acuminate, 1 cm long, spurred; spur curved, 2.5 cm long. Capsules erect, ellipsoid, narrowed at both ends, beaked, 2.5 cm long, glabrous; seeds numerous.

Fl. and Fr.: May – Nov.

Distribution: Naduvattam and Kundah (Nilgiris), Pulney hills, Kolli hills.

Impatiens gardneriana Wight, Icon, Pl. Ind. Orient. 3 (4): 1 t. 1050. 1846; Hook.f. & Thomson, J. Proc. Linn. Soc. Bot. 4: 121. 1859; Hook.f., F1. Brit. India 1: 445. 1874 and Rec. Bot. Surv. India 4: 46. 1906; Gamble, F1. Pres. Madras 1: 141. 1915, Ramamoorthy in Saldanha & Nicolson, F1. Hassan district 402. F. 78C. 1978; Yoganarasimhan *et al*, Fl. Chickmaglur District: 61. 1982; Vajravelu in Nair & Henry, F1. Tamil Nadu 1: 53. 1983; Sharma *et al.*, F1. Karnataka 38. 1984; Murthy & Yoganarasimhan, F1. Coorg: 87. 1990; Saldanha, F1. Karnataka 2: 257. 1996;

Vivekananthan *et al.*, in Hajra *et al.*, F1. India 4: 150. 1997; Ramaswamy *et al.*, F1. Shimoga District 107. 2001; Rathakrishnan *et al.*, in Daniel, F1. Kerala 1: 541. 2005; Bhaskar, Tax. Mon. *Impatiens* of Western Ghats 116. 2012 (Plate 1.3B).

Herb, 25–60 cm high. Stem quadrangular, glabrous. Leaves opposite at base, ternate at apex, ovate, elliptic to lanceolate, 3–8 x 2–8 cm, cunate to attenuate at base, apiculate-serrate at margins, acuminate at apex, hairy above, and only on nerves below; basal leaves petiolate; petioles 1–3 cm long, glabrous; upper leaves subsessile; petioles up to 3 mm long, hairy. Flowers solitary or binate, axillary, 2 cm across, dark pink; bracts triangular, 0.5 mm, acuminate at apex, scaly; pedicels 2.5–4 cm long, hairy. Lateral sepals ovate, 6 x 2 mm, concave, caudate at apex, hairy dorsally, 5-nerved, pale green. Standard petal broadly obovate, 8 x 8 mm, apiculate at apex, dorsally keeled and hairy, pink; wing petals 1.2 x 1 cm, bilobed; basal lobe oblong-lanceolate to obovate, 8 x 4.5 mm, obcordate at apex; distal lobe oblong-oblanceolate to oblanceolate, 8 x 5 mm, obcordate at apex; lip boat-shaped, 8 x 4 mm, 2.5 mm deep, caudate at apex, hairy only in middle; spur 2–3 cm long, notched to rounded at apex, sparsely hairy. Column 3 x 2 mm; filaments 3 x 0.75 mm, pink; anthers 0.75 x 0.5 mm, pink. Pistil 2.5 x 0.75 mm; ovary oblong-lanceolate, glabrous. Capsules oblong, 1.3 x 0.3 cm, curved, glabrous; pedicels 3–4.5 cm long; seeds ovoid, 2.5 x 1.5 mm, densely hairy; apical hairs spirally coiled and without thickening; basal ones not coiled but with reticulate thickenings.

Fl. and Fr.: July – Dec.

Distribution: Nilgiri region.

Impatiens goughii Wight, III. Ind. Bot. 1: 160. 1840 & Icon. t.1603. 1850; Hook. f., F1. Brit. India 1: 452. 1874; Bhaskar, Tax. Mon. *Impatiens* of Western Ghats 241. 2012. *I. microtheca* Hook.f. in Hooker's Icon. Pl. 30: t. 2910.1910. (Plate 1.3C).

Annual herbs, 5–30 cm high; stems simple or branched. Leaves opposite, ovate, subacute at apex, crenate-serrulate along margins, 1.5–3.5 x 0.8–1.5cm, glabrous, rarely hairy below; petioles up to 1.5 cm long. Flowers in terminal subumbels, pale pink; peduncle 2–5 cm long; pedicels up to 1.5 cm long; bracts persistent, minute, lanceolate. Lateral sepals falcate, 2 mm long. Lip cymbiform, spur shorter than wings, straight. Standard obovate, mucronate at the retuse apex. Wings bilobed; basal lobes oblong, rounded at apex; distal lobes longer than basal ones; dorsal auricles produced into the spur. Capsules ellipsoid, 3–5 mm long, 6–12 seeded; seeds almond-shaped, chestnut-coloured, granular with hairs.

Fl. and Fr: May – Dec.

Distribution: Nilgiris, Pulneys, Anamalais, Kannikatti, Naterikal, Kakachi, Mancholai, Shivagiri, Patchkumatchi hills (Tirunelveli), Coonur (Sim's park) and Ooty Botanic Gardens.

Impatiens grandis Heyne ex Wall. in Roxb., Fl. Ind. 2: 464.1824; Hook.f., Fl. Brit. India 1: 463. 1874; Bhaskar, Tax. Mon. *Impatiens* of Western Ghats 246. 2012. *I. hookeriana* Arn. in Hook. Companion Bot. Mag 1: 324. 1834; Hook. f., Fl. Brit. India 1: 463. 1874. (Plate 1.3D)

Herbs, robust, often shrubby, erect, 1–2.75 m high; branches terete, glabrous with prominent leaf- scars. Leaves confined to apex, incurved, bristly serrate, 8–21 x 3–8 cm, thick, prominently nerved; petiole 2–9.5 cm long with 2 pedicelled glands near apex. Flowers large in subumbellate clusters of 3–4 cm on long peduncles, pale rose or white with purple marking. Lateral sepals obovate. Lip saccate, obtuse or prolonged into very stout 4 cm long, conical straight spur. Standard orbicular, cuneate, notched. Wings bilobed, spreading; basal lobes oblong-obovate; distal lobe larger than the basal ones, rounded, notched. Capsules ellipsoid, obscurely 4 -5 angled, glabrous, 1.5–2 cm long; seeds glabrous.

Fl. and Fr.: Sept. – March.

Distribution: Kannikatti, Naterikal, Kakachi, Mancholai, Shivagiri, Patchakumatchi hills and Poopari. It is often cultivated in Ooty and Coonur Botanic Gardens.

Impatiens henslowiana Arn. in Companion Bot. Mag. 1: 322.1836; Hook.f., Fl. Brit. India 1: 458. 1874; Bhaskar, Tax. Mon. *Impatiens* of Western Ghats 223. 2012. *I. albida* Wight in Madras J. Lit. Sci. Ser. 1. 5: 7, t. 1. 1837 & Icon. t. 743. 1844; (Plate 1.3E)

Suffruticose shrubs, up to 1.5 m high; stems simple or branched, succulent when young, covered with scars of fallen leaves. Leaves alternate or spiral, elliptic-lanceolate, serrate, 3.5–14 x 1.2–5 cm, subglabrous to pilose. Flowers axillary, solitary or in fascicles, white sometime pale purple; pedicels slender, 3–10 cm long, glabrous or pilose, lateral sepal ovate, acuminate. Lip navicular; spur 4.5–6 cm long, slender, incurved, pubescent up to middle. Standard sub-orbicular, emarginated with an erect spur on back. Wings bilobed. Capsules large, fusiform, 3–4.5 cm long, glabrous or pilose; seeds almond- shaped, 3mm long, rusty- brown, furrowed.

Fl. and Fr.: March – April.

Distribution: Endemic to Western Ghats of Kerala, Tamil Nadu Kanniyakumari, Madurai, Nilgiri, Tirunelveli, Kodaikanal.

Impatiens herbicola Hook. f. in Bull. Misc. Inform. 1911: 354. 1911; Gamble, F1. Pres. Madras 141.1915; Bhaskar, Tax. Mon. *Impatiens* of Western Ghats 139. 2012 (Plate 1.3F).

Succulent herbs, 20–40 cm high. Leaves opposite, sessile or shortly petioled, linear-lanceolate, rounded or cordate at base, acuminate at apex, serrate, 3–6 x 0.4–0.6 cm. Flowers white or pink, 5 cm across, axillary, solitary or in fascicles; pedicels 5–15 mm long, 2–3 cm long in fruit, erect or deflexed. Lateral sepals linear, acuminate, 4–5 mm long. Lip cymbiform, 4–5 mm long; spur absent or provided with a punch-like structure at base. Standard orbicular, 4 mm across, keeled at dorsal side, apiculate. Wings 7–8 mm long, clawed, bilobed; basal lobes somewhat oblong; distal lobes rotund or semilunate; dorsal auricle prominent. Capsules ovoid, gibbous, 7–10 mm long, beaked, few-seeded; seeds globose, compressed, 2- 5 mm, black, shining.

Fl. and Fr.: Aug. – Oct.

Distribution: Endemic to Western Ghats of Tamil Nadu and Kerala. Coimbatore, Cumbam valley.

Impatiens hookeriana Heyne ex Wall. in Roxb., Fl. Ind. 2: 464.1824; Hook.f., Fl. Brit. India 1: 463. 1874. *I. hookeriana* Arn. in Hook. Companion Bot. Mag 1: 324. 1834; Hook. f., Fl. Brit. India 1: 463. 1874. (Plate 1.3H)

Herbs, robust, often shrubby, erect, 1–2.75 m high; branches terete, glabrous with prominent leaf-scars. Leaves confined to apex, incurved, bristly serrate, 8–21 x 3–8 cm, thick, prominently nerved; petiole 2–9.5 cm long with 2 pedicelled glands near apex. Flowers large in subumbellate clusters of 3–4 cm on long peduncles, pale rose or white with purple marking. Lateral sepals obovate. Lip saccate, obtuse or prolonged into very stout 4 cm long, conical straight spur. Standard orbicular, cuneate, notched. Wings bilobed, spreading; basal lobes oblong-obovate; distal lobe larger than the basal ones, rounded, notched. Capsules ellipsoid, obscurely 4–5 angled, glabrous, 1.5–2 cm long; seeds glabrous.

Fl. and Fr.: Sept. – March.

Distribution: Kannikatti, Naterikal, Kakachi, Mancholai, Shivagiri, Patchakumatchi hills and Poopari. It is often cultivated in Ooty and Coonur Botanic Gardens.

Note: *Impatiens hookeriana* was described by Heyne in 1834 by its special morphological characters. But, recently the species has been merged with *I. grandis* due to its few similar characters. The present study revealed that, *I. hookeriana* have a distinct character and it has been separated from *I. grandis* due to its several dissimilar sexual charecters. *I. hookeriana* has been revalidated and separated from the *I. grandis* and similar results were submitted for the recommendation.

Impatiens inconspicua Benth. ex Wight & Arn., Prodr. F1. Ind. Orientt. 139. 1834; Wight, Icon.t. 1970. 1845; Hook. f., F1. Brit. India 1: 447. 1874; Bhaskar, Tax. Mon. *Impatiens* of Western Ghats 141. 2012. *I. inconspicua* Benth. ex Wight & Arn. var. *filiformis* (Wight & Arn.) Hook. f., F1. Brit. India 1: 448. 1874. *I. inconspicua* Benth. ex. Wight & Arn. var. *ramosissima* (Dalz.) Hook.f., F1. Brit. India 1: 448. 1874. *I. pusilla* var. *nematostachys* Hook. f. in Bull. Misc. Inform. 1910: 294. 1910.

Small annuals, 10–32 cm high, much branched from base; stems erect, rigid, filiform. Leaves variable, ovate or ovate-lanceolate, rounded or acute at base, serrate, 2–5 x 0.5–1.5 cm, coriaceous, glabrous above, glaucous beneath; upper usually petioled, lower subsessile; stipules setaceous or absent. Flowers small, usually 2–3, fascicled, pink; pedicels with a line of pubescence, axillary, shorter than leaves. Lateral sepals linear-subulate. Lip saccate, acuminate; spur absent or very minute, wings with distal lobes broadly rhomboid, obovate, clawed; basal lobes small. Capsules 1 cm long, turgid in middle, acute at both ends; seeds few, globose, little compressed, black, shining.

Fl. and Fr.: July–November

Distribution: Pykara, Gudalur, Kundahs, Doddabetta, Nilgiris

Impatiens jerdoniae Wight in Madras J. Lit. Sci. Ser. 1.5: 8. 1837 and Icon. 4 (4): 15, t. 1602. 1850; Hook. f., F1. Brit. India 1: 460. 1874. excl. var. *parasitica* (Beddome) Hook. f. in Rec. Bot. Surv. India 4: 45. 1906; Rama Rao, Fl. Pl. Travancore: 60. 1914; Gamble, Fl. Madras 1: 139 (90). 1915; Manilal, Fl. Silent Valley: 40. 1988;

Figure 1.9: ***Impatiens henslowiana*** **Arn.**

Figure 1.10: ***Impatiens hookeriana*** **Heyne ex Wallich.**

Figure 1.11: ***Impatiens jerdoniae*** **Wight.**

Figure 1.12: ***Impatiens laticornis*** **Fischer.**

Figure 1.13: ***Impatiens lawsonii*** **Hook. f.**

Vivek. *et al.*, in Hajra *et al.*, Fl. India 4: 159. 1997; Bhaskar, Tax. Mon. *Impatiens* of Western Ghats 100. 2012. (Plate 1.3G)

Herbs, succulent, epiphytic on tree trunks rarely on rocks; stems tumoid, glabrous, green or purple. Leaves few, alternate, elliptic, acute, serrate, 5–8 cm long, deep green; nerves few with glandular cilia at base and on top of stout petiole. Flowers solitary or binate, 3.5–5 cm long; peduncles 2.5–8 cm long, green, stout. Lateral sepals lanceolate, acuminate, green. Lip rounded, overlapping, laterally compressed, scarlet, spur ventricose, wrinkled, saccate, very large. Standard yellow, orbicular, apiculate, narrowly keeled at back, wings bilobed; lobes rounded, overlapping, concealed under lip. Capsules ellipsoid.

Fl. and Fr.: Aug.–Dec.

Distribution: Sispara of Nilgiris.

Impatiens laticornis C. Fischer in Bull. Misc. Inform. 1930: 154. 1930 & Gamble, Fl. Pres. Madras 1: 101.1957; Bhaskar. Tax. Mon. *Impatiens* of Western Ghats 67. 2012.

Herbs, erect, acaulescent, stoloniferous; rhizome globose. Leaves 1–2, erect, thick, suborbicular or ovate, cordate at base, obtuse at apex, crenate along margins, 1.5–4 cm x 1.25–2.5 cm; pilose above, glabrous below; primary nerves 1–2, ascending; petiole glabrous. Scapes solitary, 6–17 cm long, glabrous, purple, 2–10-flowered, bracts semi-amplexicaulous, concave, oblong, obtuse, 0.3–0.6 cm long; pedicels 1–1.7 cm long. Flowers 2 cm long, white with yellow orange hairs or pink with magenta hairs. Lateral sepals ovate or subcordate, obtuse, 6–8 mm long. Spur or lip curved, bullose, 2 mm across, wing 3-lobed, pilose; distal lobes ovate-oblong, rotundate, 4–5 mm long; middle lobes sub-orbicular, 8–9 mm across; basal lobes oblong, obtuse, 9–13 mm long. Capsules ellipsoid.

Fl. and Fr.: Sept.

Distribution: Endemic to Nilgiri, in Kundahs on wet rocks in deep shade and also on tree trunks.

Impatiens latifolia L., Sp. Pl.: 937. 1753; Wight & Arn. Prodr. F1. Ind. Orient.: 138.1834; Hook.f. & Thomson, J. Proc. Linn. Soc. Bot. 4: 124.1859; Dalzell & Gibson, Bombay F1.: 44. 1861; Hook.f., F1. Brit. India 1: 450. 1874, excl. syn. *I. cuspidata* Arn. 1835 and Rec. Bot. Surv. India 4: 47. 1906; Cooke, F1. Bombay 1: 174.1901; Gamble, F1. Pres. Madras 1: 141. 1915; Blatter, J. Bombay Nat. Hist. Soc. 33. 313. 1933; Sharma *et al.*, F1. Karnataka 38. 1984; Almeida, F1. Maharashtra 1: 193. 1996; Saldanha, F1. Karnataka 2: 254. 1996; Vivekananthan *et al.*, in Hajra *et al.*, F1. India 4: 169. 1997; Mudaliar & Prasad in Singh & Karthikeyan, F1. Maharashtra State Dicotyl. 1: 455. 2000; Rathakrishnan *et al.*, in Daniel, F1. Kerala 1: 546. 2005; Bhaskar, Tax. Mon. *Impatiens* of Western Ghats 198. 2012 (Plate 1.4 A)

Shrub, up to 1–1.5 m high. Stem branched, woody at base, glabrous. Leaves alternate, subopposite or whorled, often crowded at apex, ovate to ovate- lanceolate, 5–11 x 2–3.5 cm, crenate–serrate at margins, glandular at lower margins, acuminate at apex, membranous; nerves 7–10 pairs; petiole 1–5 cm long. Flowers solitary, binate

Plate 1.3

***A: Impatiens fruticosa* Leschen. ex DC.; *B: Impatiens gardneriana* Wight; *C: Impatiens goughii* Wight; *D: Impatiens grandis* Heyne ex Wallich; *E: Impatiens henslowiana* Arn.; *F: Impatiens herbicola* Hook. f.; *G: Impatiens jerdoniae* Wight; *H: Impatiens hookeriana* Heyne ex Wallich.**

Plate 1.4

A: Impatiens latifolia L.; *B: Impatiens leptura* Hook. f.; *C: Impatiens leschenaultii* Wallich ex Wight and Arn.; *D: Impatiens maculata* Wight; *E: Impatiens minor* (DC.) Bennet; *F: Impatiens munnarensis* Barnes.

or in fascicles, pink to violet, purple or white; pedicels 2–6 cm long, deflexed in fruits. Lateral sepals ovate, 4–5 mm long, cuspidate at apex. Standard petal orbicular, bilobed at apex, dorsally keeled; wing petals 1.5–2 cm long, unequally bilobed; basal lobe smaller than distal lobe, subrounded; distal lobe ovate or dolaniform; dorsal auricle rounded; lip cymbiform, acuminate at apex; spur incurved or straight, 1.2–2 cm long, white or greenish. Capsules ellipsoid, 1–2.4 cm long, glabrous; seeds few, subglobose or obovoid, sub-compressed, 4 mm across, rugose or papillose.

Fl. and Fr.: May – Dec.

Distribution: Coonur and Doddabetta in Nilgiris, Pulney hills in Tamil Nadu.

Impatiens lawsonii Hook. f. in Rec. Bot. Surv. India 4: 45.1906; Bhaskar Tax. Mon. *Impatiens* of Western Ghats 68. 2012.

. Epiphytes, scapigerous, tuberous herbs, up to 15 cm high. Leaves radical, ovate, cuneate, obtuse or rounded, sparsely hairy above; petioles up to 2 cm long. Flowers white, 1 cm across, 4–6 flowers in straight scapes, bracteate; bracts 1.5 mm long, persistant; pedicels filiform, in fruit 2 mm long, glabrous. Lateral sepals ovate, acuminate, 3–4 nerved. Lip funnel-shaped; spur conical. Standard hood like, enclosing stamens. Wings 3-lobed, 1.5 cm long; dorsal auricle short, spiniferous, pale pink with yellow tinge at centre, whitish outside. Capsules ovate, raised in centre, glabrous; seeds powdery, minute, hairy.

Fl. and Fr.: Aug. – Sept.

Distribution: Endemic to Nilgiris, in Kundahs on wet rocks.

Impatiens lenta Hook. f. in Hooker's Icon. P1. Ser. 4.10: t. 2933.1913; Gamble, Fl. Pres. Madras 140. 1915; Bhaskar, Tax. Mon. *Impatiens* of Western Ghats 152. 2012.

Herbs, 20 cm high, glabrous; stem flaccid, simple, erect; internodes elongate. Leaves opposite, entire or remotely denticulate, 2–3 cm long; lower ones smaller, ovate, shortly petiolate; upper sessile or subsessile, oblong or ovate-oblong, truncate or cordate at base, acute at apex; midnerve strong below, lateral nerves obscure; glands at base of petiole minute, setaceous, deflexed. Flowers up to 2 cm across, pale violet; pedicels solitary, longer than leaves in fruits. Lateral sepals 2, narrowly linear, capillary at apex, 3-nerved. Lip cymbiform, acuminate, 0.3–1 cm long; spur pink, up to 1.6 cm long. Standard small, orbicular, 4–5mm broad, cuspidate at apex, dorsally keeled. Wings bilobed, long stipitate, 1.2 cm long; basal lobes small, placed on stipe, narrowly lanceolate, erect, 3 mm long; distal lobes large, obovate or broadly axe-shaped, erect, narrow at base, rounded at apex; dorsal auricle small, opposite basal lobes. Filaments subulate; anthers minute, narrow, acuminate. Capsules shortly stipitate narrowly ellipsoid, 1–1.2 cm, compressed acutely long beaked 1–1.2 cm long, many seeded; seeds minute, subglobose, 2 mm across, shining black.

Fl. and Fr.: Aug.–Sept.

Distribution: Endemic to Nilgiri.

Impatiens leptura Hook. f., F1. Brit. India 1: 467. 1875; C.E.C. Fischer in Bull. Misc. Inform. 1934: 391–393. 1934 & Gamble, Fl. Pres. Madras, Add.: 1871. 1935; Vivek. *et al.*, in Hajra *et al.*, Fl. India 4: 173. 1997; Bhaskar Tax. Mon. *Impatiens* of Western

Ghats 153. 2012. *I. cordata* sensu Hook. f. in Rec. Bot. Surv. India 4: 48. 1906 p.p., non Wight 1837; (Plate 1.4B).

Herbs, erect, about 60 cm high, unbranched; stems angled, greenish. Leaves alternate, ovate to elliptic-lanceolate, acute or acuminate at apex, cuneate at base, crenate along margins with short cilia in sinuses, 4.5–10 x 2.5–3.0 cm, midnerve slightly prominent below, lateral nerves in 5 pairs, rising almost at right angles to mid nerve, glabrous with short and stiff hairs on nerves above, pale and subglaucous below; petioles 1- 7 cm long. Peduncle axillary, 4–6.5 cm long, glabrous, 2–3 flowered; bracts ovate, 4 mm long, glabrous, green, deciduous; pedicel 1.5–2.5 cm long, fascicled. Flowers 2.5 cm across, pinkish. Lateral sepals broadly ovate, acute, 4.5–5.5 mm long, green. Lip bowl shaped, acute, 1–1.3 cm long; spur slender, tapering, incurved. Standard broadly oblong, obtuse. Distal lobes of wings 6–7 mm long, green becoming pinkish at edges. Stamens yellow. Capsules ellipsoid, beaked, 1.5 cm long; seeds orbicular, pilose when young and glabrous when mature.

Fl. and Fr.: May – Sept.

Distribution: Endemic to Coimbatore.

Impatiens leschenaultii Wallich ex Wight & Arn., Prodr. Fl. Ind. Orient. 131. 1834; Wight, Icon. t 970.1845; Hook. f. Fl. Brit. India 1: 450. 1874; Bhaskar, Tax. Mon. *Impatiens* of Western Ghats 200. 2012. *Balsamina leschenaultii* DC., Prodr. 1: 586.1824. (Plate 1.4C)

Suffruticose shrubs, up to 1.5 m high. Leaves opposite, alternate or whorled, ovate- lanceolate, acute or acuminate, 2.5–10.5 x 1.2–4 cm, crenate, sometimes ciliate on sinus; nerves 4 or 5 pairs; petioles up to 5 cm long, glandular on upper half. Flowers solitary, binate or in fascicles, pink or white; pedicels up to 8 cm long. Lateral sepals ovate- lanceolate, 3 mm long, yellow with brownish tinge. Lip cymbiform, 5–7 mm dorsal side. Wings bilobed; basal lobes smaller than distal ones. Capsules ellipsoid, beaked, 1.8 cm long, glabrous; seeds 3–5, obovoid, 3mm long, dark purplish-brown, rugose or papillose.

Fl. and Fr.: Almost throughout the year.

Distribution: Beijrim (Kodaikanal), Anaimudi, Doddabetta (Nilgiris), Pumbarai (Pulney).

Impatiens levingei Gamble ex Hook. f. in Rec. Surv. India 4 (3): 45. 1906; Gamble, Fl. Pres. Madras 138. 1915; Fyson, Fl. South Ind. Hill St. 1: 83. t. 56. 1932; Sebast. in Bull. Bot. Surv. India 2 (1-2): 1-7. 1960; Bhaskar, Tax. Mon. *Impatiens* of Western Ghats 70. 2012.

Perennial ground herbs, tuberous; tubers moniliform, stoloniferous; roots fibrous. Leaves radical, fleshy or leathery, orbicular to reniform, margin entire or distantly short toothed, nerves flabellate, glabrouse, leaf base cordate, clasping or overlapping, petiole 5 cm long. Flowers in racemose scapes, deep magenta or carmine red coloured flowers and cylindrical spur. Spur tip abruptly curved upwards, seeds minute, brownish, with dwarf bulbous dome-like hairs all over, bands spiral reticulate.

Fl. and Fr.: Oct. – Dec.

Distribution: Lamb's Rock Shola, Nilgiris.

Impatiens ligulata Beddome in Madras J. Lit. Sci. Ser. 2. 4: 67, t. 7, f. 6. 1859 & Icon. t. 149. 1868–74; Hook. f., F1. Brit. India 1: 446. 1874; and in Rec. Bot. Surv. India 4: 46. 1906; Rama Rao, Fl. Pl. Travancore: 59. 1914; Gamble, Fl. Madras: 140. 1915; Vivek. *et al.*, in Hajra *et al.*, Fl. India 4: 174. 1997; Bhaskar, Tax. Mon. *Impatiens* of Western Ghats 153. 2012.

Erect herbs, 30–60 cm high; branches slender, spreading. Leaves opposite, subsessile, linear-lanceolate, mucronate, serrate, 7.5–10 cm long, hispid above; stipules setaceous or absent. Flowers solitary or in fascicles, pink or white; pedicels slender, 3.5 cm long. Lateral sepals linear-lanceolate. Lip saccate, spurs stout, 2 cm long, undulate. Standard small with a pubescent line on dorsal side. Wings bilobed; distal lobes slender; dorsal auricle filiform, descending in to spur. Capsules ellipsoid; seeds 12, shining.

Fl. and Fr.: July – Oct.

Distribution: Hills of Anamalai above 1000 m.

Impatiens maculata Wight in Madras J. Lit. Sci. Ser. 1.5: 12.1837; Hook. f., Fl. Brit. India 1: 465. 1875; Bhaskar, Tax. Mon. *Impatiens* of Western Ghats 264. 2012. (Plate 1.4D)

Tall herbs or shrubs, up to 2 m high, slender, erect; stems and branches angled or ribbed, laxly hairy or glabrate, speckled with pinkish blotches. Leaves alternate, ovate lanceolate, acuminate, serrate, 7.5–12.5 x 3–5 cm, hairy on both surfaces; petiole 2.5–5 cm long, glanduliferous, often with 2 or more long- stipitate glands. Flowers in racemes, laxly many flowered, pink or white, 2.5 cm long; bracts ovate, persistent. Lateral sepals ovate, acute. Lip conoid; spur slender, curved upwards, longer than petals. Standard obtuse, shorter than lateral sepals. Wings bilobed; distal lobes suborbicular, very large; basal ones minute, concealed. Capsules ellipsoid, 5- angled, beaked, 8 mm long, glabrous; seeds sub-orbicular, shortly hairy.

Fl. and Fr.: May–Janu.

Distribution: Kakachi (Tirunelveli), Kundahs (Nilgiris), Waverly estate, Attakatti-Valparai (Anamalais), Sholaiyar submergible area (Coimbatore), Cumbum Valley (Theni) and Bodimetta.

Impatiens minor (DC.) Bennet in Indian J. Forest. 2: 283.1979; Almeida, FI. Savantwadi 1: 78.1990; Lakshminarasimhan & Sharma, Fl. Nasik District: 144.1991; Deshpande *et al.*, Fl. Mahabaleshwar 1: 116.1993; Almeida, Fl. Maharashtra: 194. 1996; Saldanha, Fl. Karnataka 2: 255, t. 36.1996; Vivekananthan *et al.*, in Hajra *et al.*, Fl. India 4: 181. 1997; Mudaliar & Prasad in Singh & Karthikeyan, Fl. Maharashtra State Dicotyl. 1: 457. 2000; Yadav & Sardesai, Fl. Kohlapur: 05.2002; Bhat, Fl. Udupi District: 88.2003; Rathakrishnan *et al.*, in Daniel, Fl. Kerala 1: 550. 2005; Bhaskar, Tax. Mon. *Impatiens* of Western Ghats 154. 2012. (Plate 1.4E)

Herb, 8–50 cm high. Stem semi-terete, branched or rarely unbranched, glabrous, green. Leaves opposite-decussate, sessile to shortly petiolate, lanceolate, oblong or ovate to elliptic, 3–14 x 1–3.8 cm, obtuse to semicordate at base, apiculate-crenate to serrate at margins, acute to acuminate at apex, glandular at base, hairy above, glabrous below; nerves 3–7 pairs; petioles up to 4mm long, glabrous. Flowers 1–3 per axil, 0.8–

Figure 1.14: ***Impatiens leschenaultii*** **Wallich ex Wight & Arn.**

Figure 1.15: ***Impatiens levingei*** **Gamble ex Hook. f.**

Figure 1.16: ***Impatiens minor*** **(DC.) Bennet**

Figure 1.17: ***Impatiens modesta*** **Wight**

1.5 cm across, pink with violet throat; bracts triangular to ovate, 0.5 x 0.3 mm, acute at apex, glabrous; pedicels 1–2.5 cm long, glabrous, deflexed in fruits. Lateral sepals linear-lanceolate, 3–6 x 0.7–1 mm, acute at apex, hairy on costa dorsally, 3-nerved, pale green. Standard petal oval to orbicular, 3–4 x 3–5 mm, acute at apex, pinkish white, dorsally keeled; keel glabrous to hairy, green; wing petals 0.8–1 x 0.5–0.7 cm, entire at margins; lobe asymmetrically obovate, 4–7 x 5–7 mm, obtuse at apex; stipe 2–3 x 0.5 mm; lip boat-shaped, 3–6 x 2 mm, 1–2 mm deep, acute at apex, glabrous; spur tubular to compressed, 1–1.3 cm long, rounded at apex, straight or curved, glabrous, pale green. Column 2 x 1 mm, curved; filaments 1.5 x 0.5 mm, pink; anthers 0.25 x 0.25 mm. Pistil 1.5 x 0.5 mm; ovary lanceoloid, glabrous. Capsules ellipsoid to fusiform, 1–1.5 x 2–4 mm, glabrous, pedicels 1.8–2.3 cm long; seeds globose, 0.5 mm, glabrous, shining, black to brown; funicle present.

Fl. and Fr.: June–Dec.

Distribution: Commonly distributed throughout Western Ghats above 1000 m. In Tamil Nadu- Devarshola (Nilgiris), Parambikolam, Shevoghiry, Cumbum Vaalley, Kodaikanal.

Impatiens modesta Wight in Madras J. Lit. Sci. Ser. 1.5: 13.1837 & Icon. 3 (3): 6.t. 968. 1845; Hook.f., Fl. Brit. India 1: 442. 1874; Bhaskar, Tax. Mon. *Impatiens* of Western Ghats 71. 2012.

Herbs, perennial, 8–30 cm high, epiphytic; root stock tuberous. Leaves radical, ovate-cordate or suborbicular, acute, crenate-serrate, 2.5–14 x 2–9 cm, sparsely hairy above, glabrous and pale shining- glaucous below. Scape slender, erect, 10–20 cm long; bracts subulate, lanceolate, thickened at tips. Flowers pinkish or white 1 cm across in 2–5 cm long racemes; pedicels filiform. Upper sepals broad, obovate or suborbicular, lateral ones narrowly lanceolate or subulate, incumbent on upper. Lip shorter than petals. Standard short, broader than long, entire. Wings 3-lobed; lobes subequal, obtuse; spur short, obtuse, straight. Capsules ellipsoid, 4–6 mm long, acute at both ends, glabrous; seeds minute with spiral hairs.

Fl. and Fr.: June–Sept.

Distribution: Pykara, Naduvattam, Andiparai shola, Shivagiri (Courtallum).

Impatiens munnarensis Barnes in Bull. Misc. Inform. 1938: 32; Vivek. in M. P. Nayar & Sastry, Red Data Book Indian Pl. 3: 60. 1990; Vivek. *et al.*, in Hajra *et al.*, Fl. India 4: 183. 1997; Bhaskar, Tax. Mon. *Impatiens* of Western Ghats 161. 2012. (Plate 1.4F).

Herbs, weak, erect or semi- erect, up to 1 m high; stems slender, sparingly branched, often procumbent below, rooting at lower nodes. Leaves variable, usually lanceolate, base broadest, shallowly cordate, tapering to an acute-apex or narrowly ovate, subacute at base, distinctly serrete along margins with apiculate teeth, setose at base, up to 5.5 cm–1.7 cm, pubescent above, glabrous and pale, sometimes glaucous below; petioles flattened, red above, decurrent on stem; stipule filiform, pointed, red, tips darker, bent towards right angles pointing downwards. Pedicels solitary or up to 4, fascicled in axils, very short in bud, in flower 1–2.5 cm long, partially reflexed in fruit. Flowers 1–1.3 cm long. Lateral sepals filiform, curved, 6mm long. Lip boat-shaped, cuspidate in front, white with purple nerves, bottom somewhat flattened with a small pit instead of spur, inside convex continued forward as a ridge with a small yellow or orange crest about middle. Standard ovate, keeled, tip mucronate. Wings white with liliac or violet markings; distal lobes obovate-triangular; basal lobes triangular; dorsal auricle bluntly conical or rounded, placed at right angles, dark- purple. Capsules fusiform, later swollen, acuminate, 1.3–1.5 cm long, few seeded; seeds black, glossy, 1.5 mm long.

Fl. and Fr.: Sept. – Dec.

Distribution: Meghamalai hills.

Note: After its type collection, no reports have been emerged to relocate the species. The earlier reports revealed that, the species has been extinct from the wild. The present investigation, rediscovered the species from new locality (Varaiadu mottai, High wavies mountains, Theni district, Tamil Nadu).

Impatiens munronii Wight, Ill. Ind. Bot. 1: 160. 1840 & Icon. t. 1049. 1846; Hook. f., FI. Brit. India 1: 456. 1876; Bhaskar, Tax. Mon. *Impatiens* of Western Ghats 227. 2012. (Plate 1.5A)

Undershrubs, up to 60 cm high, softly hairy; stems branched, woody, nodose, almost woolly. Leaves alternate, towards ends of branches, elliptic, acuminate, serrulate, 3–8 x 1.5–3 cm, membranous, hairy on both sides, glandular along margins; petioles 1.5–5 cm long, hairy. Flowers axillary, solitary, rose; pedicels 1.5–3 cm long, slender, hairy. Lateral sepals ovate, as large as wings, hirsute. Lip with spur trumpet-shaped, 1.5 cm long, incurved, hirsute. Standard orbicular, keeled on dorsal side, hirsute. Wings short, bipartite to middle. Capsules hairy, seeds glabrous.

Fl. and Fr.: Feb.–May.

Distribution: Endemic to Sispara Nilgiris.

Impatiens neo-barnesii C. Fischer in Bull. Misc. Inform. 330. 1930 & Gamble, Fl. Pres. Madras 1869. 1935. *I.barnesii* C. Fischer in Bull. Misc. Inform. 153-154. 1930 non Hook. f. 1909; Bhaskar, Tax. Mon. *Impatiens* of Western Ghats 73. 2012.

Herbs, epiphytic, acaulescent, delicate, pendulous; bulbs small, globose. Leaves 1–2, ovate or elliptic-ovate, rounded at base, more or less unequal, acute or subacute at apex, remotely denticulate along margins, 2–8 x 1.3–3.6 cm, sparsely pilose above, glabrous. Scapes delicate, glabrous, 4–5 mm long, few- flowered; bracts lanceolate or ovate, subacute; pedicels 1.5 cm long. Flowers cream-coloured or almost white, 2–2.5 cm long. Lateral sepals ovate-falcate, 5 mm long. Lip ovate-oblong or suborbicular, rotund, base shortly saccate at base, 1–1.2 cm long. Petals delicate, wings bilobed, 1.75–2.3 cm long; upper lobes oblong, obtuse, 7 mm long; distal lobes oblong-lanceolate, obtuse or shortly emarginated, 1.5 cm long. Capsules ellipsoid, acute, 8 mm long.

Fl. and Fr.: July–Sept.

Distribution: Endemic to Nilgiris, in Kundahs, Devabetta.

Impatiens nilgirica C. Fischer in Bull. Misc. Inform. 1931: 41 – 42. f. A-C. 1931 and in

Gamble, Fl. Pres. Madras 1869. 1935; Bhaskar, Tax. Mon. *Impatiens* of Western Ghats 74. 2012.

Herbs, erect, acaulescent; bulbs small, globose. Leaves 1–2, radical, suborbicular, prominently cordate at base, lobes rounded, often imbricate, distantly apiculate-crenate along margins, 2.5–4 cm broad, fleshy; primary nerves 7–9, fan-shaped, obscure above, distinct below; petiole 3–9 cm long, purplish. Scapes 1–2, 20 cm long, 4–10 flowered, purplish; bracts broadly ovate or suborbicular, concave, 6 mm long; pedicels 1 cm long, reflexed apex slightly cordate, 4 mm long. Lip 1 cm long, navicular, subacute; spur short, broadly obtuse. Wings 3-lobed, 2 cm long, base of lower two-lobed, pilose within; distal lobes broadly obovate, oblique, retuse, 5–6 mm long; middle lobe narrower, obovate-oblong, apex rounded, 8 mm long; basal lobes further narrower, often smaller, oblong, obtuse. Capsules ellipsoid, 1–2 cm long; seeds minute, ovate - oblong, compressed, glabrous.

Fl. and Fr.: Sept.

Distribution: Rare, strict endemic to Kundahs above 2300 m in Nilgiris.

Impatiens omissa Hook. f. in Rec. Bot. Surv. India 4 (3): 48. 1906; Gamble, Fl. Pres. Madras 144. 1915; Fyson, Fl. South Ind. Hill Stn. 1: 95. 1932; Barnes in J. Indian Bot. Soc. 18 (3): 104. 1939; Shetty & Vivek. in Bull. Bot. Soc. India 3: 29. 1971; Bhaskar, Tax. Mon. *Impatiens* of Western Ghats 250. 2012. (Plate 1.5B)

Small slender, erect branched herbs, up to 20 cm tall; stems slender, more or less quadrangular, channelled, red, roughly hairy just above nodes, upper parts glabrous, nodes swollen near branching. Leaves 1.5 cm long, opposite or alternate, ovate, deeply serrate, or coarsely crenate-serrate, acute, sessile or shortly petioled, pale in colour, glabrous below, sparsely hairy above; petioles up to 0.5 cm long. Flowers 3–5, in short or long racemes or umbels, 8–15 mm across, white with bright crimson blotches on the smaller lobe; peduncles 2.5–5 cm long, sticky, much longer than leaves, erect or spreading; bracts minute; pedicels slender, up to 3 cm long; sepals small, white crimson-streaked; lip spathulate or boat-shaped, greenish, tips red, spur distinct, 5 mm long, blunt or conical. Standard suborbicular, obtusely ridged, with a small mucro behind retuse tip, vaulted. Wings up to 1.6 cm, bilobed, upper terminal lobes large, triangular, base white, pale crimson tinged, basal lobe hidden within standard, dorsal auricle 2–4 mm long, filiform, sharp tipped or acuminate, bright orange or yellowish. Capsules up to 5 mm long, ellipsoid, swollen in middle; seeds numerous, broadly ovate, flattened, 2 mm long, dark brown, villous, hairs white, numerous all over, stout.

Fl. and Fr.: Sept.–Dec.

Distribution: Endemic to Anamalais, Palni hills.

Impatiens oppositifolia L., Sp. P1.: 937.1753; Lam, Encycl. 1: 363. 1785; Wight & Arn., Prodr. F1. Ind. Orient. 1: 139. 1834; Wight, Icon. P1. Ind. Orient. 3 (2): 7, t. 883. 1844–1845; Hook.f. & Thomson, J. Proc. Linn. Soc. Bot. 4: 120. 1859; Dalzell & Gibson, Bombay F1.: 43. 1861; Hook.f., F1. Brit. India 1: 448. 1874 & Rec. Bot. Surv. India 4: 46.1906; Cooke, F1. Bombay 1: 172. 1901; Gamble, F1. Madras 1:

141.1915; Blatter, J. Bombay Nat. Hist. Soc. 33: 312. 1933: Vartak, Enum. Plant. Gomantak 32.1966; Santapau, F1. Khandala 30. 1967; Ramamoorthy in Saldanha & Nicolson, F1. Hassan District 403. 1978; Vajravelu in Nair & Henry, F1. Tamil Nadu 1: 55. 1983; Sharma *et al.*, F1. Karnataka 38. 1984; Grey – Wilson in Dassanayake, Rev. Handb. F1. Ceylon 5: 110, F. 2X – Z. 1985; Rao, F1. Goa 1: 60.1985; Kulkarni, F1. Sindhudurg 60. 1988; Almeida, F1. Savantwadi 1: 78. 1990; Deshpande *et al.*, F1. Mahabaleshwar 1: 117. 1993; Kothari & Moorthy, F1. Raigad District 47. 1993; Saldanha, F1. Karnataka 2: 256. 1996; Vivekananthan *et al.*, in Hajra *et al.*, F1. India 4: 187.1997; Mudaliar & Prasad in Singh & Karthikeyan, F1. Maharashtra State Dicotyl. 1: 459. 2000; Ramaswany *et al.*, F1. Shimoga District 109. 2001; Yadav & Sardesai, F1. Kolhapur District 95. 2002; Bhat, F1. Udupi District 88.2003; Radhakrishnan *et al.*, in Daniel, F1. Kerala 1: 552. 2005; Bhaskar, Tax. Mon. *Impatiens* of Western Ghats 162. 2012.

Herb, 15–50 cm high. Stem often branched, semicircular at base, quadrangular above, glabrous, red, rarely green above. Leaves opposite–decussate, sessile to shortly petiolate, elliptic-lanceolate, linear-lanceolate, linear-oblanceolate or oblong to spathulate, 3.5–7 x 0.8–1.9 cm, cuneate or truncate to subcordate at base, apiculate-crenate at margins, acute to acute-apiculate at apex, hairy mostly on nerves above, glabrous below; nerves 4 or 5 pairs; petioles decurrent on stem and terminating in glands, up to 4 mm long, glabrous. Flowers 2 or 3 per axil, 1–1.6 cm across, lilac to pink with violet throat; bracts linear, 1.5 x 0.5 mm, acuminate at apex, glabrous; pedicels 1–2.6 cm long, with 2 rows of hairs, rarely glabrous, deflexed in fruits. Lateral sepals linear, 4–5 x 1 mm, acute at apex, translucent, glabrous. Standard petals broadly ovate, 3–6 x 4–6 mm, concave emarginate to obcordate at apex, pink, dorsally keeled; keel mucronate; mucro 0.5 mm long, wing petals 0.9–1.3 x 0.4–0.7 cm, bilobed; basal lobe ovate, 3–5 x 2–3 mm; distal lobe 7–9 x 4–6 mm, asymmetrically obovate, elevated at base, obtuse at apex. Lip conical to saccate, 5–6 x 2–3mm, 1–3 mm deep, acuminate at apex, glabrous; spur straight or hooked, 2–4 mm long rounded, notched to bulged at apex, glabrous, pink. Column 3 x 1.5 mm, curved; filaments 2.5 x 0.5 mm, pink; anthers 0.5 x 0.5 mm, white to pink. Pistil 2 x 0.5 mm; ovary oblong lanceoloid, curved at apex, glabrous. Capsule asymmetrically ellipsoid or lanceoloid to oblanceoloid, 0.8–1.5 x 0.4–0.5 cm, glabrous; pedicels 2–3 cm long, seeds globose, 0.5 mm, glabrous shining, black to brown; funicle present.

Fl. and Fr.: Feb.–Dec.

Distribution: Widely distributed in northern parts of Western Ghats, Kodaikanal.

Impatiens orchioides Beddome in Madras J. Lit. Sci. Ser. 3. 44: 1864 and Icon. t. 152. 186. 1874; Hook. f., Fl. Brit. India 1: 443. 1874; Bhaskar, Tax. Mon. *Impatiens* of Western Ghats 75. 2012.

Herbs, epiphytic, scapigerous; roots tuberous. Leaves radical, ovate-orbicular, deeply cordate at base, obtuse or acuminate at apex, bristly crenate along margins, 1.5–5.5 x 1–6 cm, hairy above, with numerous weak hairs, almost glabrous beneath; petioles 2–16 cm long. Scapes 5–19 cm long, 6–10 flowered at apex; bracts ovate; pedicels very slender, 1 -1.5 cm long. Flowers 8mm across, reddish-brown or pale brown. Lateral sepals ovate, acute, saccate, spurless. Standard ovate, obtuse, not vaulted. Wings with a dimidiate oblong blade produced into two linear, pendulous lobes. Capsules constricted at both ends, glabrous; seeds numerous, minute, scorbiculate, brown.

Fl. and Fr.: July–Sept.

Distribution: Strict endemic to Nilgiris, in Kundahs, Devabetta.

Impatiens parasitica Beddome in Madras J. Lit. Sci. Ser. 2. 20: 66. t. 7. f. 2. 1859 & Icon. t. 140. 1868 – 74; Bhaskar, Tax. Mon. *Impatiens* of Western Ghats 102. 2012. *I. jerdoniae* Wight var. *parasitica* (Beddome) Hook. f., Fl. Brit. India 1: 460. 1874. (Plate 1.5C)

Herbs, succulent, perennial, stems moniliform, 10–25 cm long, green or purple, internodes swollen. Leaves crowded at ends of branches, elliptic, acute, serrate, with glandular cilia at base, 3–7 x 2–3 cm; petioles 2–6 cm long. Peduncles axillary, up to 3.5 cm long, bearing 1–3 flowers; pedicels 5 cm long. Lateral sepal linear, green with a foliaceous crest. Lip saccate; spur 5 mm across the mouth, recurved at tip. Standard dorsally winged, green. Wings yellow, bilobed, concealed under lip. Capsules gibbous, glabrous; seeds numerous, small.

Fl. and Fr.: May–Dec.

Distribution: Endemic to Anamalais.

Impatiens parvifolia Beddome in Madras J. Lit. Sci. Ser. 2. 20: 66. t. 7, f. 1. 1859; & Icon. 29, t. 142. 1868–74; Hook. f., Fl. Brit. India 1: 453. 1874; Bhaskar, Tax. Mon. *Impatiens* of Western Ghats 251. 2012. *I. omissa* Hook. f. in Rec. Bot. Surv. India 4: 48. 1906.

Herbs 12–25 cm long; stems simple or branched. Leaves opposite and alternate, sessile or petioled, ovate or elliptic, crenate- serrete, 0.8–2 x 0.5–1.3 cm; petioles up to 7 mm long. Flowers axillary, solitary or in racemes, white or pale pink with crimson blotches; peduncles up to 5 cm long. Lateral sepals small, falcate. Lip cymbiform; spur green, short, blunt. Standard sub-orbicular, keeled. Wings up to 1.2 cm long, bilobed; basal lobe small; distal lobes triangular, thrice larger than basal ones; dorsal auricle minute. Capsules ellipsoid; seeds ovoid, villous.

Fl. and Fr.: Sept.–Dec.

Distribution: On wet rocks and on slopes of Anamalai hills, at 2333–2695 m.

Impatiens pendula Heyne in Wall. Cat. 4744. 1831 ex Wight. & Arn. Prodr. Fl. Ind. Orient.137. 1834; Hook. f. & Thomson in J. Linn. Soc. Lond. 4: 133. 1860; Drury, Handb. Indian Fl. 1: 186. 1864; Bhaskar, Tax. Mon. *Impatiens* of Western Ghats 207. 2012.

Small, dwarf, erect or procumbent annual bushy herbs, 10–15 cm tall, rooting at lower nodes; stem very much condensed with closely packed leaves with short internodal lengths, terete or angled above first cotyledonary leaves, hairy in opposite-decussate fashion, the lines of hairs decurrent from the side of base of petioles. Leaves opposite, decussate below including cotyledonary leaves, alternate above or subopposite, first cotyledonary leaves spherical, petiolate, entire, upper leaves upright, short petiolate, petiole about 1 cm long, hairy above, continuous with hairs on stem; lamina oval or rhomboid, leathery, often reddish-green, sharply and distantly serrate or remotely bristle serrated, denticels red, up to 2 x 1.5 cm, acute, long hairy above, hairs also reddish, continuous with hairy petiole, glabrous beneath, veins impressed on upper surface. Flowers solitary or 2 in the leaf axils, axillary, 1 cm across, white with red or maroon colour combination, buds reddish, peduncle absent, bracteate, bracts in bud fusing towards stem, linear lanceolate; a small bud is invariably produced in axil of old flower bud, pedicels 5–8 mm long, red hairy in a line, in portion facing opposite petiole; sepals 2, small, red, one fourth the length of flower bud, linear, acute, glabrous, reddish. Lip boat-shaped, spurless, with an yellow eye at the centre, long bristly hairy in the margin, like antennae of insects, glabrous outside, tip mucronate. Standard prominent, longer than all parts of flower in bud, but same length as the wing petals in the opened flower, acute, margin white, tip mucronate; keeled on back, reddish, red within including the long sharp beak, wing petals bilobed, distal larger than basal lobe, basal lobe half the size of distal lobe, but most striking character is the hairyness or velvety or villous nature on the inside part of basal lobe and distal lobe, purplish red inside mouth, white at margin or distal part, dorsal auricle present, short; pollen white, 4-colpate, exine reticulate. Capsules very small, ventricose, acuminate, densely hairy; seeds very few, brownish, globose, rugose.

Fl. and Fr.: Sept.–Oct.

Distribution: Nilgiris.

Impatiens phoenicea Beddome in Madras J. Lit. Sci. Ser. 2. 19: 175. 1858 & Icon. 30. t. 148. 1868 -1874; Hook. f., Fl. Brit. India 1: 466. 1875; Bhaskar, Tax. Mon. *Impatiens* of Western Ghats 265. 2012 (Plate 1.5D).

Herbs, suffruticose, erect, 0.3–1 m high, branched or not, glabrous. Leaves alternate, lanceolate, attenuate at both ends, bristly crenate-serrete, incurved, 5–15 x 2.5–3.5 cm long. Peduncles axillary as long as leaves, slender, erect; bracts at base of pedicels, ovate, cordate, persistent. Flowers 2.5 cm long, brilliant scarlet, yellow in centre. Lateral sepals obliquely ovate. Lip with spur blood- crimson, trumpet-shaped, incurved with swollen tip. Standard scarlet, broad, ovate. Wings bilobed, scarlet, yellow at base, small; basal lobes cordate, overlaping the larger oblong, distal lobes. Capsules 4 – 5 seeded; seeds compressed, furrowed, 3 mm long, light brown hairy.

Fl. and Fr.: July – Jan.

Distribution: Pulney hills in Kodaikanal, but rarely occurs in Anaimudi.

Impatiens pseudo-acaulis Bhaskar, Tax. Mon. *Impatiens* of Western Ghats 77. 2012. *I. scapiflora* var. *psedo-acaulis* Bhaskar & Razi in Indian J. Forestry 1 (1): 68. 1978.

Scapigerous perennial herbs, tuberous. Leaves radical, petiole 5–8 cm long, lamina reniform, deeply cordate or auricled base, margins shallowly toothed, apex round, 5–8 nerved, invariably glabrous. Flowers in scapes, nearly 3.75 cm across, pink; scapes 15–25 cm long, many-flowered; bracts ovate, 3 mm long; pedicels 1.5 cm long, glabrous; sepals inequilateral, 4–5 nerved, furunculate, apex obtuse; lip with a long slender spur, >3 cm long, straight or slightly curved. Standard broader than long, vaulted; Wings 3-lobed, basal or posterior lobe 0.9 cm long, anterior lobe 1.3 cm long, middle lobe 1 cm long and widest, venation reticulate, dorsal auricle absent. Capsules ovate-oblong, many seeded; seeds brown, with hairs all over the surface, hairs short filamentous with closely packed single helical banding.

Fl. and Fr.: March–Sept.

Distribution: Naduvattam, Kundahs, Mukurthi of Nilgiris.

Impatiens rufescens Benth. ex Wight & Arn., Prodr. Fl. Ind. Orient. 138. 1834; Wight, Icon.t. 963. 1845; C. Fisher in Gamble, Fl. Pres. Madras 1871. 1935; Bhaskar, Tax. Mon. *Impatiens* of Western Ghats 169. 2012. *I. tomentosa* Heyne ex Wight & Arn. var. *rufescens* (Benth. ex Wight & Arn.) Hook. f., Fl. Brit. India 1: 449.1874.

Herbs, 10–25 cm high; stems erect, glabrous. Leaves opposite, shortly petioled, elliptic or cordate-obovate, acute or acuminate at apex, serrate, hispid above, hairy on nerves beneath. Flowers axillary, solitary or in pairs, 1.8 cm across, rose- pink or purplish; pedicels about the length of leaves, villous. Lateral sepals linear, smaller than wings. Lip saccate, without spur. Standard orbicular, apiculate. Wings bilobed; basal lobes rounded, distal lobes larger than basal ones, capsules ovoid.

Fl. and Fr.: July – Nov.

Distribution: Nilgiris, Agastyamalai hills.

Impatiens scabriuscula B. Heyne ex Wall. in Roxb., Fl. Ind. 2: 464.1824; Wight & Arn, Prodr. Fl. Ind. Orient. 2: 136. 1834; Hook.f. & Thomson, J. Proc. Linn. Soc. Bot. 4: 131. 1859; Dalzell & Gibson, Bombay Fl.: 44. 1861; Beddome, Icon. Pl. Ind. Orient. 29, t.144.1868–1874; Hook.f., Fl. Brit. India 1: 454. 1874 and Rec. Bot. Surv. India 4: 46. 1906; Cooke, Fl. Bombay 1: 174. 1901; Gamble, Fl. Pres. Madras 1: 142. 1915; Blatter, J. Bombay Nat. Hist. Soc. 33: 314. 1933; Yoganarasimhan *et al.*, Fl. Chikmagalur District: 60.1982; Vajravelu in Nair & Henry, Fl. Tamil Nadu 1: 55. 1983; Sharma *et al.*, Fl. Karnataka: 39.1984; Almeida, Fl. Maharashtra 1: 195. 1996; Saldanha, Fl. Karnataka 2: 257. 1996; Vivekananthan *et al.*, in Hajra *et al.*, Fl. India 4: 207. 1997; Mudaliar & Prasad in Singh & Karthikeyan, Fl. Maharashtra State Dicotyl. 1: 462. 2000; Rathakrishnan *et al.*, in Daniel, Fl. Kerala 1: 556. 2005; Bhaskar, Tax. Mon. *Impatiens* of Western Ghats 210. 2012. (Plate 1.5E)

Herb, 20–60 cm high. Stem subterete to terete, sparsely and minutely hairy at base, densely hairy above, branched or not; branches alternate. Leaves alternate, linear- elliptic to oblanceolate, 1–1.5 x 0.4–1.5 cm, attenuate or rarely obtuse and oblique at base, apiculate, crenate to serrate at margins, acuminate at apex, hairy above and on nerves below; petioles 0.5–2 cm long, hairy. Flowers 2–4 per axil, 0.9–1.2 cm across, pink; bracts ovate, 0.75 x 0.5 mm, acute at apex, dorsally hairy; pedicels 0.6–1 cm long, hairy, deflexed in fruits. Lateral sepals 1.25 mm long, acute to acuminate at apex, dorsally hairy, light green. Standard petal broadly obovate, 8 x 6 mm, concave, retuse- mucronate at apex, hairy outside; mucro horned, 2 mm long, hairy; wing petals 1–1.4 x 0.4–0.6 cm, unequally bilobed; basal lobe triangular, 4–5 x 2–4 mm; distal lobe asymmetrically obovate, 6–8 x 4–5 mm, acute at apex, papillate at base

Figure 1.18: ***Impatiens munnarensis*** **Barnes**

Figure 1.19: ***Impatiens munronii*** **Wight**

Figure 1.21: ***Impatiens phoenicea*** **Beddome**

Figure 1.22: ***Impatiens tangachee*** **Beddome**

Figure 1.20: ***Impatiens nilgirica*** **Fischer**

towards inner side; auricle 1.5 mm long, rounded at apex; lip saccate, 5–9 x 4–5 mm, 2–4 mm deep, hairy, light pink; spur absent. Column 4 x 1.5 mm; filaments 3 x 0.5 mm; anthers 0.5 x 0.5 mm. Pistil 3 x 0.5 mm; ovary lanceoloid to elliptic- lanceoloid, minutely hairy. Capsules broadly ellipsoid, 0.8–1 x 0.5–0.7 cm, villous; pedicels 1–1.3 cm long; seeds globose, 1mm, brown, shortly papillate; papillae dark brown.

Fl. and Fr.: July–Jan.

Distribution: Naduvattam, Sholur in Nilgiris.

Impatiens scapiflora B. Heyne ex Roxb., Fl. Ind. 2: 464.1824; Wight & Arn. Prodr. Fl. Ind. Orient. 137. 1834; Hook.f. & Thomson, J. Proc. Linn. Soc. Bot. 4: 118. 1859; Hook.f., Fl. Brit. India 1: 443. 1874 & Rec. Bot. Surv. India 4: 44. 1906; Gamble, Fl. Pres. Madras 1: 138. 1915; Blatter, J. Bombay Nat. Hist. Soc. 33: 310. 1933; Ramamoorthy in Saldanha & Nicolson, Fl. Hassan District: 403.1978; Yoganarasimhan *et al.*, Fl. Chikmagalur District: 60.1982; Vajravelu in Nair & Henry, Fl. Tamil Nadu 1: 56. 1983; Sharma *et al.*, Fl. Karnataka: 39.1984; Almeida, Fl. Maharashtra 1: 196. 1996; Saldanha, Fl. Karnataka 2: 257. 1996; Vivekananthan *et al.*, in Hajra *et al.*, Fl. India 4: 208. 1997; Mudaliar & Prasad in Singh & Karthikeyan, Fl. Maharashtra State Dicotyl. 1: 461. 2000; Ramaswamy *et al.*, Fl. Shimoga District: 109.2001; Rathakrishnan *et al.*, in Daniel, Fl. Kerala 1: 557. 2005; Bhaskar, Tax. Mon. *Impatiens* of Western Ghats 80. 2012. (Plate 1.5F).

Terrestrial or epiphytic on moist, scapigerous, tuberous or rhizomatous herb, 20–45 cm high; tubers oblong, 1–1.5 cm across. Leaves radical, 2–5 per tuber, reniform, ovate to obovate, 2.5–20 x 2–15 cm, cordate to rounded at base, crenate to distantly serrate at margins, acute to acuminate at apex, pubescent above, glabrouse below. Inflorescence raceme, 20–45 cm long, 4–16 flowered. Flowers 2.5–5 cm across, pink; bracts ovate, 2–8 x 1–5 mm, acute to rounded at apex, slightly concave, thick, fleshy, glabrous, deflexed in fruits. Lateral sepals asymmetrically ovate, 3–5 x 1–3 mm, acute at apex, glabrous, pale green, 5–7 nerved. Standard petal broadly reniform, 0.6–1 x 1–1.5 cm, forming a hood above column, humped at base dorsally, emarginate to obcordate at apex, glabrous, pink outside, white inside; wing petals 2.3–3.0 x 1.7–2.6 cm, 3-lobed; basal lobe oblong, 1.1–1.8 x 0.3–0.6 cm, rounded at apex, curved, straight or coiled, glabrous. Column 5 x 3 mm, bent forward; filaments 5 x 3 mm; anthers 2 x1 mm. Pistil 5 x 2 mm; ovary lanceoloid, glabrous. Capsules lanceoloid, 1–1.5 x 0.5–0.8 cm, glabrous; pedicels 4–6 cm long; seeds oblongoid to ovoid – oblongoid, 1.5 x 0.5 mm, reticulately nerved, hairy all over; hairs spirally coiled, forming a cone like structure.

Fl. and Fr: May – Oct.

Distribution: Nilgiri, Cumbum Valley.

Impatiens shevaroyensis Bhaskar, Tax. Mon. *Impatiens* of Western Ghats 83. 2012.

Scapigerous perennial herbs, tuberous. Leaves radical, petiole 6–14 cm long, lamina 3 x 2 cm to 7.5 x 5 cm, orbicular, deeply cordate or auricled at base, margin shallowly toothed, apex round, 3–8 nerved, sparsely hairy above, glabrous beneath. Flowers in racemose scapes, scapes 18–22 cm long, many flowered, deep pink, 2 x

1.75 cm; bracts ovate, 3mm long; pedicels 2.8 cm long, glabrous; sepals inequilateral, 4–5 nerved, 3 mm x 2.5 mm, apex obtuse; lip 6.5 mm long on top, with a long slender spur, spur 's'-shaped, up to 3 cm long. Standard longer than broad, 0.8 cm x 0.5 cm, cupulate, deep pink outside and white inside, tip round, wings 3-lobed, the basal lobe half the size of other two, usually overlapping or clasping on middle lobe, 1.5–2.1 cm along distal lobe, proximal lobe the smallest, 6 mm long, anticous lobe 6 x 3.5–6.0 mm, middle lobe broadest, 4–5 mm x 6 mm, base of all lobes white, dorsal auricle absent. Pollen white. Capsules ovate-oblong, 1 cm long when mature, red-tinged on a side, many seeded; seeds brown, 1 mm long, with hairs all over the surface. Hairs filamentous, with closely packed single standard helical banding.

Fl. and Fr.: August – September.

Distribution: Endemic to North-East monsoon zone of South India, in Shevaroy hills in Yercaud of Tamil Nadu.

Impatiens stocksii Hook. f. & Thomson. in. J. Linn. Soc. Lond. 4: 119. 1860; Drury, Handb. Indian Fl. 1: 191. 1964; Hook. f., Fl. Brit. India 1: 442. 1874; Bhaskar, Tax. Mon. *Impatiens* of Western Ghats 85. 2012

Epiphytic scapigerous herbs, scarcely reaching 5 cm tall, tuberous. Leaves 3–4, radical, petiole up to 2 cm long, lamina 1.5 x 1.3 cm, orbicular or broadly ovate, crenate, apex round, apiculate, pubscent above. Flowers in small stout scapes scarcely reaching 4 cm long, bent downwards due to weight of flowers and fruits; pedicels about 1 cm, stout and thick; bracts ovate, acute; sepals similar to bracts, persistent; lip spurless. Standard more or less rhomboid, as broad as long, wings 3-lobed, distal lobe longest, 0.5 cm long, dorsal auricle absent. Pollen white, 3-colpate, radial, exine reticulate, lumen broad, muric, slightly irregular or wavy. Capsules ellipsoid; seeds minute, linear oblong, brownish, long hairy at either ends of seed, bands spiral.

Fl. and Fr.: July – Sept.

Distribution: Mahendragiri hills, Kanyakumari.

Impatiens tangachee Beddome in Madras J. Lit. Ser. 2.20: 66. 1859 & Icon. 30. t. 147. 1868 – 1874; Hook. f., Fl. Brit. India 1: 467. 1875; Bhaskar, Tax. Mon. *Impatiens* of Western Ghats 268. 2012. (Plate 1.6A)

Herbs 20–40 cm high; stems forked, subprocumbent, rooting at lower nodes, glabrous. Leaves crowded towards tip, obscurely whorled, alternate, subsessile, elliptic- acuminate at apex, serrate along margins, 3–10 x 1–1.2 cm; nerves few, slender. Peduncle axillary, stout, 4–16 cm high, exceeding leaves; bracts ovate, coriaceous, persistant. Flowers 4–8, in racemes or subumbels, bright rose; pedicels erect, 1–2.5 cm long. Lateral sepals orbicular–oblong. Standard short, broad. Wings bilobed; basal lobes broadly truncate; distal lobes truncate with erose apex. Capsules gibbous, glabrous, 8 mm long, seeds many, small.

Fl. and Fr.: June–Dec.

Distribution: Avalanchi (Nilgiris), Periya shola, Stream Tiger Shola (Pulnis).

Impatiens tenella B. Heyne ex Wight & Arn., Prod. Fl. Ind. Orient.: 140. 1834; Hook.f. & Thomson, J. Proc. Linn. Soc. Bot. 4: 123.1859; Hook.f., Fl. Brit. India 1: 447. 1874

& Rec. Bot. Surv. India 4: 46. 1906; Gamble, Fl. Madras 1: 141. 1915; Vajravelu in Nair & Henry, Fl. Tamil Nadu 1: 56.1983; Sharma *et al.*, Fl. Karnataka: 39.1984; Vivekananthan *et al.*, in Hajra *et al.*, Fl. India 4: 215. 1997; Bhaskar, Tax. Mon. *Impatiens* of Western Ghats 175. 2012.

Herb, 20–60 cm high. Stem semi-terete, swollen at nodes; branches opposite, glabrous, reddish pink. Leaves opposite, lanceolate, 4–5.5 x 1.5–2 cm, cuneate at base, apiculate-serrate margins, acute to acuminate at apex, pubescent above, glabrous below, shortly petiolate or sessile; nerves obscure above, prominent below; petioles decurrent on stem and terminating in glands, 2 mm long. Flowers binate, axillary, 1 cm across, dark pink, lilac to white; bracts linear, 2 x 0.5 mm, acute to acuminate at apex, glabrous; pedicels slender, 2.3–2.7 cm long, pubescent with 2 rows of hairs, deflexed in fruits. Lateral sepals linear-lanceolate to oblanceolate, 0.5 x 0.75 mm, acute to acuminate at apex, glabrous, distinctly 1- nerved. Standard petal broadly ovate, 4 x 3.5 mm, concave, mucronate at apex, white to pink. Wing petals 1 x 0.4 cm, auricled near base, bilobed; basal lobe ovate to subulate, 3 x 1 mm, obtuse at apex; distal lobe asymmetrically obovate, 6 x 4 mm, rounded to obtuse at apex, auricle 1 mm long, rounded at apex; lip saccate, 8 x 3 mm, 3mm deep, acute at apex; spur basal, tubular, 7mm long, rounded at apex, slightly curved, glabrous. Column 3 x 1.5 mm, curved; filaments 0.2 x 0.5 mm; ovary linear- oblongoid, curved at apex, glabrous. Capsules ellipsoid or oblongoid to oblanceoloid, 1.2 x 0.5 cm, glabrous; pedicels 3 – 4 cm long; seeds subglobose, 2.5 x 2 mm, laterally compressed, glabrous, shining, dark brown to black; funicle prominent.

Fl. and Fr.: July – Nov.

Distribution: Endemic to Nilgiris in Naduvattam.

Impatiens tirunelvelliensis L. Joseph & Bhaskar in Bhaskar, Tax. Mon. *Impatiens* of Western Ghats 87.2012.

Epiphytic scapigerous seasonal herbs with corms hidden inside mosses, 18–22 cm height. Leaves 2–3, radical; orbicular with dark veins, base cordate, sparsely hairy above and glabrous beneath, petiole 6–8 cm long, faintly crenate, size varying 3–5 cm across. Flowers in erect scapes, scape 18–22 cm long, flowers pink with purplish hairs at the mouth, 1.5 cm to 1.7 cm across; 2 lateral sepals minute, 3 mm long, ovate, lip and spur white, lip 5 mm long and flat, spur 10 mm long, terete or cylindrical, straight or slightly bent in the middle part; Standard broad with lateral wings, 5 mm long, 8 mm wide, wings 3-lobed, dorsal auricle absent, basal lobe 7 mm long, middle lobe 5 mm long, distal lobe 15 mm long. Capsule ovoid, many seeded, seeds minute, oblong, brownish, hairy all over, hairs very short or dwarf with spiral banding pattern; pollen pink, spherical or triangular, 3-colpate with reticulate exine sculpturing.

Fl. and Fr.: July – Sept.

Distribution: Endemic to Tirunelvelli hills.

Impatiens tomentosa B. Heyne ex Wight & Arn., Prodr. Fl. Ind. Orient.2: 139.1834; Wight, Icon. Pl. Ind. Orient. 3 (1): 2, t. 749. 1844; Hook.f. & Thomson, J. Proc. Linn. Soc. Bot. 4: 121. 1859; Dalzell & Gibson, Bombay Fl.: 43.1861; Hooker, Fl. Brit.

India 1: 449.1874, excl. var. *rufescens* and Rec. Bot. Surv. India 4: 46. 1906; Cooke, Fl. Bombay 1: 173.1901; Gamble, Fl. Pres. Madras 1: 141.1915; Blatter, J. Bombay Nat. Hist. Soc. 33: 312. 1933; Vajravelu in Nair & Henry, Fl. Tamil Nadu 1: 56.1983; Kulkarni, Fl. Sindhudurg: 61.1988; Deshpande *et al.*, Fl. Mahabaleshwar 1: 119.1993; Vinekananthan *et al.*, in Hajra *et al.*, Fl., India 4: 217. 1997; Mudaliar & Prasad in Singh & Karthikeyan, Fl. Maharashtra State Dicotyl. 1: 443. 2000; Yadav & Sardesai, Fl. Kolhapur District: 95. 2002; Rathakrishnan *et al.*, in Daniel, Fl. Kerala1: 559. 2005; Bhaskar, Tax. Mon. *Impatiens* of Western Ghats 179. 2012. (Plate 1.6B)

Herb, 25–30 cm high. Stem branched; branches semi-qudrangular, densely hairy above, glabrous below, pinkish red. Leaves opposite-decussate, shortly petiolate at base, subsessile at apex, oblong to oblanceolate, 3–4.5 x 1.5–2 cm, truncate to obcordate at base, serrete at margins, acute to acuminate at apex, pubescent above, hairy on nerves below; petioles decurrent and terminating in glands, 2 mm long. Flowers 4 or 5 per axil, 1.5 cm across, dark pink with purple throat; bracts linear, 2mm long, dorsally hairy on nerve in middle; pedicels 2-3 mm long, hairy with 2 rows of hairs. Lateral sepals 2, linear, 9 x 1 mm, acute to acuminate at apex, slightly keeled, hairy on keel and side facing downwards, pink. Standard petal suborbicular, 9 x 8 mm, pink, dorsally keeled; keel mucronate at apex, hairy, green, wing petals 2.2 x 0.8 cm, auricled at base, 2- lobed; basal lobe triangular, 6 x 4 mm, acute to obtuse at apex; distal lobe asymmetrically obovate, 1.3 x 0.8 cm, elevated at base towards inner side, obtuse at apex; lip conical, 11 x 4 mm, 6mm deep, acuminate at apex; hairy, pink; spur cylindric, 3mm long, hooked, notched at apex, hairy green. Column 5 x 3 mm, bent on one side; filaments 4.5 x 1 mm, pink; anthers 0.5 x 1 mm, pink. Pistil 3.5 x 1 mm; ovary oblong-lanceoloid, glabrous. Capsules asymmetrically ellipsoid to elliptic- lanceoloid, glabrous; pedicels 2–3.5 cm long; seeds oblongoid, 2.5 x 1.5 mm, glabrous, shining, brown to brownish black; funicle prominent.

Fl. and Fr.: June – Nov.

Distribution: Coimbatore, Madurai, Nilgiri, Tirunelveli, Theni.

Impatiens travancorica Beddome, Icon. 29. t. 142. 1868 – 1874; Hook.f., Fl. Brit. India 1: 464. 1874; Bhaskar, Tax. Mon. *Impatiens* of Western Ghats 252. 2012. (Plate 1.6C)

Herbs, annual, succulent, up to 30 cm high; stems usually simple, rarely branched, glabrous. Leaves crowded and rosulate towards apex, elliptic, sub-acute, crenate, ciliate in crenatures; petiole as long or longer than lamina. Peduncles 1–3, subterminal, erect, 2.5 to 5 cm long, 2–4 flowered; bracts green, subulate, lanceolate, spreading. Flowers umbelled, white, streaked with red with pink patches near base, membranous. Lateral sepals obliquely ovate, acuminate. Lip boat-shaped; spur short, stout, obtuse or absent. Standard small, concave, distal lobes of wings larger, sessile.

Fl. and Fr.: September–Jan.

Distribution: Mahendragiri hill, Kalaar, Kadalaar valleys, Tirunelveli ranges.

Plate 1.5

A: Impatiens munronii **Wight;** ***B: Impatiens omissa*** **Hook. f.;** ***C: Impatiens parasitica*** **Beddome;** ***D: Impatiens phoenicea*** **Beddome;** ***E: Impatiens scabriuscula*** **B. Heyne ex Wallich;** ***F: Impatiens scapiflora*** **B. Heyne ex Roxb.**

Plate 1.6

A: Impatiens tangachee Beddome; *B: Impatiens tomentosa* B. Heyne ex Wight & Arn.; *C: Impatiens travancorica* Beddome; *D: Impatiens trichocarpa* Hook. f.; *E: Impatiens verticillata* Wight; *F: Impatiens viridiflora* Wight.

Impatiens trichocarpa Hook. f. in Hooker's Icon. Pl. Ser. 4. 10: t. 2914. 1913; Gamble, Fl, Pres. Madras 142. 1915; Bhaskar, Tax. Mon. *Impatiens* of Western Ghats 218. 2012. (Plate 1.6D)

Annuals, 20–60 cm high, branching, appressed hairy. Leaves alternate, subsessile, elliptic-lanceolate, acute, serrulate along margins, 4–7 x 2.5–3.5 cm, coriaceous. Flowers solitary or binate; pedicels 1–1.5 cm long; glabrous or puberulous. Lateral sepals ovate, acute. Lip cymbiform, 8 mm long, pubescent; spur shorter than lip, robust, incurved. Standard orbicular, 7 mm across, cucullate, keeled on dorsal side ending into a mucro. Wings 1.2 cm long, bilobed; basal lobe rounded; distal lobes larger than basal ones, semilunate or dolabriform; dorsal auricle prominent, decurved. Capsules ellipsoid, 1–1.2 cm long, turgid, tomentose; seeds globose, 2 mm, papillose.

Fl. and Fr.: Aug.–Dec.

Distribution: Endemic to Nilgiris, but also reported from Tirunelveli.

Impatiens umbellata B. Heyne ex Roxb., Fl. Ind. 2: 464. 1824; Hook. f., FI. Brit. India 1: 461.1874; Bhaskar, Tax. Mon. *Impatiens* of Western Ghats 253. 2012. (Fig. 24).

Herbs, annual, stems simple, rarely branched, 10 – 20 cm high, naked below. Leaves rosulate, ovate or elliptic, sometimes orbicular, obtuse or acute, broadly crenate, 2.5–5cm, often dark-brownish to purplish beneath. Peduncles subterminal, erect, 1–6 flowered, shorter than leaves. Lateral sepals broad, ovate. Lip boat-shaped; spur 2.5 cm long, slender, incurved. Standard orbicular. Wings with subequal lobes, obtuse. Capsules ellipsoid, turgid, glabrous; seeds many, echinate.

Fl. and Fr.: Aug. – Oct.

Distribution: Endemic to Coutrallum in Tirunelveli.

Impatiens uncinata Wight in Madras J. Lit. Sci. Ser. 1. 5: 11. t.6. 1837 & Icon. 3 (1): 2. t. 747. 1844; Hook. f., Fl. Brit. India 1: 465. 1875; Bhaskar, Tax. Mon. *Impatiens* of Western Ghats 254. 2012.

Herbs, slender, up to 30 cm high, erect; stems sparingly branched, leafy. Leaves alternate, ovate or suborbicular, acute or cordate at base, acute or acuminate at apex, crenate to bristly serrate along margins, 3–7 x 2–4 cm, membranous, often ciliate at base, glabrous or hairy on nerves above. Petioles 1.5–4 cm long, glanduliferous at apex. Flowers 4–8 on short racemes, 2cm across, pink; penducles axillary, solitary, up to 8 cm long; pedicels 1–2 cm long. Lateral sepals obliquely ovate, pointed. Lip sub-campanulate; spur inflated in middle, hooked at tip, short, stout. Standard elliptic. Wings bilobed, divaricate; distal lobes obovate or falcate; basal lobes oblong, smaller than distal ones. Capsules ellipsoid, attenuate below, beaked above, glabrous, 8 mm long; seeds few, globose, glabrous.

Fl. and Fr.: May – Jan.

Distribution: Kannikatti, Naterikal, Kalivayalpil, Kakachi of Agasthymalai.

Impatiens verticillata Wight in Madras J. Lit. Sci. Ser. 1. 5: 15. 1837; Hook. f., Fl. Brit. India 1: 452. 1874; Bhaskar, Tax. Mon. *Impatiens* of Western Ghats 256. 2012. (Plate 1.6E).

Herbs or undershrubs up to 60 cm high; stems branching in lower half, swollen at joints. Leaves opposite or in whorls of 3–6, elliptic or elliptic-lanceolate, acuminate, crenate-serrulate along margins, 6–12 x 1.2–2.2 cm; petioles 0.5–1.5 cm long. Flowers in 3 to many flowered subumbels, scarlet; peduncles up to 4 cm long; pedicels 0.7–2 cm long. Lateral sepals linear-lanceolate, 0.8–1 cm long. Lip cymbiform; spur slender, curved, 2.5–3.5 cm long. Standard cucullate, 1 x 1.2 cm. Wings bilobed, clawed, up to 2.8 cm long; lobes oblong-obovate, with hairs on dorsal keels. Capsules fusiform, 0.8 -1.2 cm long, many-seeded; seeds 2.5 mm long, brown, hairy at pointed end.

Fl. and Fr.: Almost throughout the year.

Distribution: Anamalais, Tirunelveli, Shivagiri, Travancore high range and in hills of Coimbatore district.

Impatiens viridiflora Wight in Madras J. Lit. Sci. Ser. 1. 5: 9. 1837; Hook. f., Fl. Brit. India 1: 460.1974; Bhaskar, Tax. Mon. *Impatiens* of Western Ghats 106. 2012. (Plate 1.6F)

Herbs, erect, glabrous, epiphytic on tree trunks, stems fleshy, thick, 2.5 cm across; branches few, short nodes. Leaves congested at ends of branches, elliptic, sometimes ovate, obovate or lanceolate, acute at apex, crenate-serrate, crenatures larger with a few glandular cilia at base, 3–5 cm long, fleshy, petiole stout. Peduncles axillary, two, rarely one. Flowers completely green, 2.5 cm along with lip. Lateral sepals reflexed or deflexed, linear-oblong, acuminate. Lip saccate, narrowed into a stout involute spur; spur wrinkled, green, laterally compressed. Standard orbicular, bilobed, strongly keeled on back. Wings fleshy, bilobed; basal lobes concealed under standard, distal lobes oblong, rounded. Capsules glabrous.

Fl. and Fr.: Sept.

Distribution: Hills of Shivagiri, Tirunelveli above 1500 m.

Impatiens viscida Wight in Madras J. Lit. Sci.Ser. 1.5: 12. 1837; Hook. f., Fl. Brit. India 1: 462. 1874; Bhaskar, Tax. Mon. *Impatiens* of Western Ghats 257. 2012

Herbs, slender, 0.5–1 m high, branching, rooting at lower nodes; stems angled. Leaves alternate, elliptic-lanceolate, serrate, 5–8 cm long, pubescent beneath, nerves on both surfaces covered with stiff hairs. Peduncles axillary, viscid, 5–10 cm long. Flowers umbelled or in short racemes, deep-pink or light-purple; pedicels viscid. Lateral sepals sub-orbicular, equalling the small standard. Lip boat-shaped; spur long, strong, incurved. Distal lobes of wings large, suborbicular. Capsules glabrous, ellipsoid, beaked, seeds globose, hairy.

Fl. and Fr.: June – January.

Distribution: Chiefly distributed in Pulney hills, Pambar shola, Silver Cascade, Pambar river, Perumalmalai Kodaikanal.

Impatiens viscosa Beddome in Madras J. Lit. Sci. Ser. 2. 10: 68. t. 7. f., 7.1859; Hook. f., Fl. Brit. India 1: 453. 1874; Bhaskar, Tax. Mon. *Impatiens* of Western Ghats 258. 2012. *I. ballandii* Beddome, Icon. 44, t. 192. 1868–74; Hook. f., Fl. Brit. India 1: 482. 1874.

Figure 1.23: ***Impatiens tenella*** **Heyne.**

Figure 1.24: ***Impatiens umbellata*** **Heyne ex Roxb.**

Figure 1.25: ***Impatiens verticillata*** **Wight**

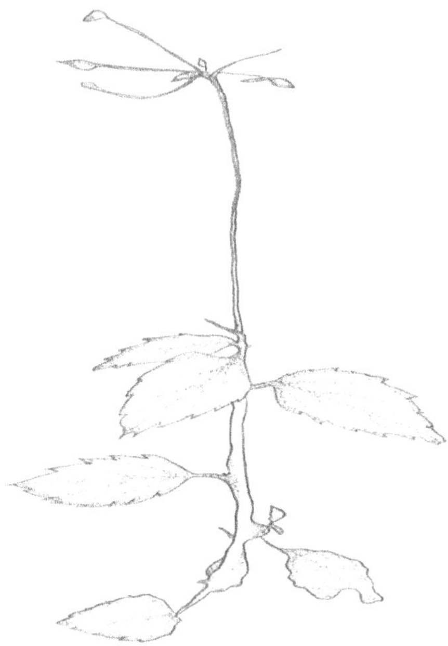

Figure 1.26: ***Impatiens viscosa*** **Beddome**

Herbs, 12–30 cm high; stems succulent, often branched, rooting below. Leaves opposite, rarely alternate, elliptic-ovate, serrate, 2–6 x 0.8–3 cm, sparsely hairy above, glabrous below, glandular at base. Flowers in axillary and terminal subumbels, pink; peduncles 2.5–7 cm long; pedicels slender, 0.5–1.5 cm long. Lateral sepals ovate-subulate. Lip cymbiform; spur as long as or longer than wings, incurved, inflated at middle, subulate at tip. Standard orbicular, concave, obscurely keeled on dorsal side. Wings bilobed; distal lobes larger than basal ones, clawed, dorsal auricle filiform, enclosed in spur. Capsules ovoid, turgid, 7–10 seeded; seeds echinate.

Fl. and Fr.: July – Dec.

Distribution: Paralai, Anamalais.

Impatiens wightiana Beddome in Madras J. Lit. Sci. Ser. 2. 20. 67. 1858 & Icon. 30. t. 146. 1868–1874; Hook. f., Fl. Brit. India 1: 467. 1875; Bhaskar, Tax. Mon. *Impatiens* of Western Ghats 269. 2012

Herbs, erect, suffruticose, branched, rooting at lower nodes. Leaves alternate, lanceolate, unequal at base, acuminate at apex, bristly serrate, 5–18 x 1.5–4.5 cm, glabrous or glabrescent, lanceolate; petiole with numerous stalked glands towards apex, up to 8 cm long. Peduncles axillary, solitary, little longer than petioles. Flowers white, mottled with pink, 1 cm across, in axillary racemes, up to 11cm long. Lateral sepals small. Lip funnel-shaped; spur short, hooked. Standard orbicular, ovate, apiculate with a green dorsal ridge. Capsules ovoid–trapezoid, acute with a pointed tip, glabrous; seeds 4–5, hairy.

Fl. and Fr.: Sept.

Distribution: Parali, Anamalais.

Impatiens yercaudensis Bhaskar, Tax. Mon. *Impatiens* of Western Ghats 180. 2012

Annul or perennial herbs, up to 60 cm tall; stems woody, stiff, terete or angled, erect, yellowish- red, often from a rooting prostrate or procumbent basal stem, internodal length short, < 2 cm in the upper part, about 2 cm long in the middle, > 2 cm below, rooted at lower nodes, whole plant including stem, leaves, pedicel, lip, spur prominently densely scabrous hairy, even the older parts of stem pubescent and flower buds, including spur, are more densely pubescent. Leaves opposite-decussate, arranged at apex, stipulate, petiolate or sessile, leaf shape varies from rhomboidal to ovate, ovate-oblong or elliptic, 1.5 x 1 cm, acute or obtuse, base acute, rounded or sub-cordate, often pale reddish, prominently impressed venation above, scabrous hairy all over upper surface, hairy only on bulged veins below surface, serrate, serratures setaceous. Flowers axillary, solitary or fascicled, in opposite-decussate pattern as in the leaves, 1.3–2.5 cm across, pale pink with deep pinkish blotches at throat, honey guides present in wings as well as lip; pedicels longer than leaves, 4.5 cm in flower, in fruits up to 5.5 cm long, deflexed in fruit, densely hairy; sepals linear-acute, hairy; lip boat-shaped, with prominent honey guides and an yellow eye at the centre inside, puberulous outside; spur slender, 2–3 cm long, slightly incurved, distinctly hairy all along, often bifid or notched at tip. Standard orbicular, acuminate or horned, hairy above, wing bilobed, larger anterior lobe 1.5 x 2 cm, smaller basal lobe deltoid, 3–5 mm long, 3mm wide, enclosed by hood-like standard, tip depressed and minutely

Plate 1.7

A: Impatiens balsamina **variety; *B: Impatiens walleriana*** **variety; *C: Impatiens walleriana*** **variety; *D: Impatiens walleriana*** **variety; *E: Impatiens walleriana*** **variety; *F: Impatiens walleriana*** **variety; G. *Impatiens repens;*** **H. *Impatiens walleriana*** **variety.**

mucronate, dorsal auricle short, round, 1 mm long deep pink or dark purple in colour, appear like two projections in the throat placed opposite to each other. Pollen light pink gradually turning to cremish yellow. Capsules 1.2–2 cm long, ellipsoid, turgid in middle, fruiting pedicel hairy, fruit glabrous; seeds many, 16–18 per fruit, smooth black shiny.

Fl. and Fr.: Nov.–Jan.

Distribution: Endemic to Shevaroy hills Yercaud.

Commercial Varieties

In addition to the wild species reported from Tamil Nadu, more than 50 commercial varieties of *Impatiens* are cultivated in the garden. *Impatiens repens*, Sri Lankan creeping balsam is also grown in several gardens in Tamil Nadu (Plate 1.7A-H).

Disucssion

The family Balsaminaceae has about 1000 species with two genera, the monotypic *Hydrocera* Blume and the large genus *Impatiens* L. (Fischer, 2004; Stevens, 2004). The genus *Hydrocera* is a monotypic and can be characterized by its free petals and berry like capsular fruits. The *Impatiens* has four lateral petals connate in pairs and five valves, loculicidal capsule. The sister group relationship between *Hydrocera* and *Impatiens* is also confirmed by recent molecular analysis (Yuan *et al.*, 2004). Among these 1000 species reported in the world, 230 species have been reported in India, of which 109 species occur in South India (Hooker, 1904; Gamble, 1915; Blatter, 1933; Fischer, 1930-1938; and Barnes, 1939.) Among them at least 90 per cent of the species are restricted to hills of South India.

Species belonging to this genus are commonly referred as Balsams or Jewel weeds. Many species are cultivated as ornamental and very few of them are used as medicine and cosmetics. *Impatiens* are primarily annual or perennial herbs and occasionally shrubby. The stems typically rather than thick and fleshy and often semitranslucent, rarely becoming thickened and woody. Leaves are fleshy and thin becoming membranous and often transparent when dry. Flowers are zygomorphic and fleshy and described with three or occasionally five sepals, of which the lower one is much larger and often coloured. Generally, it is characterized by a nectary tipped spur. Corolla consists of five petals, a large and sepaloid standard in the upper position and four lateral wings. The explosive nature of the seeds is an important factor for the derivation of family. The fruits are capsules with a fleshy pericarp, the outer larger cells are highly turgid and thus a great strain is put upon the whole. The dehiscence mechanism is septifragal and is affected by a touch when the fruit is ripen. Abounded seeds are usually found and seeds of few South Indian Balsams exhibit dormancy for 7–8 months to overcome dry months from December – May (Ramasubbu, 2009, Ramasubbu *et al.*, 2011, 2013). Also, the tubers in the case of Scapigerous and other tuberous species exhibit the same.

In Species Plantaram, Linnaeus (1753) described seven species of *Impatiens* including *I. balsamina* which had been cultivated for several hundred of years.

Taxonomical studies in *Impatiens* in India were initially started by Wight who recognized 16 species from South India and this was followed by R. H. Beddome (1869-1874) who recognized 14 species and others who also made substantial contributions. The major contribution to the Indian species of *Impatiens* was by J. D. Hooker (1904–1906) and described 43 species of *Impatiens* from Indian subcontinent and listed around 135 species known from British India. The studies on *Impatiens* were contributed by Gamble and Fischer (1930–38) and Barnes (1938, 1939) who concentrated mainly on South Western Ghats. Eight new species of *Impatiens* were published by Fischer from Southern Ghats and it was followed by a number of later works in floristic analysis. They reported 109 species of *Impatiens* in South India are restricted to 5 endemic centers, *viz.*, Agasthiyamalai, Anamalai – high ranges, Pulney hills, Nilgiri – Wayanad – Silent Valley – Coorg and Shimoga.

In the present study, a total of 74 species including one *Hydrocera* are documented during the systematic studies on family Balsaminaceae in Tamil Nadu. Among the 74 species reported in the different forest regions of Tamil Nadu, 64 herbs and 9 shrubs are included. In the total 73 species of *Impatiens,* 15 scapigerous, 6 epiphytes and 52 terrestrial plants are documented. The number of species of *Impatiens* distributed in Maharashtra was 23 (including 2 varieties), in Karnataka and Goa 50 (including 11 varieties), in Kerala 51 (including 2 varieties), and in Andhra Pradesh 6 (including 4 varities). But, the number of species reported in Tamil Nadu is comparatively than the remaining states of South India. There are 73 species of *Impatiens* and one species of *Hydrocera* reported from different forest regiongs of Tamil nadu (Table 1.1).

Table 1.1: Distribution of of *Impatiens* in different States of Peninsular India.

Sl.No.	*State*	*Species*	*Varieties*	*Total No. of Taxa Reported*
1.	Maharashtra	21	2	23
2.	Karnataka and Goa	39	11	50
3.	Kerala	51	2	53
4.	Tamil Nadu	74	3	77
5.	Andhra Pradesh	2	4	6

The family Balsaminaceae has wide range of distribution in Tamil Nadu. *Hydrocera triflora* is restricted to the marshy and aquatic plains of Thanjavur district of Tamil Nadu. Among the 30 districts of Tamil Nadu, *Impatiens* is distributed in 10 districts. Among the 10 districts, Nilgiri and Coimbatore have more number of species, followed by Dindigul, Tirunelveli, Theni, Kanyakumari (Table 1.2). Nilgiri, Coimbatore, Dindigul, Tirunelveli are mostly occupied with hilly regions. Due to the high altitude habit, forest dwelling nature and orophytic behavior, the *Impatiens* distributed in more number in the above reported districts.

Impatiens hookeriana was described by Heyne in 1834 by its special morphological characters. But, recently the species has been merged with *Impatiens grandis* due to its few similar characters. The present study reveled that, *Impatiens hookeriana* have a distinct character and it has been separated from *I. grandis* due to its several dissimilar sexual charecters. *I. hookeriana* has been revalidated and separated from the *I. grandis* and similar results were submitted for the recommendation.

Table 1.2: Distribution of *Impatiens* L. in Top 10 Districts of Tamil Nadu.

Sl.No.	*Name of the District*	*No. of Species Reported*
1.	Nilgiris	39
2.	Coimbatore	23
3.	Dindigul	19
4.	Tirunelveli	19
5.	Theni	12
6.	Kanyakumari	6
7.	Namakkal	2
8.	Dharmapuri	1
9.	Thanjavur	1
10.	Virudhanagar	1

Impatiens munnarensis, a herbaceous species described by Barnes in 1938 from its type locality of top station in Munnar, Idukki district, Kerala, India. After its type collection, no reports have been emerged to relocate such beautiful balsam species. The earlier reports revealed that, the species has been extinct from the wild. The present investigation, rediscovered the species from new locality (Varaiadu mottai, High wavies mountains, Theni district, Tamil Nadu). The species was studied carefully and confirmed as *I. munnarensis* by various monograph published by several researchers. The current research work is being not only rediscovery of species but also the new distributional record to the Tamil Nadu state.

From the past, there were several Botanical and non-botanical researchers involved in the survey, exploration and documentation of the family Balsaminaceae in Tamil Nadu. Even then, the proper taxonomical documents of *Impatiens* have not been raised. In the present study, we documented 73 species of *Impatiens* only from the political boundaries of Tamil Nadu. In the current research work, we found that most of the specie of balsams blooming from August –October (Figure 1.27). Due to impact of south west monsoon, the balsams acquiring favourable condition for flowering and fruiting. Very few balsams alone flowering throughout year and most of them completing their life cycle within 3-6 months.

From the present study, we conclude that the high range is the suitable habitat for the growth of *Impatiens* species. The high range is one of the botanically richest areas in the Western Ghats with respect to the species of balsams. The highly undulating mountainous configuration and the formation of a well marked upland shola forest with very high rainfall provide suitable habitat for *Impatiens* (Barnes, 1939). Many species of *Impatiens* are found restricted to specific altitudinal zones and most of the exclusive endemics are found in restricted isolated pockets in the high altitudes (2000m asl) in Western Ghats. According to Bhaskar (1981), the high rate of endemism in this genus is due to its orophytic nature, extreme adaptation to some particular microclimate. This perplexing genus has two main centers of distribution in India; the Himalayas in the north and the Western Ghats in the south and each area has a natural home for 100 species each. It generally occurs at high altitudes (1500m asl), where there is active monsoon (rainfall), optimum temperature (20° C) and sunlight. The hilly ranges of Western Ghats, phytogeographically an important topographical unit in southern India consists of areas with intermittent rain,

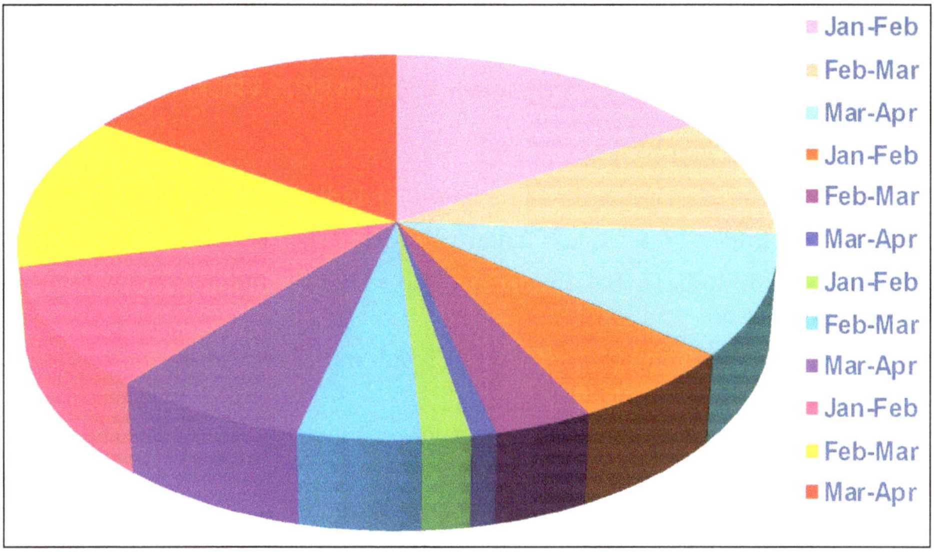

Figure 1.27: Flowering Period of *Impatients* Distributed in Tamil Nadu.

innumerable valleys, grassy slopes, perennial streams and water falls form an appropriate habitat for the balsams. Most of the Indian *Impatiens*, particularly in the Western Ghats are narrow endemics and confined to specialized ecological habitats. Their colonizing abilities and regenerative capacities are very poor due to human interference. Hence they become extinction prone. When the habitat is disturbed or destroyed, these endemics may become rare, vulnerable, endangered or even extinct. Several factors, therefore, may be responsible for the distribution and subsequent establishment of balsams in the Western Ghats. The current research work will serve as a better monograph for the identification of wild *Impatiens* in the Western Ghats of Tamil Nadu. The details provided in the current research work will helpful for the better understanding of the family Balsaminaceae.

Acknowledgement

We thank Dr. A.G. Pandurangan and Dr. A.K. Sreekala, JNTBGRI, Kerala for their support and encouragement. We also thank Deputy Director, BSI (Southern Circle), Coimbatore for the permission to consult the herbarium specimens.

References

Anil Kumar, N., Narayanan, M.K.R, Sujanpal, P., Meera Raj, R., Sujana, K. A. and Mithunlal. 2011. *Impatiens veerapazhasii* (Balsaminaceae), a new scapigerous balsam of Wayanad, Western Ghats, India. J. Bot. Res. Inst. Texas **5 (1):** 153-158.

Arnott, G. A. W. 1834. New species of Indian Balsaminaceae. Hooker's Comp. Bot. Mag.**1:** 320-325.

Augustine, J., Sasidharan, N. and Sivadsan, M. 1999. Balsams of Periyar Tiger Reserve, Southern Western Ghats, Kerala. In: Sivadasan, M. and Mathew, P. (eds.) Biodiversity, Taxonomy of flowering plants, Mentor books, Calicut. pp 275-292.

Blatter, E. 1933. Revision of the Flora of Bombay Presidency: Balsaminaceae. J. Bombay Nat. Hist. Soc. **36 (2):** 307-320.

Barnes, E. 1938. New or little known plants from South India: VIII. Kew Bull. **1938:** 32-34.

Barnes, E.1939. The species of Geraniaceae occurring on the Travancore High ranges including the descriptions of a new balsam. J. Indian Bot. Soc. **18 (3):** 95-105.

Beddome, R. H. 1869-1874. Icones Plantarum Indiae Orientalis 1: t. 140-153.

Bhaskar, V. 1975. Studies in Balsaminaceae. Ph. D. Thesis submitted to University of Mysore. Mysore. pp 496.

Bhaskar, V.1981. The genus *Impatiens* L. in South India: Endemism and affinities. Indian Foreste **107:** 368-376.

Bhaskar, V. 2006. *Impatiens clavata* Bhaskar, a new scapigerous balsam (Balsaminaceae) from Bisle Ghat, Western Ghats, South India. Curr. Sci. **91 (9):** 1138-1140.

Bhaskar, V. 2012. *Impatiens* L. (Balsaminaceae) of Western Ghats, South India the key to the genus for endemism, Centre for Plant Taxonomic studies, Bangalore, pp 31-261.

Candolle, A. P. de. 1824. Prodromus systematis naturalis regni vegetabilis, sive enumeration contracta irdinum, generum, specie rumque plantarum hucusque cognitarum juxtra method naturalis norma digesta, Paris 1: 685-688.

Chandrabose, M. 1978-79. A new species of *Impatiens* Linn. (Balsaminaceae) from South India *(I. chandrasekharanii)*. J. Bombay Nat. Hist. Soc. **75:** 901-902.

Chandrabose, M., Chandrasekaran, V. and Nair, N. C. 1984. A new species of *Impatiens* L. (Balsaminaceae) from South India. J. Bombay Nat. Hist. Soc., **81 (3):** 676-677.

Dessai, J.R.N. and Janarthanam, M. K. 2011. The genus *Impatiens* (Balsaminaceae) in the northern parts of central Western Ghats. Rheeda **21 (1):** 23-80.

Dessai, J.R.N., Joseph, L. and Janarthanam, M. K. 2009. A new species of epiphytic *Impatiens* (Balsaminaceae) from the Western Ghats, India. Taiwania **54 (2):** 149: 151.

Fischer, C. E. C. 1930–1938. New or little known plants from South India. Bull. Misc. Inf. Kew **1930:** 150, 330; **1931:** 41; **1934:** 389-394; **1935:** 92-97 and 157-160; **1938:** 32-37.

Fischer, E. 2004. Balsaminaceae. In: The families and genera of Vascular plants–Dicotyledons: Celastrales, Oxalidase, Rosales, Cornales, Ericales vol. 6. (Kubitzki, K. ed.), Spinger-Verlag, Berlin, Germany, pp 20-25.

Fyson, P. F. 1915. The Flora of the Nilgiri and Pulney hill tops. Today and Tomorrows Printers and Publishers, 1: 59-71; 2: t. 47-55.

Fyson, P. F. 1932. The flora of the South Indian hill stations: Ootacamund, Coonoor, Kotagiri, Kodai, Yercaud and the country around. Today and Tomorrows Printers and Publishers, Vol.1. New Delhi.

Gamble, J. S. 1915. Flora of the Presidency of Madras, Bishen Singh Mahendra Pal Singh Calcutta.1: 134-145.

Ganeshaiah, K. N. 2012. Plants of Western Ghats. Vol. 2. National Bioresource development Board, Department of Biotechnology, New Delhi, pp 42-46.

Grey -Wilson, C. 1980. *Impatiens* of Africa. Balkema, A. A., Rotterrdam, Netherlands.

Hooker, J. D. and Thomson, T. 1859-60. Praecursores ad Floram Indicum: Balsaminaceae. J. Linn.Soc. Lond. (Bot.) **4:** 106-157.

Hooker, J. D.1904-06. An epitome of the British Indian species of *Impatiens* L. Rec. Bot. Sur. India, **3:** 106-157; **4 (1):** 1-10; **4 (2):** 11-35 (1905); **4 (3):** 37-58 (1906).

IUCN, 2001. IUCN Red list categories and criteria: 2001.1. Version 3

IUCN, 1987. Centre of plant diversity: A guide and strategy for their conservation. IUCN Publications, UK, pp 167.

Jain, S. K. 1963. Flora and vegetation of India – an outline. Dept. of Environ., New Delhi. pp. 1-24.

Jain, S. K. and Sastry, A. R. K. 1980. Threatened plants of India. A state of the art report, New Delhi.

Kumar, M. and Sequiera, S. 1996. *Impatiens sivarajanii*–a new species of Balsaminaceae from Silent Valley National Park, Kerala, India. Rheedea, **6 (2):** 51-54.

Kumar, M. and Sequiera, S. 1999. Observation on the epiphytic flora of Silent valley. In: Silent valley – Whispers of Reason (Manoharan, T. M. *et al.*, eds.) Kerala Forest Dept. and Kerala Forest Research Institute. pp. 252–256.

Kumar, M. and Sequiera, S. 2001. Two new species of *Impatiens* (Balsaminaceae) from India. Sida **19 (4):** 795-801.

Linnaeus, C. 1753. Species Plantarum. 1st ed. Stockhom.

Mathew, K. M. 1981. Illustrations on the Flora of the Tamil Nadu Carnatic.The Rapinet Herbarium, St. Joseph College, Tiruchurapalli.

Mathew, K. M. 1991. An excursion Flora of Central Tamil Nadu, Oxford and IBH Publ. Co. Pvt.Ltd., New Delhi. pp. 62-64.

Mathew, K. M. 1999. The Flora of the Palni Hills, South India. The Rapinat Herbarium, St. Joseph College, Tiruchurapalli. pp. 1880.

Mohanan, M. and Henry, A. N. 1994. The Flora of Thiruvananthapuram. Bot. Surv. Ind.Calcutta.

Nair, N. C. 1991. Endemism on the Western Ghats with special reference to *Impatiens* L. In: Proceedings of the Symposium on rare, endangered and endemic plants of the Western Ghats (ed. C. Karunakaran). Kerala Forest Department, Kerala. Spl. Publ. No. III. pp. 93-102.

Narayanan, M. K.R., Anil Kumar N., Meera R. R., Sivadasan M. and Alfarhan A. H. 2011. A new scapigerous species of *Impatiens* (Balsaminaceae) from India. Bangl. J. Plant Taxon., **18 (2):** 141-148.

Nayar, M. P. 1980. Endemic flora of Peninsular India and its significance. Bull. Bot. Surv. India. **22:** 2-23.

Nayar, M. P. 1997. Biodiversity challenges in Kerala and science of conservation biology. In: P. Pushpangadan and K. S. S. Nair (eds.), Biodiversity of Tropical Forests: the Kerala Scenario. SETC, Kerala, Trivandrum.

Nayar, T. S., Razia Begham and Rajkumar, G. 2006. Flowering plants of Kerala– Hand book. Tropical Botanic Garden Reserch Institute, Palode, Kerala.

Pandurangan, A. G. 1996. Systematics studies on Balsaminaceae in Peninsular India, Final Project submitted to Kerala Forest Department, Thiruvananthapuram, Kerala.

Pandurangan, A. G. and Nair, V. J. 1995. *Impatiens kulamavuensis,* a new species of Balsaminaceae from India. Novon **5:** 57-58.

Pandurangan, A.G. and Sreekala, A. K. 2008. Studies on reproductive biology of selected rare, endemic and horticulturally promising balsams from the Western Ghats, Final Project report submitted to Department of Science and Technology (SERC), New Delhi.

Rajalal, R., Pandurangan, A. G. and Pushpangadam, P.1996. Systematic studies of Balsaminaceae in Peninsular India. J. Swamy Bot. Club **13:** 59-62.

Ramasubbu, R. 2009. Reproductive biology of three endemic and endangered balsams (*Impatiens coelotropis* Fischer, *I.phoenicea* Bedd. and *I. platyadena* Fishcer) of Western Ghtas. Ph.D. thesis submitted to the University of Kerala, Thiruvananthapuram, Kerala, India.

Ramasubbu, R., Sreekala, A.K., Pandurangan, A.G. and Kulloli, S.K., 2011. Reproductive Ecology of *Impatiens platyadena* Fischer- A critically endangered balsam of Western Ghats. Curr. Sci.- **100 (10):** 1551-1554.

Ramasubbu, R. 2013. Studies on diversity, endemism and rarity in the genus *Impatiens* in Southern Western Ghats. Proc. Second Indian Biodiversity Congress, pp. 45.

Rao, R. V. S., Ayyangar, K. R. and Sampathkumar, R.1986. On the karyological characteristic of some members of Balsaminaceae. Cytologia **51:** 251-260.

Rathakrishnan, N. C., Vivekananthan, K. and Nair, V. J. 2005. In: The flora of Kerala. Vol.1. Daniel, P. (ed.). Botanical Survey of India. Kolkatta. PP 524-565.

Roxburgh, W. 1824. Flora Indica. Vol. 2. Serampore.

Shimizu, T. 2005. Seed morphology and classification of *Impatiens* (Balsaminaceae). Blumea, **50:** 447-456

Stevens, P. F. 2004. Angiosperm phylogeny website: http: /www.mobot.org/mobot/ research/apweb/.

Subramanyam, K. and Nayar, M. P.1974. Vegetation and phytogeography of the Western Ghats. Mani, M. S. (ed.), Ecology and Biogeography in India, The Hague.

Turrill, W. B. 1953. Pioneer plant Geography. The Hague.

Vajravelu, E.1983. Balsaminaceae In: Nair, N. C. and Henry, A. N. (ed.), Flora of Tamil Nadu, India. Ser. 1: Analysis. Vol.1. Bot. Surv. Ind., Coimbatore.

Vajravelu, E. and Daniel, P. 1983. Enumeration- Peninsular India. (Jain, S. K. and Sastry, A. R. K. eds.) Materials for a catalogue of threatened plants of India. Botanical Survey of India, Calcutta.

Viswanathan, M. B. and Manikandan, U. 2003. A new species of Balsaminaceae, *Impatiens tirunelvelica* from peninsular India. Bull. Bot. Surv. India, **45 (1-4):** 189-194.

Vivekananthan, K; Rathakrishnan, N. C; Swaminathan M. S. and Ghara, L. K. 1997. *Balsaminaceae.* In: (Hajra, P. K., Nair, V. J. and Daniel, P. eds) Flora of India, vol 4. Botanical Survey of India, Calcutta. pp. 95-229.

Wallich, N. 1831. A numerical list of dried plants in the East India Company's museum, London.

Wight, R1838-1853. Icones Plantarum Indiae Orientalis, Hafner Publishing Company, New Delhi (1963 Repr. ed)

Song, Y., Yuan, Y-M and Kupfer, P. 2003. Chromosomal evolution in Balsaminaceae with cytological observations on 45 species from South Asia. Carylogia **56 (4):** 463-481.

Yuan, Y. M., Song, K., Geuten, E., Rahelivololona, S., Wohlhauser, E., Fischer, E., Smets, and Kufer, P. 2004. Phylogeny and Biogeography of Balsaminaceae inferred from ITS sequence data. Taxon **53**: 391-403.

Chapter 2

Diversity and Distribution of Macrofungi in Kodagu Region (Western Ghats): A Preliminary Account

N.C. Karun[1], *K.R. Sridhar*[1]* *and K.A. Anu Appaiah*[2]

[1]*Department of Biosciences, Mangalore University, Mangalagangotri, Mangalore – 574 199, Karnataka, India*

[2]*Department of Food Microbiology, Central Food Technological Institute, Mysore – 570 020, Karnataka, India*

ABSTRACT

This chapter provides preliminary observations on the diversity, distribution and ecology of macrofungi in selected biomes (sacred groves, reserve forests and agroforests) of Kodagu region of the Western Ghats. One-time survey of three biomes during monsoon season (August-October, 2011) yielded 25 species with a highest of 23 species in agroforests and the rest two biomes possess 12 species each. A brief description based on the field and laboratory studies along with substrates and distribution in the Western Ghats are given. None of the macrofungi was frequent in all biomes and nine species were restricted to agroforests and one species each confined to sacred groves and reserve forests. Among the macrofungi recovered, a maximum of 17 species are known from the Western Ghats of Kerala followed by four species from Karnataka and one species from

* Corresponding Author: E-mail: sirikr@yahoo.com

Tamil Nadu. Eight species were edible, five species were medicinal and one species was ectomycorrhizal, and all except for *Tricholoma* sp. were recovered from agroforests shows the potential of this biome of Kodagu region in cultivation and harnessing economically valuable macrofungi. Strategies and necessity of further exploration of macrofungi in different ecological niches of the Western Ghats are highlighted.

Keywords: *Macrofungi, Mushrooms, Woodrot fungi, Western Ghats, Sacred groves, Reserve forests, Agroforests, Plantations, Diversity, Economic value.*

Introduction

Fungi constitute the third essential functional component after flora and fauna as decomposers, symbionts and pathogens representing highest global diversity between 1.5 and 3 million species based on continuous rise of reports on new species and in turn fungus/plant ratios in tropics (Hawksworth, 2001, 2012). Macrofungi especially mushrooms represent up to 41,000 species, of which about 850 species are recorded from different parts of India especially in Himalayan and Western Ghat region (Manoharachary *et al.*, 2006). Like Himalayas, the Western Ghats of India is also one of the most valuable natural repositories of macro- and microfungi of the world. It runs 1,600 km (~160,000 km^2) through the western part of Deccan Plateau in Gujarat, Maharashtra, Goa, Karnataka, Tamil Nadu and Kerala states passing through complex riverine system encompassing a wide range of hills and mountains (up to 1,200 m asl).

Some studies are available on the macrofungi in different parts of the Western Ghats. Thomas *et al.* (2001) have reported 13 species belonging to Bolbitiaceae from Kerala. In 25 sacred groves, 23 coffee agroforests and 10 reserve forest sites in Karnataka, a total of 163 macrofungi were recovered and identified only up to genus level (Bhagwath *et al.*, 2005; Brown *et al.*, 2006). Nearly 195 species and 28 species of agarics have been reported from the Nilgiri Biosphere Reserve of Tamil Nadu and Kerala region (Natarajan *et al.*, 2005a). Up to 25 species of ectomycorrhizal fungi in dipterocarp evergreen Kadamakkal Reserve Forest of Kodagu (Karnataka State) were documented by Natarajan *et al.* (2005b). Twenty-five species of *Hygrocybe* were described from Kerala by Leelavathy *et al.* (2006). Riviere *et al.* (2007) compared the diversity of ectomycorrhizal fungi of the rain forests of Africa and Western Ghats of India based on mitochondrial large subunit rRNA. Surveys in semi-evergreen and moist-deciduous forests of Karnataka yielded 315 species of macrofungi (Swapna *et al.*, 2008). Twenty-two species of *Lepiota*, nine species of *Leucoagaricus* and 15 species of *Leucocoprinus* were recorded from Kerala (Kumar and Manimohan, 2009a, 2009b). Mohanan (2011) documented 550 species (belonging to 51 families) from various parts of Kerala. Ranadive *et al.* (2011) gave a checklist of 256 species of Aphyllophorales from Maharashtra. A total of 68 species of mushrooms were recovered from Thiribhuvanamala *et al.* (2011) from Tamil Nadu and Kerala. Survey of Tamil Nadu and Karnataka yielded 56 species of woodrot fungi (Selvam *et al.*, 2012). However, compared to the diversity of flora and fauna and varied ecological niches exists in the vast area of Western Ghats, our knowledge on the macrofungal diversity, distribution and ecological functions is limited. In this report, occurrence, diversity and distribution of macrofungi from four protected biomes (two each of sacred groves and reserve forests) and two cultivated landscapes (coffee agroforests) of Kodagu region of the Western Ghats based on one-time survey are reported with a brief description of each species along with distribution in different parts of the Western Ghats.

Study Area and Sampling

Kodagu District (formerly known as Coorg) in Karnataka State is situated between 11°56, 12°52 N and 75°22, 76°11 E (Pascal and Meher-Homji 1986). The mean temperature during the coldest months (December-January) in the study area selected ranges between 8°C and 20°C, approximate rainfall ranges from 200-500 cm/annum and dry season persists for about 3-4 months (February-May; 28-32°C). The landscape of Kodagu consists of a large number of sacred groves, reserve forests, tree-covered plantations and treeless cultivated lands (Bhagwath *et al.*, 2005). For example, sacred groves (known as 'Devarakadu') are common in Kodagu region occupied about 2 per cent of the landscape accounting one grove for every 300 ha, which vary in area from a fraction of hectare up to a few tens of hectares (Kushalappa and Bhagwat 2001; Bhagwath *et al.*, 2005; Brown *et al.*, 2006). The reserve forests occupied up to 30 per cent of the area, while the shade-grown plantations occupied about 60 per cent *(e.g.*, coffee) and the treeless cultivated lands occupied 8 per cent *(e.g.*, paddy).

Macrofungi were assessed in three biomes: i) Sacred groves (Maggula and Bettoli); ii) Reserve forests (Sampaje and Makutta); iii) Coffee agroforests (V'Badaga and B'Shettigeri) (Figure 2.1). Sacred groves studied have least human interference and encompass economically valuable plant species those were maintained traditionally by the local communities as a measure of biodiversity preservation and restoration [*Artocarpus hirsutus* (wild jack), *Canarium strictum* (black dhupa), *Caryota urens* (toddy palm), *Dysoxylum malabaricum* (white cedar), *Holigarna beddomei* (kadu geru), *H. nigra* (kadu geru), *Lagerstroemia microcarpa* (nandi), *Mangifera indica* (kadu mavu), *Syzygium cumini* (nerale), *Toona ciliata* (red cedar), and *Vateria indica* (white dhupa)]. Usually edible mushrooms like *Lentinus* and *Pleurotus* are common in standing dead, woody litters and logs of *C. strictum*, *D. malabaricum*, *M. indica* and *V. indica*. The reserve forests studied are located in hilly zones with homogenous evergreen tree vegetation with high rainfall. The agroforests studied consists of shade-grown coffee plantations with common native *(Artocarpus hirsutus*, *Dalbergia latifolia*, *Syzygium cumini* and *Toona ciliata*) and exotic *(Acacia mangium*, *Acrocarpus fraxinifolius*, *Erythrina subumbrans* and *Grevelia robusta*) tree species.

Opportunistic mode of sampling was performed in three biomes during August (sacred groves), September (reserve forests) and October (agroforests) 2011. In each sampling, an area of 50 m^2 was intensely surveyed for the presence of fruit bodies of woodrots and mushrooms. Representative fruit bodies were sampled from different substrates like soil, leaf/bark/twig/woody litter, rotting logs/stubs, termite mounds and standing dead trees. The snaps of each mushroom was taken on the spot, fruit bodies were assessed for general features like type, colour, dimension, extent of fruit bodies (solitary or gregarious) and substrates. Part of fragile mushrooms possessing

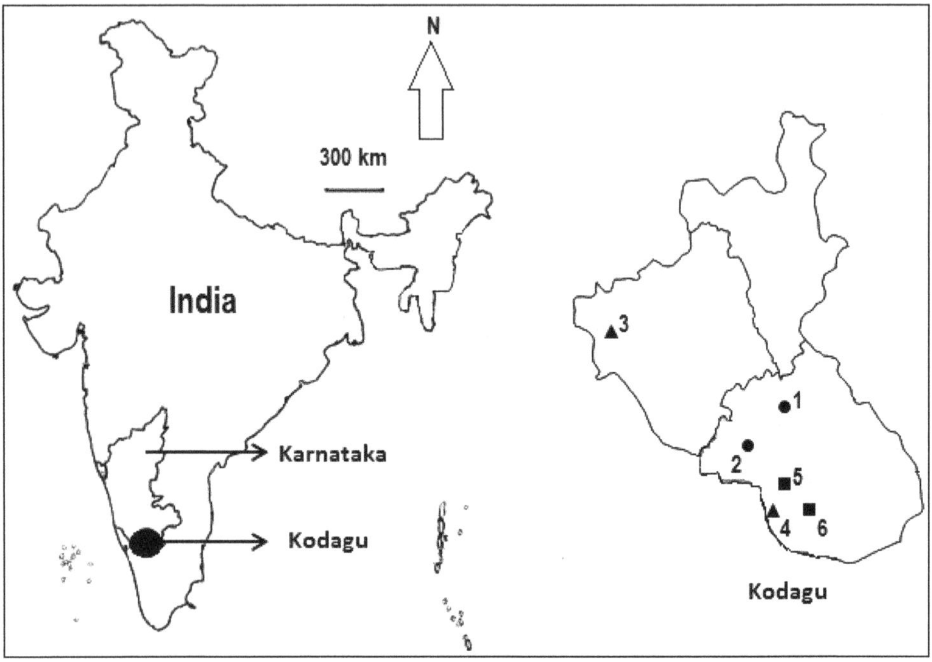

Figure 2.1: Map of the Study Area and Biomes Studied in Kodagu District of Karnataka State of the Western Ghats

Sacred grove: 1. Maggula, 2. Bettoli; Forest reserve: 3. Sampaje, 4. Makutta; Agroforest: 5. V'Badaga, 6. B'Shettigeri.

spores was fixed in lactophenol or aniline-blue-lactophenol in labeled vials. Sturdy fruit bodies of mushrooms were collected in labelled polythene bags and brought to the laboratory for further observations. Spores of each mushroom were evaluated based on the light microscopic observations. Each mushroom was fixed (formalin-ethanol-water) and deposited in the herbarium of the Department of Biosciences, Mangalore University (# MUBSNCKKRSMF-002-026) (Figures 2.2–2.5). A brief morphological description of each fungus is given after confirming the identity based on the monographs (Kibby, 1979; Purkayastha and Chandra, 1985; Jordan, 1995; Mohanan, 2011) along with their substrates and distribution.

Diversity and Distribution

One-time survey in sacred groves, reserve forests and agroforests in Kodagu region yielded 25 macrofungi (Table 2.1; Figures 2.2–2.5). Agroforests consists of a maximum of 23 species and the rest (sacred groves and reserve forests) possess 12 species each. Eight species were seen in all biomes *(Auricularia auricula-judae, Cookeina tricholoma, Daldinia concentrica, Ganoderma applanatum, Lentinus dicholamellatus, Lenzites vespacea, Termitomyces tylerianus* and *Xylaria longipes*), nine species were restricted to agroforests *(Auricularia mescenterica, Clitocybe infundibuliformis, Crepidotus variabilis, Exidia glandulosa, Gymnopilus junonius, Lentinus patulus, Pleurotus pulmonarius, Ramaria*

Table 2.1: Distribution of Macrofungi in Three Biomes of Kodagu Region of the Western Ghats.

Species (Figures 2.2–2.5)	*Sacred-Groves*	*Reserve Forests*	*Agro-forests*	*Economic Value*
Auricularia auricula-judae (Bull.) Quél. (2a)	++	++	+++	Edible and medicinal
A. mescenterica (Dicks.) Pers. (2b and c)	–	–	++	
Clitocybe infundibuliformis Quél. (2d and e)	–	–	+	Edible
Cookeina tricholoma (Mont.) Kuntze (2h)	+++	+++	+	
Crepidotus variabilis (Pers.) P. Kumm. (2f and g)	–	–	+++	
Cyathus striatus (Huds.) Willd. (2i)	+	–	+++	
Daldinia concentrica (Bolton) Ces. & De Not. (3a)	++	++	++	Medicinal
Dictyophora cinnabarina W.S. Lee (2j)	–	+	+	
Exidia glandulosa (Bull.) Fr. (3b)	–	–	+	
Ganoderma applanatum (Pers.) Pat. (3c and d)	++	++	++	Medicinal
Geastrum sp. (3g)	+	–	++	
Gymnopilus junonius (Fr.) P.D. Orton (3e and f)	–	–	+	Edible
Ileodictyon gracile Berk. (3h)	–	++	++	
Lentinus dicholamellatus Manim. (4a and b)	+	+	++	Edible
L. patulus Lév. (4e)	–	–	+++	
Lenzites vespacea (Pers.) Pat. (4c and d)	++	++	++	
Panus conchatus (Bull.) Fr. (4g and h)	–	++	++	
Pleurotus pulmonarius (Fr.) Quél. (4a and b)	–	–	++	Edible
Polyporus sp. (4c and d)	+	–	–	
Ramaria sp. (4f)	–	–	++	
Scleroderma verrucosum (Bull.) Pers. (5e)	++	–	++	Edible and Ectomycorrhizal
Termitomyces tylerianus Otieno (5f and g)	++	++	+++	Edible and medicinal
Tricholoma sp. (5i and j)	–	++	–	Edible
Trogia infundibuliformis Berk. and Broome (5h)	–	–	++	
Xylaria longipes Nitschke (5k)	++	++	++	Medicinal
Total species	12	12	23	

–: Not found; +: Rare; ++: Common; +++: Frequent.

sp. and *Trogia infundibuliformis*) and one species each confined to the sacred groves (*Polyporus* sp.) and reserve forests (*Tricholoma* sp.). None of the macrofungi was frequent in all biomes. Only *Cookeina tricholoma* was frequent in sacred groves as well as reserve forests, while *A. auricula-judae*, *C. variabilis*, *C. striatus*, *L. patulus* and *T. tylerianus* were frequent in agroforests. Based on published records, a maximum of 17 species are known from the Western Ghats of Kerala followed by four species in Karnataka and one species in Tamil Nadu. One species each of *Geastrum, Polyporus, Ramaria* and *Tricholoma* was not identifiable up to species level and seem to be new species to the Western Ghats needs further enquiry. Among the macrofungi recorded, eight species were edible, five species were medicinal and one species was ectomycorrhizal based on literature and traditional knowledge (Purkayastha and Chandra, 1985; Jordan, 1995; Puttaraju *et al.*, 2006; Das, 2010; Jha *et al.*, 2011; Mohanan, 2011). All of them (except for *Polyprus* sp. and *Tricholoma* sp.) were recovered from the agroforests shows the potential of this biome in Kodagu region for cultivation and harnessing economically important macrofungi.

Description

Auricularia auricula-judae (Bull.) Quél. (Figure 2.2a) (Auriculariaceae – Basidiomycotina) (# MUBSNCKKRSMF-002)

Non-gilled, small to medium, fleshy, neither encrusting nor bracket-like, purple brown to grey brown, fruit bodies sessile, ear-like, grow in clusters, frequent, annual, lignicolous, odour and taste not distinctive, edible and medicinal.

Fruit body. Upper surface purple brown with grey tinge when damp, shiny, lower surface grey brown, shiny, smooth becoming wrinkled, inverted cup-/irregular bracket (flabelliform)/fan-shaped, often fused with adjacent fruit bodies and narrowly attached: 4.4 (2.1-6.2) cm diam. (n=10). Flesh purple brown, gelatinous, tough and rubbery when damp, and hard and brittle on drying.

Spores. Hyaline, allantoid or cylindrical and smooth: 12.9 (11.1-13.9) × 5.2 (4.7-5.5) µm (n=10).

Substrate. Rotting logs and bark of *Acacia mangium, Acrocarpus fraxinifolius, Artocarpus hirsutus, Careya arborea, Olea dioica* and *Toona ciliata*.

Occurrence. Found in all the locations of sacred groves, reserve forests and agroforests.

Distribution. Karnataka (Swapna *et al.*, 2008), Tamil Nadu (Johnsy *et al.*, 2011; Sargunum *et al.*, 2012) and Kerala (Mohanan, 2011).

Auricularia mesentrica (Bull.) Quél. (Figure 2.2a) (Auriculariaceae – Basidiomycotina) (# MUBSNCKKRSMF-003)

Non-gilled, small to medium, fleshy, neither encrusting nor bracket-like, purple brown to grey brown, fruit bodies sessile, ear-like, grow in tires, annual, lignicolous, infrequent, odour and taste are not distinctive and inedible.

Fruit body. Upper surface brownish-grey, zoned, more pallid at margin, hairy, lobed margin, initially disc-shaped and becoming expanded into densely tiered brackets. Under surface reddish purple, wrinkled, white and pruinose: 3.5 (1.1-4.8) cm diam. (n=14). Flesh brown, gelatinous, elastic when damp, hard and brittle when dry.

Spores. Hyaline, smooth and ellipsoidal: 8.5 (6.6-9.7) × 5.4 (4.6-5.9) µm (n=10).

Substrate. Rotting logs and bark of *Memecylon umbellatum* and *Syzygium cumini*.

Occurrence. Coffee agroforest in B'Shettigeri.

Distribution. Kerala (Mohanan, 2011).

Clitocybe infundibuliformis Quél. (Figures 2.2d and e) (Tricholomataceae – Basidiomycotina) (# MUBSNCKKRSMF-004)

Gilled, medium-size agaric, fleshy, stipe central, foamy white, solitary or in small troops, annual, rare, humicolous/lignicolous, odour of almonds, taste not distinctive and edible.

Cap. Whitish to cream, smooth and silky, infundibuliform with typical wavy margin: 1.8-6.6 cm (n=5). Flesh whitish, thick and soft.

Gills. Creamish, deeply decurrent, narrow and crowded.

Spore. Hyaline, smooth and ellipsoidal: 5.9 (4-7.9) × 3.2 (2.6-4) µm (n=14).

Stipes. Concolorous with cap, smooth, equal and slightly swollen at base: 0.7-1.8 cm tall × 0.2-0.4 cm thick (n=5). Flesh creamish, tougher than in cap and partly stuffed or hollow.

Substrate. On the decomposing coir of *Cocos nucifera.*

Occurrence. Coffee agroforest in V'Badaga.

Distribution. Not known.

Cookeina tricholoma (Mont.) Kuntze (Figure 2.2h) (Sarcocyphaceae – Ascomycotina) (# MUBSNCKKRSMF-005)

Small to medium, top shaped, fleshy, orange, spiny, in trooping groups/clusters, frequent, annual, lignicolous, odour and taste not distinctive and inedible.

Fruit body. Orange coloured, fading to pinkish-orange with age, deeply cup- to funnel-shaped: 2.1 (1.3- 2.4) × 2.8 (2.2-3.6) cm (n=10), sparse long whitish hairs throughout, more hairs towards margin, without hairs on stipe, inner hymenial surface orange coloured, smooth and bears sporangium.

Asci. Cylindrical and equal: 320.9 (289.3-341.9) × 19.9 (18.4-21) µm.

Spores. Hyaline, smooth and broadly fusiform: 28.9 (26.8-31.6) × 14.1 (11.8-16.3) µm.

Stipe. Smooth, equal and hallow: 1.5 (0.4-2.8) × 0.4 (0.3-0.5) cm (n=10).

Substrate. On rotting twigs of *Canarium strictum, Dysoxylum malabaricum, Euodia lunu-ankenda, Toona ciliate* and *Vateria indica.*

Occurrence. Found in all the locations of sacred groves, reserve forests and agroforests.

Distribution. Kerala (Mohanan, 2011).

Crepidotus variabilis (Pers.) P. Kumm. (Figures 2.2f and g) (Crepidotaceae – Basidiomycotina) (# MUBSNCKKRSMF-006)

Gilled, small, fleshy, sessile, whitish to creamish, irregular, in trooping groups/ clusters, annual, rare, lignicolous, odour and taste not distinctive and inedible.

Fruit body. Upper surface whitish become creamish with age, kidney or reniform or irregularly fan-shaped or lobed and faintly striate towards the margin: 1.7 (0.7-2) cm (n=9). Flesh white, thin, soft and brittle.

Gills. Whitish becoming pink-violet, decurrent and fairly crowded.

Spores. Pinkish-brown, minutely warty and elongated to ellipsoidal: 7.5 (5.3-7.9) × 4.7 (3.3-5.3) µm (n=10).

Stipe. lateral, rudimentary or absent.

Substrate. On the rotting burnt *Bambusa* sp. and rotting twigs of *Jatropha curcas*.

Occurrence. Coffee agroforest in B'Shettigeri.

Distribution. Not known.

Cyathus striatus (Huds.) Willd. (Figure 2.2i) (Nidulariaceae – Basidiomycotina) (# MUBSNCKKRSMF-007)

Non-gilled, small, cone-/trumpet-shaped, leathery, blackish-greyish brown, hairy, in small clusters, rare, annual, lignicolous, odour and taste not distinctive and inedible.

Fruit body. Cone- or trumpet-shaped peridium: 0.9 (0.8-0.9) × 0.8 (0.7-0.9) cm (n=10). Outer surface brownish, shaggy and covered in fine felty hairs. Inner surface greyish-brown to greyish-silver, vertically striated and covered by operculum (whitish membrane), which ruptures at maturity to expose several (20-30) greyish-white lens shaped 'eggs' (peridioles) (2 mm diam.), immersed in mucilage at the base and attached by fine mycelial strands. Flesh soft, paper-thin and resistant.

Spores. Hyaline, smooth and ellipsoidal: 6.8 (5.3-7.9) × 3.7 (2.9-4) µm (n=10).

Substrate. On rotting twigs of *Aporusa lindleyana, Euodia lunu-ankenda, Erythrina subumbrans* and on soil rich with woody debris.

Occurrence. Sacred grove in Bettoli and coffee agroforest in B'Shettigeri.

Distribution. Karnataka (Swapna *et al.*, 2008) and Kerala (Mohanan, 2011).

Daldinia concentrica (Bolton) Ces. & De Not. (Figure 2.3a) (Xylariaceae – Ascomycotina) (# MUBSNCKKRSMF-008)

Small to medium, irregularly sub-spherical, tough and woody, reddish-brown to purple brown, sessile, annual, frequent, lignicolous, solitary to gregarious, odour and taste not distinctive, inedible and medicinal.

Fruit body. Stroma reddish-purple brown, becoming black with age by shedding black spores, sub-spherical, sometimes shiny, ostiolate: 5.1 (1.8-7.8) cm diam. (n=12). Perithecia black, sub-spherical, embedded in stromatal tissue in single layer below the surface. Flesh blackish-grey with zebra-banded silver sheen on cut surface, hard, brittle and fibrous.

Asci. Baseball bat-shaped, long and 8-spored: 141.5 (113.1-173.6) × 9.2 (6.8-10.5) µm (n=10).

Spores. Black, smooth and ellipsoidal with longitudinal slit 12.5 (11.6-15.1) × 5.6 (5.3-6.1) µm (n=10).

Figure 2.2: Photographs of Macrofungi of Kodagu Region.

a: *Auricularia auricula-judae* on wood; b and c: *A. mescenterica* on wood; d and e: *Clitocybe infundibuliformis* on coconut surface (coir); f and g: *Crepidotus variabilis* on wood; h: *Cookeina tricholoma* on bark and wood; i: *Cyathus striatus* on twigs and embedded in soil; j: *Dictyophora cinnabarina* on soil.

Figure 2.3: Photographs of Macrofungi of Kodagu Region.

a: *Daldinia concentrica* on wood; b: *Exidia glandulosa* on wood; c and d: *Ganoderma applanatum* on wood; e and f: *Gymnopilus junonius* on burnt soil; g: *Geastrum* sp. on leaves, twigs and soil; h: *Ileodictyon gracile* on twigs.

Substrate. Rotting logs and bark of *Acacia mangium, Acrocarpus fraxinifolius, Artocarpus hirsutus, Careya arborea* and *Toona ciliata.*

Occurrence. Found in all the locations of sacred groves, reserve forests and agroforests.

Distribution. Karnataka (Swapna *et al.*, 2008) and Kerala (Mohanan, 2011).

Dictyophora cinnabarina W.S. Lee (Figure 2.2j) (Phallaceae – Basidiomycotina) (# MUBSNCKKRSMF-009)

Non-gilled, medium, fleshy, phallic, white stipe with brownish head and orange net, solitary, annual, rare, silvicolous, odour foetid/stringent, taste not distinctive and inedible.

Fruit body. White egg, sub-spherical, resistant to rubbery outer membrane with gelatinous matrix separated from the embryonic spore mass and stipe by an inner membrane. Attached to soil by rhizomorphs (3-4) and on maturity the egg ruptures and the spore mass are carried rapidly upward on the receptacle in tip of stipe. Stipe whitish, honey-combed, cylindrical, spongy, hollow: 13.1 (12.9-13.3) × 1.7 (1.6-1.7) cm (n=3) and leaving membranous volva around the base of stipe. Volva white to creamish or yellowish brown, thickly gelatinous, soft: 3.6 (3.5-3.8) cm diam. (n=3), surmounted by a greyish yellow to brownish yellow conical head carrying the spore mass and orange indusium hanging out from under the lower margin. The head is strongly chambered, bears a reticulate mesh of raised ribs with apical disc perforated in centre and surrounded throughout with dark olive-green to metallic green mucilaginous gleba (spore mass): 2.2 (2.1-2.3) × 2.2 (2-2.3) cm (n=3). Indusium orange to yellowish orange, netted, partly elastic and flaring at bottom: 6.9 (6.7-7.1) cm height (n=3).

Spores. smooth, ellipsoidal and hyaline. 3.3 (2.6-3.9) × 1.3 (1.1-1.6) µm (n=10).

Substrate. Soil below the trees of *Lagerstroemia microcarpa.*

Occurrence. Reserve forest in Sampaje and coffee agroforest in V'Badaga.

Distribution. Karnataka (Abrar *et al.*, 2007; Swapna *et al.*, 2008) and Kerala (Mohanan, 2011).

Exidia glandulosa (Bull.) Fr. (Figure 2.3b) (Tremellaceae – Basidiomycotina) (# MUBSNCKKRSMF-010)

Non-gilled, small to medium, gelatinous, blackish brown to blackish ash, sessile, contorted disc-shaped, grow in groups, annual, lignicolous, rare, odour and taste not distinctive and inedible.

Fruit body. Brownish when damp, ash black when dry, expanses of tissues appear like brain folding and fused with adjacent sporophores, smooth, shiny with glandular warts and sessile. Under surface creamish, wrinkled and wholly attached to the decaying log: 2.8 (1.6- 4.7) cm diam. (n=5). Flesh brownish, gelatinous/soft when damp, dry on hard, become thin and membranous.

Spores. Hyaline, smooth and ellipsoidal: 8.1 (6.6-9.2) × 4.7 (3.9-5.3) µm (n=10).

Substrate. On the rotting logs of *Artocarpus heterophyllus.*

Occurrence. Coffee agroforest in B'Shettigeri.

Distribution. Not known.

Ganoderma applanatum (Pers.) Pat. (Figures 2.3c and d) (Ganodermataceae – Basidiomycotina) (# MUBSNCKKRSMF-011)

Non-gilled, large to very large, woody bracket, sessile, grey-brown, grow in limited overlapping tiers, perennial, lignicolous, common on wood, odour fungoid, taste bitter, inedible and medicinal.

Fruit body. Greyish-brown, often discoloured into reddish brown or cocoa brown from deposited spores, generally flattened, radially wavy or wrinkled, concentrically grooved and zoned, broadly attached, sessile: 18.8 (12.6-24.9) cm diam. (n=5). Flesh dark brick, thicker than the tube region, very tough and fibrous.

Pores. White, bruising brown and circular in annual layers.

Spores. Brown, smooth ovoid to broadly ellipsoidal and flattened at one end with hyaline germ pore: 7.8 (7.7-7.9) × 5.4 (5.3-5.9) µm (n=14).

Substrate. Rotting trunks of *Artocarpus heterophyllus* and *A. hirsutus*.

Occurrence. Found in all the locations of sacred groves, reserve forests and agroforests.

Distribution. Maharashtra (Ranadive *et al.*, 2011) and Kerala (Mohanan, 2011).

***Geastrum* sp.** (Figure 2.3g) (Geastraceae – Basidiomycotina) (# MUBSNCKKRSMF-012)

Non-gilled, small, sessile, creamish-white, rounded/bulb on a star shaped outer wall, grow in trooping groups, annual, humicolous, infrequent, odour and taste not distinctive and inedible.

Fruit body. at first brownish, coarsely scaly, bulb shaped, 0.6 (0.4-0.7) cm diam. (n=5), sessile, thick and brittle. Outer peridium splitting at maturity into 5 pointed starfish-like rays which reflex back to reveal the creamy-white inner fleshy layer. Inner peridium (sporesac, gleba and spore mass) creamish-white, sub-spherical, thin, papery, sessile and opening by a fringed apical pore. 1.4 (1.1-1.8) cm diam. (n=7). Gleba (spore mass) at first pallid, firm and becoming brown with powdery deposits.

Spores. spherical, warty and brown, 4-5 µm diam. (n=10).

Substrate. On decomposing leaf litter and twigs of *Acacia mangium* and *Sapium insigne*.

Occurrence. Sacred grove in Maggula and coffee agroforest in V'Badaga.

Distribution. Not known.

Gymnopilus junonius (Fr.) P.D. Orton (Figures 2.3e and f) (Strophariaceae – Basidiomycotina) (# MUBSNCKKRSMF-013)

Gilled, medium to large agaric, orange to yellow in dense caespitose clusters, annual, carbonicolous, rare, odour acidic or bamboo shoot, taste not distinctive and edible.

Cap. Initially dark orange fades to yellow with age, appressed fibrillose-squamulose, convex, becoming expanded, plane and wavy and depression often deepens to the stipe. Flesh concolorous, firm and medium: 5.2 (2.8-8.2) cm diam. (n= 10).

Gills. Orange brown to creamish yellow, adnate, broad and crowded.

Spores. Brownish orange, finely warty and ellipsoid: 8 (7.9-8.3) × 5.4 (5.3-5.9) µm (n=14).

Stipe. Ring absent, pallid ochraceous-brown, more or less equal or slightly bulged towards the base and fibrillose-striate with scattered squamules: 4.9 (2.4-8.4) × 0.9 (0.35-1.6) cm (n=10). Flesh yellowish, firm, stuffed and full.

Substrate. On soil, roots and stumps of burnt decaying *Bambusa* sp.

Occurrence. Coffee agroforestry in B'Shettigeri.

Distribution. Kerala (Mohanan, 2011).

Ileodictyon gracile Berk. (Figure 2.3h) (Phallaceae – Basidiomycotina) (# MUBSNCKKRSMF-014)

Non-gilled, small, sessile, white, lattice-like sphere extruding from a partially submerged egg, grow in troops or solitary, annual, humicolous, infrequent, odour foul/foetid, taste not distinctive and inedible.

Fruit body. Initially bone-white, egg-shaped/sub-spherical: 0.7 (0.6-0.8) × 0.5 (0.4-0.6) cm (n=6), attached to the substrate by a mycelial cord, outer membrane delicate, finely elastic, and closing the compressed lattice surrounded the white gleba (spore mass). At maturity the egg ruptures and the spore mass carried upwards on the inside surface of the bright white lattice: 2.7 cm (1.5- 3.3) × 1.3 cm (1.0-1.6) diam. (n=6).

Spores. Hyaline, smooth and ellipsoidal: 4.4 (3.3-4.6) × 1.6 (1.3-1.8) µm (n=10).

Substrate. Rotting sheath of inflorescence of *Cocos nucifera* and rotting twigs and leaf litter of *Eupatorium odoratum* and *Pongamia pinnata*.

Occurrence. Reserve forest of Makutta and coffee agroforest of V'Badaga.

Distribution. Kerala (Mohanan, 2011).

Lentinus dicholamellatus Manim. (Figures 2.4a and b) (Polyporaceae – Basidiomycotina) (# MUBSNCKKRSMF-015)

Gilled, large, irregular infundibuliform bracket with lateral stipe, creamish with brownish squamules/tinge, solitary or in groups, annual, lignicolous, frequent, odour mealy, taste not distinctive and edible with second preference.

Cap. Upper surface at first pale yellow changing to brownish orange and densely dotted with very fine appressed or erect squamules. Initially convex with a depressed centre becoming deeply infundibuliform, finely striate, radially fibrillose, margin incurved/inrolled and lobate: 19.3 (25.6- 14.2) cm diam. (n=3). Flesh creamish, thin, elastic and leathery.

Gills. Yellowish-white, deeply decurrent and crowded.

Spores. Hyaline, smooth and ellipsoidal: 7.7 (7.2-7.9) × 3.8 (3.4-3.9) µm (n=10).

Stipe. Concolorous with cap, often dotted with erect pointed squamules, central/eccentric/lateral, almost equal and solid: 4.1 (3.1-4.7) × 1.3 (1.2-1.3) cm (n=3).

Substrate. Rotting logs of *Canarium strictum, Mangifera indica* and *Vateria indica.*

Occurrence. In Bettoli sacred grove, reserve forest in Makkuta and coffee agroforest in V'Badaga.

Distribution. Kerala (Mohanan, 2011).

Lentinus patulus Lév. (Figure 2.4e) (Polyporaceae – Basidiomycotina) (# MUBSNCKKRSMF-016)

Gilled, small to medium, infundibuliform with central stipe, creamish-white with brownish squamules, in small clusters, annual, lignicolous, infrequent, odour mealy, taste not distinctive and inedible.

Cap. White to cream, with fine brown to blackish squamules more towards the centre, at first convex with depressed centre, finally attain infundibuliform, finely sulcate, striate towards margin, margin regular and thin: 3.8 (2.7-4.9) cm (n=10). Flesh white, thin and at first weakly elastic becoming more leathery on drying.

Gills. White becoming creamish buff, decurrent, narrow and moderately crowded.

Spores. Hyaline, smooth and ellipsoidal: 8 (6.6-9.2) × 4.6 (4-5.3) µm (n=14).

Stipe. Ring absent, concolorous with cap, possess fine blackish fibrillose squamules, typically central to eccentric, short, more or less equal, arising from rhizomorph: 1.7 (0.9-3.8) × 0.3 (0.2-0.6) cm (n=10). Flesh white, firm, tough and thin.

Substrate. On dead stub of *Mangifera indica.*

Occurrence. Coffee agroforest in B'Shettigeri.

Distribution. Kerala (Mohanan, 2011).

Lenzites vespacea (Pers.) Pat. (Figures 2.4c and d) (Polyporaceae – Basidiomycotina) (# MUBSNCKKRSMF-017)

Non-gilled, Large, semi-circular or fan-shaped bracket, creamish-white to brownish-orange, solitary or in limited tires, perennial, lignicolous, common, odour and taste not distinctive and inedible.

Fruit body. Upper surface creamish brown with concentric zones (brown, tan and grey), downy or finely hairy, concentrically grooved and radially wrinkled, sometimes surface shows algal growth, narrowly attached and sessile: 15 (10.4-23.4) cm diam. × 1-1.2 cm thick (n=3). Flesh creamish, leathery and tough.

Pores. Whitish-buff and straw/gill/maize-like in a radial arrangement with branching.

Spores. Hyaline, broadly ellipsoidal and smooth: 7.8 (5.3-9.2) × 5.7 (5.3-6.6) µm (n=11).

Substrate. Rotting logs of *Acacia mangium, Acrocarpus fraxinifolius, Artocarpus heterophyllus, A. hirsutus, Lagerstroemia microcarpa, Syzygium cumini, Tectona grandis* and *Toona ciliata.*

Occurrence. Found in all the locations of sacred groves, reserve forests and agroforests.

Distribution. Maharashtra (Ranadive *et al.*, 2011) and Kerala (Mohanan, 2011).

Panus conchatus (Bull.) Fr. (Figures 2.4g and h) (Polyporaceae – Basidiomycotina) (# MUBSNCKKRSMF-018)

Gilled, medium, trumphet-shaped agaric, infundibuliform with central stipe, brownish-orange, velvety, solitary or in small clusters, annual, lignicolous, infrequent, odour and taste not distinctive and inedible.

Cap. Upper surface at first purple red fading to brownish-orange with age, applanate to depressed, surface striate, velvety, smooth, thin, leathery, appressed squamules at centre, margin inrolled, lobed and hairy throughout: 8.8 (6.5-10.2) cm diam. (n=5). Lower hymenial surface at first violet or purple fading to creamish, possess a series of irregular, blunt gill-like wrinkles and folds and decurrent.

Stipe. Purple-brown, thickly velvety/pubescent throughout, fading with age, tough, long, solid and more or less equal: 4.6 (3.4-5.8) × 0.9 (0.6-1.1) (n=5). Flesh creamish-brown, tough, leathery and thin.

Spores. Hyaline, smooth and ellipsoidal: 5.3 (4.6-5.9) × 2.8 (2.6-3.2) µm (n=12).

Substrate. Rotting logs of *Delonix regia, Mangifera indica* and *Memecylon umbellatum.*

Occurrence. Reserve forest in Sampaje and coffee agroforestry in B'Shettigeri.

Distribution. Kerala (Mohanan, 2011).

Pleurotus pulmonarius (Fr.) Quél. (Figures 2.4a and b) (Pleurotaceae – Basidiomycotina) (# MUBSNCKKRSMF-019)

Gilled, small to medium, fleshy, stipe lateral or absent, whitish becoming creamish, irregular, fan- or bracket-shaped, in caespitose tufts or groups, annual, infrequent, lignicolous, mealy odour and taste, and edible.

Cap. Whitish becoming creamish-yellow with age, at first shallowly convex then flattened, reniform to irregularly flabelliform, undulating and wavy, lobed as well as split margin, narrowing at point of attachment, fleshy and smooth: 2.1 (0.8-3.8) cm diam. (n=6). Flesh white, slightly fibrous and moderate.

Gills. White, becoming creamish, decurrent, narrow and close.

Spores. Hyaline, smooth and ellipsoidal or cylindrical: 8.1 (7.9-8.6) × 4.9 (4-5.3) µm (n=12).

Stipe. White, lateral or eccentric, very short, slightly woolly, stout and hard: 0.5 (0.3-0.7) × 0.4 (0.3-0.5) cm (n=6). Flesh white, firm and full.

Substrate. Rotting twigs of *Erythrina subumbrans* and *Jatropha curcas.*

Occurrence. Coffee agroforestry in B'Shettigeri.

Distribution. Not known.

Figure 2.4: Photographs of Macrofungi of Kodagu Region.

a and b: *Lentinus dicholamellatus* on wood; c and d: *Lenzites vespacea* on stub; e: *Lentinus patulus* on stub; f: *Ramaria* sp. on soil; g and h: *Panus conchatus* on twig.

Figure 2.5: Photographs of Macrofungi of Kodagu Region.

a and b: *Pleurotus pulmonarius* on wood; c and d: *Polyporus* sp. on twig; e: *Scleroderma verrucosum* on soil; f and g: *Termitomyces tylerianus* on termite soil; h: *Trogia infundibuliformis* on twig; i and j: *Tricholoma* sp. on soil with woody debris; k: *Xylaria longipes* on wood embedded in soil.

***Polyporus* sp.** (Figures 2.4c and d) (Polyporaceae – Basidiomycotina) (# MUBSNCKKRSMF-020)

Porous, small, fan- or kidney-shaped, creamy-brown, solitary, annual, lignicolous, rare, odour and taste not distinctive and inedible.

Cap. Camouflaged, upper surface creamy with coffee-brown concentric bands, kidney- or fan shaped, fleshy, shiny, smooth and with regular margin: 2.6-2.8 cm diam. (n=2). Flesh white, thick and fleshy.

Stipe. Lateral, short, ivory-colour and tough.

Pores. Whitish, irregularly angular, decurrrent and 3.5-4 mm deep.

Spores. White, fusiform and smooth: 10.7 (9.2-13.2) × 5.4 (5.3-5.9) µm (n=6).

Substrate. Rotting twigs of *Ficus religiosa.*

Occurrence. Sacred grove in Maggula.

Distribution. Not known.

***Ramaria* sp** (Figure 2.4f) (Gomphaceae – Basidiomycotina) (# MUBSNCKKRSMF-021)

Non-gilled, small to medium, coral-like and antler-tip-like, lilac to creamish, solitary, or in small groups, annual, carbonicolous, rare, odour and taste not distinctive and inedible.

Fruit body. Lilaceous to creamish, fragile with loose limited branching and coral-like with pointed forked tips (antler-like). Branches filiform arising from long slender stipe and dusted with smokes of grey spores and minutely downy at the base: 5.4 (2.5-9.6) × 1.8 (1.1-2.5) cm diam. (n=10). Flesh creamish and elastic.

Spores. Fusiform, hyaline and smooth: 10.3 (7.9-13.2) × 5.6 (5.3-7.9) µm (n=10)

Stipe. Long, slender, fibrous, tough, smooth and equal: 4.6 (2.2-8.3) × 0.2 (0.2-0.3) cm (n=10).

Substrate. On laterite soil and clay roof tiles.

Occurrence. Clay tiled roofs of coffee agroforest in B'Shettigeri.

Distribution. Not known.

Scleroderma verrucosum (Bull.) Pers. (Figure 2.5e) (Sclerodermataceae – Basidiomycotina) (# MUBSNCKKRSMF-022)

Non-gilled, medium, knobbly, yellowish-brown, solitary or in small groups, annual, silvicolous, rare, odour and taste not distinctive, edible and ectomycorrhizal.

Fruit body. Yellowish-brown, knobbly (round headed hammer-like) and tapered below into a thick grooved hard long stipe-like base: 5.7 (3.8-7.9) × 2.4 (1.8-3.3) cm (n=10). Attached to the soil by a dense mass of mycelial cord, wall of head/hymenial portion thin, smooth, brittle and sparsely scaly throughout: 4.3 (3.2-6.6) cm diam. (n-10). On cracking opens at apex to form an irregular slit for the release/dispersal of spores.

Spores. Spherical, dark brown and spiny: 11.1 (9.2-13.2) × 10.8 (9.2-13.2) µm (n=10).

Substrate. On soil below the trees of *Canarium strictum*, *Holigarna nigra* and *Vateria indica.*

Occurrence. Sacred grove in Bettoli and coffee agroforest of V'Badaga.

Distribution. Kerala (Mohanan, 2011).

Termitomyces tylerianus Otieno (Figures 2.5f and g) (Lyophyllaceae – Basidiomycotina) (# MUBSNCKKRSMF-023)

Gilled, medium to large agaric, grayish-white, solitary or occasionally in small groups (3-5), annual, particolous, common, odour and taste mealy, edible and medicinal.

Cap. Upper surface greyish-white to slightly grayish-yellow, convex to plane, with a prominent umbo/perforatorium and pointed to broadly conical (sharply to bluntly umbonate): 9.5 (4.9-14.4) cm diam. (n=10). Margin straight, smooth, entire and splitting with age. Flesh white to creamish, thick, smooth and fleshy.

Gills. White, free, crowded and regular.

Spores. Hyaline, smooth and broadly ellipsoidal: 7.2 (5.9-9.2) × 3.8 (3.3-4.7) µm (n=10).

Stipe. White, long, fibrillose, cylindrical with a swollen base and tapering downwards to a long firm pseudorrhiza: 14.6 (10.2-17.6) × 2.3 (1.2- 4.5) cm (n=10).

Substrate. Termite mounds and soils of grasslands, path of paddy fields, playgrounds, sacred groves, forests and coffee plantations.

Occurrence. Found in all the locations of sacred groves, reserve forests and agroforests.

Distribution. Kerala (Mohanan, 2011).

***Tricholoma* sp.** (Figures 2.5i and j) (Tricholomataceae – Basidiomycotina) (# MUBSNCKKRSMF-024)

Gilled, medium to large agaric, creamish-brown to potato-brown, caespitose clusters, annual, silvicolous and lignicolous, rare, odour and taste mealy and edible.

Cap. Greyish-brown with deposition of fine powder paling almost wholly after rain to potato/sapota-brown. Initially hemispherical, becoming campanulate, margin inrolled, smooth and dry: 3.9 (5.5-2.8) cm diam. (n=10). Flesh white and medium.

Gills. Initially whitish becoming slightly creamish, emarginate and distant.

Spores. Hyaline, smooth and oval-ellipsoidal: 6.9 (5.3-7.9) × 4.4 (4-5.3) µm (n=12).

Stipe. Long: 9.2 (7.2-11.3) cm tall (n=10), pallid, decorated with whitish-brown speckled fibrils below ring-zone, bulbose at base: 2.02 (1-2.7) cm diam. (n=10), tapering towards apex: 1.3 (0.7-1.9) cm diam. (n=10) and ring absent. Flesh white, becoming yellowish with age and stuffed.

Substrate. On soil rich with debris of *Caryota urens*.

Occurrence. In reserve forest in Sampaje.

Distribution. Not known.

Trogia infundibuliformis Berk. & Broome (Figure 2.5h) (Marasmiaceae – Basidiomycotina) (# MUBSNCKKRSMF-025)

Vein-like ridges, small, funnel-shaped, brownish to orange, solitary or in small groups, annual, lignicolous, infrequent, odour fungoid, taste not distinctive and inedible.

Cap. Brown to orange, fading with age, deeply infundibuliform, fleshy, thin, smooth, shiny, surface deeply sulcate and margin wavy and irregular: 2.6 (1.4-3.3) cm diam. (n=6). Gills absent, lower hymenial surface orange, possess vein-like ridges and decurrent.

Stipe. Brownish fading with age, smooth, hollow and more or less equal but slightly bulged at the base or point of attachment to wood: 2.2 (1.5-2.6) × 0.3 (0.2-0.4) cm (n=6). Flesh brownish orange, tough and thin.

Spores. Hyaline, smooth and ellipsoidal: 7.8 (7.2-8.5) × 4.1 (3.3-4.6) μm (n=10).

Substrate. On decaying twigs of *Coffea robusta.*

Occurrence. Sacred grove in Maggula and coffee agroforestry in V'Badaga.

Distribution. Kerala (Kumar and Manimohan, 2009c; Mohanan, 2011).

Xylaria longipes Nitschke (Figure 2.5k) (Xylariaceae – Ascomycotina) (# MUBSNCKKRSMF-026)

Club-shaped mature fruit body, develops on wood, project above ground, usually gregarious, common, possess stipe, tough or woody, annual, odour and taste not distinctive, and medicinal.

Fruit body. Upper stromatal surface black, warty, often somewhat bent over and narrowing slightly into a brownish-black sterile. Stipe cylindrical or baseball bat-shaped or clavate and smooth or slightly downy. Perithecia black, sub-spherical, fully embedded in stromatal tissue and arranged in a single dense layer just below the surface towards the apex: 4.8 (2.8-6.1) × 0.7 (0.4-0.8) cm (n=10). Flesh white and hard.

Asci. Cylindrical: 155.6 (147.3-163.1) × 7.5 (5.5-9) μm. Spores dark brown, smooth, ellipsoid to reniform, flattened on one side with distinct cleft, non-septate and uniseriate: 13.3 (12.5-14.5) × 5.9 (5.5-6) μm (n=10).

Substrate. Rotting logs and stumps of *Acrocarpus fraxinifolius*, *Artocarpus hirsutus*, *Erythrina subumbrans*, *Memecylon umbellatum* and *Syzygium cumini.*

Occurrence. Found in all the locations of sacred groves, reserve forests and agroforests.

Distribution. Kerala (Mohanan, 2011).

Discussion and Outlook

Two megadiversity hotspots in India, the Western Ghats and Himalayas are the treasures of a variety of life forms and always fascinating for exploration. Mushrooms especially edible forms are of immense economic value as alternative food source against plant and animal food sources. They are versatile as source of adequate protein, fibre and a variety of bioactive compounds and possess the capacity to decrease cholesterol level in blood. Besides food source, several macrofungi serve as medicinal and some are of agricultural importance as ectomycorrhizal fungi. Interestingly, a wide range of agarics has been reported from the Nilgiri Biosphere Reserve (Natarajan *et al.*, 2005a) and ectomycorrhizal fungi from dipterocarp forest stands of Kodagu region (Natarajan *et al.*, 2005b).

The current study projected several edible, medicinal and one ectomycorrhizal fungi in one-time survey. Sacred groves and reserve forests provide fairly undisturbed natural habitats for a variety of macrofungi. These biomes are the ideal sites for successful conservation of macrofungi as natural repositories (Brown *et al.*, 2006). Domestication of desired macrofungi can be achieved in agroforest consisting of shade-grown commercial plantations (*e.g.* coffee, cardamom, cocoa and areca), monocrops (*e.g.* rubber) and tree-less field crops (*e.g.* paddy, sugarcane and banana). Brown *et al.* (2006) studied the impact of fragmentation and disturbance on the diversity and distribution of macrofungi in tropical rain forests of Kodagu region. They recorded macrofungi during three wet seasons in 25 sacred groves, 10 reserve forests and 23 coffee plantations. Based on this long-term study, they concluded that habitat degradation is an important threat to macrofungal diversity than fragmentation. Although sacred groves occupied less area, they are valuable in macrofungal conservation due to unique habitat with higher abundance of sporocarp as well as morphospecies compared to reserve forests and coffee-agroforests. On the contrary, our one-time study revealed the highest species richness in coffee-agroforests than sacred groves and reserve forests. However, there may be some difference regarding the abundance of sporocarps of macrofungi in these habitats and long-term studies may reveal more improved perception. In shade-grown coffee-agroforests, in addition to native trees, several exotic tree species are common (*e.g. Acacia mangium, Acrocarpus fraxinifolius, Artocarpus hirsutus, Dalbergia latifolia, Grevelia robusta, Syzygium cumini* and *Toona ciliata*). Besides, small tree species are also common (*e.g. Erythrina subumbrans* and *Gliricidia sepium*). It is likely, regular pruning of such shade trees as litter source for coffee plantation and misting during dry seasons might be responsible for occurrence of macrofungi throughout the year. As human disturbances have to be avoided in sacred groves and reserve forests, agroforests constitute ideal habitats to domesticate edible, medicinal and ectomycorrhizal fungi for economic growth. Similar to flora and fauna, macrofungi also consititute another important link in forest ecosystems as detritus decomposers. But, knowledge on the diversity, distribution of macrofungi and impact of forest degradation/fragmentation on macrofungi is insufficient. Rehabilitation of many forest communities needs considerable time especially macrofungi prefers old-growth forests than the new forests for their perpetuation (Norden and Appelqvist, 2001). For instance, distinct succession pattern of ectomycorrhizal fungi was seen in old (11–17 years) and young (3–7 years) dipterocarp forests in Kodagu region (Natarajan *et al.*, 2005b). Besides plant detritus and pure forest stands (*e.g.* dipterocarp and bamboo groves), there are several ecological niches deserve meticulous attention to explore macrofungi in the Western

Ghats: coprophilus fungi on the dung of wild animals (*e.g. Amanita, Entoloma* and *Xylaria*), medicinally valuable entomopathogens (*e.g., Cordyceps* and *Ophiocordyceps*) and termite mounds for edible mushrooms (*e.g., Termitomyces*).

Each geographical location has specific characteristic features and yield of optimum number of macrofungi depends on the time and frequency of field survey (Halme and Kotiaho, 2012). For each biome in a location, it is necessary to earmark substrate, time and frequency of sampling for maximum recovery of macrofungi. In addition, phase-wise long-term studies on macrofungi are warranted for a comprehensive and total understanding of the diversity in different parts of the Western Ghats to improve our knowledge on the species richness, functional diversity, conservation strategies and economic benefits. Traditional knowledge of local people and tribals should not be ignored to fill the gaps in our knowledge on the value of macrofungi.

Acknowledgement

Authors are thankful to Mangalore University for permission to carry out this study in the Department of Biosciences.

References

Abrar, S., Swapna, S. and Krishnappa, M. 2007. *Dictyophora cinnabarina*. Curr. Sci. **92:** 1219-1220.

Bhagwat, S., Kushalappa, C., Williams, P. and Brown, N. 2005. The role of informal protected areas in maintaining biodiversity in the Western Ghats of India. Ecology and Society **10:** article # 8, 1-40: http: //www.ecologyandsociety.org/vol10/iss1/art8/

Brown, N., Bhagwat, S. and Watkinson, S. 2006. Macrofungal diversity in fragmented and disturbed forests of the Western Ghats of India. J. Appl. Ecol. **43:** 11-17.

Das, K. 2010. Diversity and conservation of wild mushrooms in Sikkim with special reference to Barsey Rhododendron Sanctuary. NeBIO **1:** 1-13.

Halme, P. and Kotiaho, J.S. 2012. The importance of timing and number of surveys in fungal biodiversity research. Biodiversity and Conservation **21:** 205–219.

Hawksworth, D.L. 2001. The magnitude of fungal diversity: the 1.5 million species estimate revisited. Mycol. Res. **105:** 1422–1432.

Hawksworth, D.L. 2012. Global species numbers of fungi: Are tropical studies and molecular approaches contributing to a more robust estimate? Biodiversity and Conservation **21:** 2425–2433.

Jha, K.S., Kumar, V. and Tripathi, N.N. 2011. Survey of ethnomedicinal macrofungi of Nagarjuna and Phulchowki areas of Kathmandu Valley, Nepal. Intern. J. Pharmaceutical Sci. Review Res. **11:** 147-151.

Johnsy, G., Sargunam, S.D., Dinesh, M.G. and Kaviyarasan, V. 2011. Nutritive value of edible wild mushrooms collected from the Western Ghats of Kanyakumari district. Botany Res. Intern. **4:** 69–74.

Jordan, M. 1995. The Biology of Higher Fungi: The Encyclopedia of Fungi of Britain and Europe. David and Charles Publishers, UK, p. 384.

Kibby, G. 1979. Mushrooms and Toadstools a Field Guide. Oxford University Press, Oxford, UK, p. 256.

Kumar, T.K. and Manimohan, P. 2009a. The genus *Lepiota* (Agaricales, Basidiomycota) in Kerala State, India. Mycotaxon **107:** 105–138.

Kumar, T.K. and Manimohan, P. 2009b. The genera *Leucoagaricus* and *Leucocoprinus* (Agaricales, Basidiomycota) in Kerala State, India. Mycotaxon **108:** 385–428.

Kumar, T.K. and Manimohan, P. 2009c. Rediscovery of *Trogia cyanea* and a record of *T. infundibuliformis* (Marasmiaceae, Agaricales) from Kerala State, India. Mycotaxon **109:** 429–436.

Kushalappa, C.G. and Bhagwat, S. 2001. Sacred groves: biodiversity, threats and conservation, In: Forest Genetic Resources: Status, Threats, and Conservation. Shaankar, U.R., Ganeshaiah, K.N. and Bawa, K.S. (Ed.), Oxford and IBH, New Delhi, pp. 21–29.

Leelavathy, K.M., Manimohan, P. and Arnolds, E.J.J. 2006. *Hygrocybe* in Kerala State, India. Persoonia **19:** 101–151.

Manoharachary, C., Sridhar, K.R., Singh, R., Adholeya, A., Suryanarayanan, T.S., Rawat, S. and Johri, B.N. 2006. Fungal biodiversity: Distribution, conservation and prospecting of fungi from India. Curr. Sci. **89:** 58–71.

Mohanan, C. 2011. Macrofungi of Kerala. Kerala Forest Research Institute, Hand Book # 27, Kerala, India, p. 597.

Natarajan, K., Narayanan, K., Ravindran, C. and Kumaresan, V. 2005a. Biodiversity of agarics from Nilgiri Biosphere Reserve, Western Ghats, India. Curr. Sci. **88:** 1890–1893.

Natarajan, K., Senthilarasu, G., Kumaresan, V. and Riviere, T. 2005b. Diversity in ectomycorrhizal fungi of a dipterocarp forest in Western Ghats. Curr. Sci. **88:** 1893–1895.

Norden, B. and Appelqvist, T. 2001. Conceptual problems of ecological continuity and its bioindicators. Biodiversity and Conservation **10:** 779–791.

Pascal, J.P., and Meher-Homji, V.M. 1986. Phytochorology of Kodagu (Coorg) district, Karnataka. J. Bombay Nat. Hist. Soc. **83:** 43–56.

Purkayastha, R.P. and Chandra, A. 1985. Manual of Indian Edible Mushrooms. Today and Tomorrow's Printers and Publishers, New Delhi, p. 267.

Puttaraju, N.G., Venkateshaiah, S.U., Dharmesh, S.M., Urs, S.M.N. and Somasundaram, R. 2006. Antioxidant activity of indigenous edible mushrooms. J. Agric. Food Chem. **54:** 9764–9772.

Ranadive, K.R., Vaidya, J.G., Jite, P.K., Ranade, V.D., Bhosale, S.R., Rabba, A.S., Hakimi, M., Deshpande, G.S., Rathod, M.M., Forutan, A., Kaur, M., Naik-Vaidya, C.D., Bapat, G.S. and Lamrood, P. 2011. Checklist of Aphyllophorales from the Western Ghats of Maharashtra State, India. Mycosphere **2:** 91–114.

Riviere, T., Diedhiou, A.G., Diabate, M., Senthilarsu, G., Natarajan, K., Verbeken, A., Buyck, B., Dreyfus, B., Bena, G. and Ba, A.M. 2007. Genetic diversity of ectomycorrhizal basidiomycetes from African and Indian tropical rain forests. Mycorrhiza **17:** 415–428.

Sargunum, S.D., Johnsy, G., Samuel, A. and Kaviyarasan, V. 2012. Mushrooms in the food culture of the Kaani tribe of Kanyakumari district. Indian J. Trad. Knowl. **11:** 150–153.

Selvam, K., Priya, M.S., Sivaraj, C. and Arungandhi, K. 2012. Identification and screening of wood rot fungi from Western Ghats area of South India. Intern. J. ChemTech Res. **4:** 379–388.

Swapna, S., Abrar, S. and Krishnappa, M. 2008. Diversity of macrofungi in semi-evergreen and moist deciduous forest of Shimoga District–Karnataka, India. J. Mycol. Plant Pathol. **38:** 21–26.

Thiribhuvanamala, G., Prakasam, V., Chandrasekar, G., Sakthivel, K., Veeralakshmi, S., Velazhahan, R. and Kalaiselvi, G. 2011. Biodiversity, conservation and utilization of mushroom flora from the Western Ghats region of India. Proceedings of the 7th International Conference on Mushroom Biology and Mushroom Products, pp. 163–172.

Thomas, K.A., Hausknecht, A. and Manimohan, P. 2001. Bolbitiaceae of Kerala State, India: New species and new and noteworthy records. *Österreichische* Zeitschrift für Pilzkunde **10:** 87–114.

Chapter 3

Western Ghats: The World Heritage Site for Biodiversity and Ecological Security

S. Paulsamy*

Department of Botany,
Kongunadu Arts and Science College, Coimbatore – 641 029

Our nation's pride, the Western Ghats are at last included in the list of World Heritage Sites by UNESCO on 01.07.2012 after the sustained efforts taken by us. The Western Ghats form a practically unbroken relief dominating the western coast of the Indian peninsula for almost 1600km with an area of about 1,60,000 sq.km. They extend from the mouth of the river Tapti to the tip of South India, with Palghat Gap, the only gap in the chain. This apparent unity makes the great heterogeneity of the environmental conditions resulting from geographical, geological and demographic differences. Besides monsoon, the thermic convectional rainfall and the reliefs (hills and valleys) are the characteristics of Western Ghats. The varied climate, altitude and diverse topography create a wide array of habitats and hence many types of vegetations like forests of different types, grasslands, sholas etc in Western Ghats that support unique sets of organisms which include species of microbes, lower plants, higher plants, lower invertebrates, butterflies, fishes, amphibians, reptiles, birds and other animals. A cluster of 39 sites which includes the Nilgiris supported for getting this status of World Heritage Site to Western Ghats due to high species richness both at flora and fauna levels. It is estimated that Western Ghats harbour

---

* E-mail: paulsami@yahoo.com

about 4000 flowering plants and 326 pteridophytes. In addition, among the various biogeographic zones of India, Western Ghats contain high degree of endemism also. Nearly 2000 species of higher plants, 92 species of amphibians, 89 species of reptiles, 15 species of birds and 12 species of mammals are endemic to Western Ghats. It has been identified that nearly 75 per cent of endemics of peninsular India are confined to the Western Ghats alone. It was surveyed that the entire Western Ghats consists of 44 rare and endangered species which were published in British Fern Gazette in 1995. Apart from biological diversity, the region boasts of high levels of cultural diversity too as many indigenous people inhabit its forests.

By having all these significances, inclusion of Western Ghats in the list of World Heritage Site is not astonishing. In spite of the great natural value, the IUCN Director, Tim Badman cautioned about the conservation challenges and the need of additional monitoring of this mountain range by the World Heritage Committee to ensure the effective conservation of many precious biological organisms. *In situ* conservation practices being followed now in the protected areas of Western Ghats are the most effective strategies for the protection of our valuable bioresources which include many red listed species of both plants and animals in this region. It is evidenced from our early study in the shola understories of Nilgiris, a first ever declared Biosphere Reserve in India in 1986 that through intensive habitat protection exerted by our

Figure 3.1: The Forest Vegetation Endowed with Huge Biodiversity in Nilgiri Biosphere Reserve.

Figure 3.2: *Anaphalis beddomei* (Asteraceae)

Figure 3.3: *Disporum leschenaultianum* (Convallariaceae)

Figure 3.4: *Ceropegia pusilla* (Asclepiadaceae)

Figure 3.5: *Smilax wightii* (Smilacaceae)

Figure 3.6: *Biophytum polyphyllum* (Oxalidaceae)

Figure 3.7: *Agrostis pilosula* (Poaceae)

Figure 3.8: *Asparagus fysonii* (Asparagaceae)

Figure 3.9: *Dumasia villosa* (Fabaceae)

Figure 3.10: *Gaultheria fragrantissima* (Ericaceae)

Figure 3.11: *Helichrysum bracteatum* (Asteraceae)

Figure 3.12: *Pogostemon wightii* (Lamiaceae)

Figure 3.13: *Strobilanthes foliosa* (Acanthaceae)

Forest Department, many plant species of red listed category and endemics improved their ecological position through stronger establishment. The perpetuation level is determined to be highly appreciable in terms of major ecological attributes such as distribution, density and basal cover. The importance value index (IVI), a synthetic ecological attribute to know the level of establishment of a plant species, estimated for certain red listed species is quite amazing as a result of *in situ* conservation.

For instance, the rare and threatened plant species, *Anaphalis beddomei* (Asteraceae) secured 1.2 as its IVI among the average of 80 species in shola understories. Similarly, the rare species, *Disporum leschenaultianum* (Convallariaceae) with 1.2 IVI, the threatened species, *Ceropegia pusilla* (Asclepiadaceae) with 1.6 IVI, and the two rare and endemic species, *Smilax wightii* (Smilacaceae) and *Biophytum polyphyllum* (Oxalidaceae) respectively with the IVI of 1.4 and 1.2 are distributed comfortably in many sholas of Nilgiris. The ecological studies on certain endemics like *Agrostis pilosula* (Poaceae), *Asparagus fysonii* (Asparagaceae), *Dumasia villosa* (Fabaceae), *Helichrysum bracteatum* (Asteraceae), *Gaultheria fragrantissima* (Ericaceae), *Pogostemon wightii* (Lamiaceae) and *Strobilanthes foliosa* (Acanthaceae) in sholas also showed the improved status of these species in their habitats by obtaining adequate IVI. Therefore, declaration of still more biomes in Western Ghats as 'protected area for biodiversity conservation' definitely will ensure conservation of our species and their germplasm as well. In addition, modern propagation technologies like clonal propagation and *in vitro* regeneration by employing tissue culture technology may also be developed for those species which have barriers in natural reproduction. Selection of micro-sites for raising such kinds of clones to be produced through these modern methods is another area to be worked out to hold that species very successfully in their habitats of Western Ghats.

In addition to the need of severe biodiversity conservation measures, strict banning must be imposed severely by both central and state governments on mining activities for exploiting physical resources, construction activities and conversion of natural habitats for plantation of cash crops which affects the natural gene flow in evolution process. As the tribals are the integral part of the ecosystems, they must not be disturbed and they should be supported for their prosperous life. Indeed, by using traditional knowledge and techniques they are the people involving in the conservation of many precious wild plants and animals very effectively even before the employing of formal scientific conservation strategies by our authorities after declaring the status of 'protected areas' for many biomes in Western Ghats. All these measures to be adopted will synergistically regulate the ecological processes like energy flow, material cycling, hydrologic cycling, trophic relations among various organisms, species diversity and finally the homeostasis in a perfect and balanced state in Western Ghats. This will definitely provide long term ecological security for all organisms inhabiting not only in Western Ghats but also in all other parts of peninsular India.

Chapter 4

Bryophytes of the Western Ghats: An Overview

A.E.D. Daniels

Bryology Laboratory, Department of Botany and Research Centre, Scott Christian College (Autonomous), Nagercoil – 629 003, Tamil Nadu

Indian bryology has contributed significantly to the knowledge on the diversity and ecology of this delicate and elusive group of plants of the country. However, knowledge on the diversity and ecology of the bryophytes of Western Ghats is far from adequate. Hence, a brief review on the studies and contributions made on this elusive group might awaken the botanists of the country to re-orient their thoughts towards bryophyte studies. In addition, a brief account on their habit and habitat might enable the readers to distinguish between the various forms seen in the said region.

Bryophytes were the pioneers to colonize terrestrial habitats from an aquatic environment and hence they are regarded as the most primitive of the land plants. They form the second largest group of terrestrial plants next to phanerogams. Their adaptation to a terrestrial mode of life is partial as water is indispensable at one stage or other in their life cycle. Hence, they are also known as the 'amphibians' of the plant kingdom. Their remarkable capacity to absorb water and turn fresh in a very short time has earned them the name the 'resurrection plants'.

The term bryophyta is derived from the Greek words *bryon* meaning a moss and *phyton* meaning a plant. It was first used by A. Braun in 1864 to include the algae, fungi, lichens and mosses. However, later on this term was applied exclusively to denote bryophytes which comprise the Hepaticae, Anthocerotae and Musci now referred to as Marchantiophyta, Anthocerotophyta and Bryophyta.

Bryophytes are more common in humid areas but are abundant during rainy seasons in drier areas too. However, they usually do not form a very conspicuous part of any vegetation. There are a few aquatic forms also namely *Riccia fluitans, Ricciocarpus natans* and *Riella* spp. *Cryptothallus* and *Buxbaumia* are the saprophytic genera of liverworts and mosses respectively.

Generally, bryophytes prefer microclimatic niches such as crevices of rocks and trees, and vicinity of small shady springs. Daniels and Kariyappa's (2007) study on the bryodiversity of various vegetation types found in the Southern Western Ghats proves this since they found evergreen forests to have the highest diversity of bryophytes than the rest of the vegetation types. They may grow on barks (corticolous) or leaves (epiphyllous or folicolous) of trees, rocks (rupicolous), pebbles and stones (saxicolous), logs (lignicolous), river banks and roadside cuts (terricolous). However, they also grow on strange substrata like old discarded/abandoned leather, rubber tyres and wooden goods, tiled and asbestos roofs, and mortar of stone and mud walls. Some bryophytes have a remarkable capacity to endure drought and are usually known as the xeromorphic forms. They are the secondary colonizers on barren rocks after lichens.

When compared to the rest of the autotrophic plants, the life cycle of bryophytes is unique comprising of two distinct heteromorphic alternating phases, namely the haploid gametophytic phase which is the dominant one and the diploid sporophytic one dependent on the former for its survival. Most bryophytes reproduce by specialised vegetative bodies known as gemmae which are produced either on leaves or specialized structures called gemma cups as in *Marchantia* spp. and *Lunularia cruciata.*

Though mosses and liverworts differ from each other in a number of features, the differences between them are not always that pronounced. However, there are certain characters with which they can be distinguished. A moss is differentiated into stem and leaves, hence a bryophyte without this differentiation is a liverwort. A bryophyte with leaves deeply lobed or segmented is a liverwort. No moss has this feature. In leafy liverworts the leaves are generally arranged in three ranks on the stem, two being lateral and the third ventral, touching or almost touching the substratum, called the underleaves, usually smaller than the lateral ones. However, exceptions are not few in which the underleaves may be absent or all the three ranks of leaves are almost of the same size and structure. Any bryophyte in which the leaves possess a nerve, single or forked, is a moss since a liverwort lacks this. Mosses with nerveless leaves may invariably be distinguished from liverworts by their very different cell structure and the absence of oil bodies. A bryophyte that possesses delicate, colourless and unicellular rhizoids is a liverwort since mosses possess multicellular rhizoids which are almost always coloured. In a moss the sporogonium is differentiated into a foot embedded in the plant body, a long and slender seta and a capsule which invariably possesses a lid called the operculum and a protective covering the calyptra. But in a liverwort the sporogonium does not show this differentiation and is enclosed by a perianth. But at maturity the seta elongates pushing the capsule out of the perianth. A hornwort, on the other hand, possesses a sporogonium which is totally different from that of a liverwort and moss. Its sporogonium resembles the antlers of an antelope with which it can be differentiated from a thalloid liverwort. These characters though recognizable in the field either with naked eyes or with the aid of a hand lens, require a microscopic examination for conclusive determination.

Bryophytes occupy an important place in the plant kingdom although they were thought to be of little economic value. They are the secondary colonizers next to lichens on barren rocks in plant succession in a xerosere and facilitate weathering of soil. Some of them are extremely good soil binders as they form large mats on forest floor and roadside cuts, thus controlling soil erosions. They are a good source of humus and hence a haven for a number of soil-borne invertebrates like earthworms and form excellent seed beds for seedlings and saplings particularly in evergreen forests. They are also very good indicators of environmental pollution. Thallus extracts of *Marchantia polymorpha* and *Lunularia cruciata* show antifungal and antibacterial properties. A number of liverworts are known to be rich sources of terpenoids such as sesqui- and diterpenoids, and also aromatic compounds. Some are very good laboratory specimens to carry out experiments on hybridization, apogamy, apospory, DNA targeting etc. and also to understand the process of evolution of land plants. A number of mosses are used in decorations in bouquets, vases, bonsais etc.

India with a large area and a variety of phytoclimatic conditions within its different biogeographical zones that contribute to the great diversity of the flora, is one of the 12 megabiodiversity countries in the world. These biogeographical zones with their diverse topography and climatic conditions have been divided into seven bryogeographical regions (Pandé, 1958). They are the Western and Eastern Himalaya, Punjab and West Rajasthan, Gangetic Plains, Central India, the Western and Eastern Ghats and the Deccan Plateau. Singh and Semwal (1995) proposed to include the Andaman and Nicobar Islands in the Bay of Bengal as the eighth one.

The Western Ghats also known as the Sahyadris in Peninsular India has been identified as one of the 25 'hotspots' in the world (Myers *et al.*, 2000). It comprises 27 per cent of the country's flora (Nayar, 1996). Taxonomy and distribution of the flowering plants and the fern and fern allies of this region have been studied with emphasis (Nair and Daniel, 1986). But knowledge on the taxonomy and distribution of bryophytes of this region is far from adequate (Daniels and Daniel, 2009a,b; Daniels *et al.*, 2010b, 2011d) and is still mostly based on the contributions made by the Colonial Europeans of the late 19th and early 20th centuries.

Bryophyte Studies: A Review

In India, the first bryophyte report was made by Rheede (1678–1703) in his monumental work Hortus Indicus Malabaricus. He described one moss as 'puem-peda' from the W. Ghats of Kerala. Robinson presumed that it was *Bryum coronatum* Schwägr., one of the commonest species in Southern India. But after going through Rheede's comments he concluded that it was more likely to be *Bryum bicolor* Dicks. (*vide* Nicolson *et al.*, 1988). Dillenius (1741) included some Indian mosses in his Historia Muscorum and later Linnaeus (1753) described some Indian mosses along with higher plants in his Species Plantarum. Buchanan-Hamilton, a British medical officer, made a collection of mosses in Nepal in 1802. Hooker studied Hamilton's collection and published it in 1808 as Musci Nepalensis. This kindled Hooker to make more serious studies on Indian mosses and in 1818–1820 he described several new species. In 1825 he made two joint publications with Greville on Indian mosses. Schwägrichen (1826–1827) included some Indian mosses in his Supplement to Hedwig's (1801) Species Muscorum.

Our present knowledge on the mosses of the W. Ghats is to a large extent based on the early works by Harvey and Hooker (1840) and Griffith (1842, 1843). Montagne (1842a) reported 66 species from the Nilgris on the W. Ghats and Müller (1853, 1854) over 100 species again from the Nilgris. Mitten (1859) described over 700 species in his Musci Indiae Orientalis which has species from the W. Ghats as well. Brotherus (1899) reported 86 species from Coorg (Karnataka) in the W. Ghats and Dixon (1909, 1914, 1921) particularly from the northern part of the W. Ghats. Potier de la Varde (1922–1925) described 40 species from the Palni Hills, W. Ghats of Madurai (Tamil Nadu). Brühl (1931) published a census of Indian mosses that comprised 2471 species including those from the W. Ghats. Foreau (1930, 1931, 1961, 1964) listed 368 species from the Palni Hills which included 95 new species and 15 varieties. Raghavan and Wadhwa (1968, 1970) studied the mosses of the Agumbe and Hulical Ranges in Shimoga District, Karnataka and listed 28 species. Wadhwa's (1969, 1971) checklist on the mosses of Peninsular India has 331 species that includes Nilgris as well. Crum (1972), in his taxonomic account of the family *Erpodiaceae*, mentioned one collection from the Travancore Hills. Srinivasan (1974) reported 64 species from South India. Chopra (1975) dealt with nearly 2000 mosses in his Taxonomy of Indian mosses where there are species from the W. Ghats as well. The only other comprehensive work that gives some information on the habitat and distribution of the south Indian mosses including the W. Ghats is Gangulee's mosses of Eastern India and adjacent regions (1969–1980). Further, Mohamed's studies (1981, 1982, 1984) which include *Rhodobryum commersonii* and *R. aubertii* in Southern India, *Bryum wightii* and related species, and a synopsis of the genus *Rhodobryum* in Asia respectively cannot be overlooked. Based on a preliminary study, Vohra *et al.* (1982) listed 83 species of mosses from the Silent Valley in Kerala. Ellis (1989) in his revision of the genus *Calymperes* in Southern India and neighbouring islands dealt with 14 species mostly from the W. Ghats. Kumar and Maniselvan (1994), Maniselvan and Kumar (1998) made a report on the moss flora of Udhagamandalam (Nilgris) and Kodaikanal (Palnis) respectively. In continuation, Maniselvan and Kumar (2000a,b) again made a report on the moss flora of Udhagamandalam and studied the cytology of some of them. Meanwhile, Dabhade (1998) who studied the mosses of Khandala and Mahabaleshwar in Maharashtra reported 87 species. Daniels and Daniel (2003a,b) added 2 species to the country from the W. Ghats. Madhusoodhanan and Nair (2004) made a contribution from the Eravikulam National Park where they ennumerated 8 species. Again Daniels and Daniel (2005, 2007) added 4 species to the Indian mainland and 9 to Peninsular India respectively. Lal's (2005) Checklist of Indian mosses and Aziz and Vohra's (2008) Indian *Pottiaceae* are others that include species from the W. Ghats. Subramanian (2008) included 218 species and described 6 novelties in the mosses of Tamil Nadu. Unfortunately, this piece of work has no key to taxa at any rank and provides only the correct name of the species without the original citation, basionym and synonym (s); descriptions are far from adequate to identify a given material; has no details on distribution or specimens studied and the illustrations do not provide the required details for identification. Even the names of all the new species are not validly published since the author does not satisfy the requirements of Art. 40 of the Code. Hence, this work cannot be considered anywhere. Latest contributions include Kariyappa and Daniels (2010), Daniels *et al.* (2011a,b,c and 2012a) and Manju *et al.* (2012).

Similarly, our present knowledge on the liverworts is also by and large based on the works of Montagne (1842b) who reported 34 species, Mitten (1861) who reported 290 species in his Hepaticae Indiae Orientalis and that of Stephani (1900–1924) and Benedix (1953). The contributions of Chopra (1938, 1943), Bharadwaj (1950, 1958, 1960), Kachroo (1969, 1970a, 1970b, 1973), Kachroo *et al.* (1977), Udar and Srivastava (1975, 1977), Srivastava and Udar (1976), Joshi and Biradar (1984), Asthana and Srivastava (1991), Nath and Asthana (1998), Asthana *et al.* (1995) and Srivastava and Sharma (2000), pertain to different areas in Southern India including the W. Ghats.

Parihar *et al.* (1994) brought out a checklist of Indian liverworts and hornworts. The Hepaticology of India (Bapna and Kachroo, 2000), the monograph on Indian *Geocalycaceae* (Srivastava and Srivastava, 2002), the *Notothylaceae* of India and Nepal (Singh, 2002) and the Indian *Cololejeunea* (Asthana and Srivastava, 2003) are some of the more recent contributions that include species from the W. Ghats.

Contributions exclusively to the liverwort flora are by Daniels (1998, 2001), Daniels and Daniel (2002, 2004, 2008, 2009a,b), Daniels and Mabel (2010), Daniels *et al.* (2010a), Daniels and Raja (2011) and Daniels *et al.* (2012a) and also Nair *et al.* (2005b, 2006) and Pócs *et al.* (2007).

Contributions that include bryophytes as a whole (mosses, liverworts and hornworts) may be Easa's (2003) documention of the bryophytes of Kerala based on earlier reports and includes 232 species of which 169 are mosses, Nair *et al.* (2005a) Bryophytes of Wayanad with 171 species, Chaudhary *et al.* (2006) Bryophyte flora of Gujarat which includes parts of the northernmost W. Ghats as well with 67 species and Chaudhary *et al.* (2008) Bryophytes from North Konkan, Maharashtra with 100 species.

Additions to the Bryoflora of the Region

About 960 genera and 20,000 species of bryophytes are so far known to plant science. In India about 2000 species of mosses belonging to 342 genera and 54 families are known to occur (Vohra and Aziz, 1997). From the West Coast and the Western Ghats 682 species of mosses with 190 endemics have been reported.

Daniels *et al.* (2010b) described a new genus *Indopottia* (*I. zanderi*) from the Silent Valley National Park, Daniels *et al.* (2011a and d) added the genus *Ochrobryum* Mitt. (*O. kurzianum* Hampe ex Besch.) and 7 species respectively, Kariyappa and Daniels (2010) added the African *Trachyphyllum dusenii* (Müll.Hal.) Broth., Daniels *et al.* (2011b) added 2 species of *Daltonia* Hook. *viz.*, *D. angustifolia* Dozy and Molk. and *D. contorta* Müll.Hal. to the bryoflora of the country from the W. Ghats. Again Daniels *et al.* (2012b) added *Aulacopilum beccarii* (Müll.Hal. ex Venturi) Mitt. and *Erpodium glaziovii* Hampe to the moss flora of Asia, and South and Southeast Asia respectively, whereas

Aulacopilum glaucum Wilson to the moss flora of India. All these additions were made from the Agasthyamalai Biosphere Reserve in the Southern W. Ghats. Moreover, Manju *et al.* (2012 a,b) described a new species *Aerobryopsis eravikulamensis* and added *Trichosteleum stigmosum* Mitt. to the moss flora of the country respectively from this region. Considering all these recent additions, the number of moss species is now 914.

Concerning liverworts and hornworts, there are about 850 species of liverworts belonging to 140 genera and 52 families (Singh, 1997). From the West Coast and the Western Ghats 121 species of liverworts (including hornworts) with 10 endemics have been reported. Daniels (2010) included 211 liverworts and 9 hornworts (given erroneously as 8 in the abstract) in his Tamil Nadu checklist. Taking into account the contributions made by earlier workers the number of liverworts is 284 and the hornworts is 23. Daniels and Daniel (2002) described 2 new species of *Riccia* L. and Daniels *et al.* (2010a) a new species of *Spruceanthus* Verd. Manju *et al.* (2011) described *Chiloscyphus chinnarensis* as new from the W. Ghats. Daniels & Daniel (2003b) added *Radula madagascariensis* Gottsche and *Leptolejeunea himalayensis* Pandé and R.N. Misra to the W. Ghats, Daniels and Daniel (2004, 2008 and 2009a,b) added *Leptolejeunea balansae* Steph. to the liverwort flora of the Indian mainland, *Frullania ceylanica* Nees, *Cololejeunea distallopapillata* (E.W. Jones) R.M. Schust. and *C. vidaliana* Tixier and the *Australian spruceanthus* (Gottsche & F. Muell.) B. Thiers and Gradst. to the country. Daniels and Felix (2009) rediscovered the hornwort *Notothylas anaporata* Udar and D.K. Singh from Shendurney Wildlife Sanctuary in Kerala. Moreover, Daniels and Mabel (2010) added *Cheilolejeunea trifaria* (Reinw. *et al.*) Mizut. and *Plagiochila grossa* Grolle and M.L. So, Daniels & Raja (2011) added the New Guinean *Thysananthus appendiculatus* Steph. & Daniels *et al.* (2011c) added *Thysananthus spathulistipus* (Reinw. *et al.*) Lindenb., to the liverwort flora of the country from the W. Ghats. Daniels *et al.* (2012a) added *Anastrophyllum aristatum* (Herzog ex N. Kitag.) A.E.D. Daniels *et al.* to the liverwort flora of the country. Thus, the number of liverworts and hornworts so far recorded in the W. Ghats is 296 and 23 respectively.

Needless to say that the need of the hour is to pursue more systematic studies on the bryophytes of the W. Ghats and should be taken up before it is too late.

Acknowledgements

Daniels thanks Drs. D.G. Long (E), R.L. Zhu (HSNU), Dr. G. Winter, Senckenberg Natural History Museum, Senckenberg, Germany, Mr. M.J. Wigginton (Peterborough) and Mr. B.J. O'Shea (London), Drs. G. Hardy (E), M. Price (G), M.A.H. Mohammed, University of Brunei Darussalam, Brunei Darussalam, D.K. Singh, P. Venu and S. Bandyopadhyay (CAL), V. Nath and A.K. Asthana (LWG), S.C. Srivastava and G. Asthana (LWU) for help with literature.

References

Asthana, A.K. and Srivastava, S.C. 1991. Indian hornworts (A taxonomic study). Bryophyt. Biblioth. **42**: 1–158.

Asthana, G. and Srivastava, S.C. 2003. Indian *Cololejeunea*. A taxonomic study. Bryophyt. Biblioth. **60**: 1–155.

Asthana, G., Srivastava, S.C. and Asthana, A.K. 1995. The genus *Cheilolejeunea* in India. Lindbergia **20**: 125–143.

Aziz, M.N. and Vohra, J.N. 2008. Pottiaceae (Musci) of India. Bishen Singh Mahendra Pal Singh, Dehra Dun, India.

Bapna, K.R. and Kachroo, P. 2000. Hepaticology in India I and II. Himanshu Publ., New Delhi, India.

Benedix, E.H. 1953. Indomalayische *Cololejeuneen*. Feddes Repert. Spec. Nov. Regni Veg. Beih. **134**: 1–88.

Bharadwaj, D.C. 1950. Studies in Indian Anthocerotaceae. I. On the morphology of *Anthoceros crispulus* (Mont.) Douin. J. Indian Bot. Soc. **29**: 145–163.

Bharadwaj, D.C. 1958. Studies in Indian Anthocerotaceae. II. The morphology of *Anthoceros* cf. *gemmulosus* (Hattori) Pandé. J. Indian Bot. Soc. **37**: 75–92.

Bharadwaj, D.C. 1960. Studies in Indian Anthocerotaceae. III. The morphology of *Anthoceros erectus* Kash. and some other species. J. Indian Bot. Soc. **39**: 568–592.

Brijithlal, N.D., Mabel, J.L. and Daniels, A.E.D. 2008. Additions to the moss flora of Peninsular India from the Neyyar Wildlife Sanctuary, Kerala, India. Indian J. Forest. **31**: 643 - 646.

Brotherus, V.F. 1899. Contribution to the bryological flora of southern India. Report on the collection made by Dr. T.L. Walker in Coorg during the cold weather of 1897–98. Rec. Bot. Surv. India **1 (12):** 311–329.

Brühl, P. 1931. A census of Indian mosses with analytical keys to the genera. Rec. Bot. Surv. India **13 (1):** 1–135.

Chaudhary, B.L., Sharma, T.P. and Sanadhya, C. 2006. Bryophyte flora of Gujarat (India). Himanshu Publ., New Delhi, India.

Chaudhary, B.L., Sharma, T.P. and Bhagora, F.S. 2008. Bryophyte Flora of North Konkan: Maharashtra (India). Himanshu Publ., New Delhi, India.

Chopra, R.S. 1938. Notes on Indian hepatics I. South India. Proc. Indian Acad. Sci. **B7:** 239 - 251.

Chopra, R.S. 1943. A census of Indian hepatics. J. Indian Bot. Soc. **22:** 237–260.

Chopra, R.S. 1975. Taxonomy of Indian mosses. CSIR, New Delhi, India.

Crum, H. 1972. A taxonomic account of the family Erpodiaceae. Nova Hedwigia **23:** 201–224.

Dabhade, G.T. 1998. Mosses of Khandala and Mahabaleshwar in Western Ghats (India). A.S. Dalvi, Sanman Co-op. Hsg. Ltd., Thane, India.

Daniels, A.E.D. 1998. Ecological adaptations of some bryophytes of the Western Ghats. J. Ecobiol. **10**: 261–270.

Daniels, A.E.D. 2001. *Cololejeunea furcilobulata* (Berrie et Jones) Schuster and *Heteroscyphus argutus* (Reinw. *et al.*) Schiffn. from Mahendragiri hills of Kanyakumari District of South India. In: Nath, V. and A.K. Asthana (eds.) Perspectives in Indian Bryology: 301 - 307. Dehra Dun, India.

Daniels, A.E.D. 2010. Checklist of the bryophytes of Tamil Nadu. Arch. Bryol. **65**: 1–118.

Daniels, A.E.D. and Daniel, P. 2002. Two new species of *Riccia* L. (Hepaticae: Marchantiales) from the Western Ghats of Tamil Nadu. Bull. Bot. Surv. India **44**: 135–140.

Daniels, A.E.D. and Daniel, P. 2003a. *Fissidens griffithii* Gangulee (Musci: Fissidentales) – An addition to the bryoflora of India. Indian J. Forest. **26**: 93–194.

Daniels, A.E.D. and Daniel, P. 2003b. Additions to the bryoflora of Peninsular India. Indian J. Forest. **26:** 389–396.

Daniels, A.E.D. and Daniel, P. 2003c. An Addition to the bryoflora of India. Bull. Bot. Surv. India **45**: 225–226.

Daniels, A.E.D. and Daniel, P. 2004. *Leptolejeunea balansae* (Hepaticae: Jungermanniales)–a new record to the bryoflora of the Indian mainland. J. Bombay Nat. Hist. Soc. **101:** 333 - 334.

Daniels, A.E.D. and Daniel, P. 2005. Additions to the moss flora of the Indian mainland. Bull. Bot. Surv. India **47**: 93–100.

Daniels, A.E.D. and Daniel, P. 2007. The mosses of the southern Western Ghats. In: Nath, V. and A.K. Asthana (ed.), Current trends in bryology: 227–243. Bishen Singh Mahendra Pal Singh, Dehra Dun, India.

Daniels, A.E.D. and Daniel, P. 2008. *Frullania ceylanica* Nees (Frullaniaceae)–New to the hepatic flora of India. Indian J. Forest. **31**: 637–639.

Daniels, A.E.D. and Daniel, P. 2009a. *Cololejeunea distallopapillata* (E.W. Jones) R.M. Schust. and *C. vidaliana* Tixier (Lejeuneaceae)–New to the liverwort flora of India. Acta Bot. Hung. **51**: 61–66.

Daniels, A.E.D. and Daniel, P. 2009b. The Australian *Spruceanthus thozetianus* (Gottsche and F. Muell.) B. Thiers & Gradst. (Hepaticae) discovered in the Western Ghats of India. Acta Bot. Hung. **51**: 283–287.

Daniels, A.E.D and Felix, R. 2009. The endemic and rare *Notothylas anaporata* Udar & D.K. Singh (Notothylaceae: Anthocerotae) rediscovered. Nelumbo **51**: 217–218.

Daniels, A.E.D., Long, D.G., Kariyappa, K.C. and Daniel, P. 2012a. *Anastrophyllum aristatum* (Herzog ex N.Kitag.) A.E.D. Daniels *et al.*, comb. et stat. nov. (Marchantiophyta: Anastrophyllaceae) from India and China. J. Bryol. **34:** 146–149.

Daniels, A.E.D. and Kariyappa, K.C. 2007. Bryophyte diversity along a gradient of human disturbance in the southern Western Ghats. Curr. Sci. **93:** 976–982.

Daniels, A.E.D., Kariyappa, K.C. and Daniel, P. 2011a. The moss genus *Ochrobryum* Mitt. (Leucobryaceae) added to the bryoflora of the Western Ghats. Nelumbo **53**: 155–160.

Daniels, A.E.D., Kariyappa, K.C. and Daniel, P. 2011b. Two species of *Daltonia* Hook. & Taylor (Hookeriaceae: Bryophyta) new to the moss flora of India. Nelumbo **53:** 133–139.

Daniels, A.E.D., Kariyappa, K.C. and Daniel, P. 2011c. The liverwort *Thysananthus spathulistipus* (Reinw. *et al.*) Lindenb. (Lejeuneaceae) long-lost in India rediscovered. Acta Bot. Hung. **53:** 283–289.

Daniels, A.E.D., Kariyappa, K.C. and Daniel, P. 2010a. A new species of *Spruceanthus* Verd. (Hepaticae: Jungermanniales) from the Western Ghats of India. Cryptog. Bryol. **31**: 121 - 123.

Daniels, A.E.D. and Mabel, J.L. 2010. Two liverworts new to the Indian bryoflora. Lindbergia **33**: 77–80.

Daniels, A.E.D., Mabel, J.L. and Daniel, P. 2012b. The Erpodiaceae (Bryophyta: Isobryales) of India. Taiwania **57**: 168–182.

Daniels, A.E.D., Mabel, J.L. and Sreeji, S. 2011d. Mosses new to the bryoflora of the Western Ghats from the Agasthyamalai Biosphere Reserve. Nelumbo **53:** 145–154.

Daniels, A.E.D., Raja, R.D.A. and Daniel, P. 2010b. *Indopottia zanderi* (Pottiaceae, Bryophyta) gen. et sp. nov. from the Western Ghats of India. J. Bryol. **32**: 216–219.

Daniels, A.E.D. and Raja, R.D.A. 2011. The New Guinean *Thysananthus appendiculatus* Steph. (Lejeuneaceae) discovered in the Western Ghats of India. Lindbergia **34:** 40–43.

Dillenius, J.J. 1741. Historia muscorum. Oxford, England.

Dixon, H.N. 1909. Mosses from Western Ghats. J. Bot. **47:** 157–164.

Dixon, H.N. 1914. Report on the mosses collected by C.E.C. Fischer and others from south India and Ceylon. Rec. Bot. Surv. India **6 (3):** 75–91.

Dixon, H.N. 1921. On a collection of mosses from Kanara District. J. Indian Bot. **2:** 174–188.

Easa, P.S. 2003. Biodiversity documentation for Kerala. Part 4: Bryophytes. KFRI Handb. **17:** 1–47. KFRI, Peechi, Kerala, India.

Ellis, L.T. 1989. A taxonomic revision of *Calymperes* in southern India and neighbouring islands. J. Bryol. **15**: 697–732.

Foreau, G. 1930. Notes on bryological geography for the Presidency of Madras. J. Madras Univ. **2:** 238–250.

Foreau, G. 1931. Notes on bryological geography for the Presidency of Madras. J. Madras Univ. **3:** 118–126.

Foreau, G. 1961. The moss flora of Palni Hills. J. Bombay Nat. Hist. Soc. **58:** 13–47.

Foreau, G. 1964. Some South Indian mosses. J. Bombay Nat. Hist. Soc. **61:** 223–226.

Gangulee, H.C. 1969–1980. Mosses of Eastern India and adjacent regions. Fasc. 1–8: 1–2142. Calcutta, India.

Griffith, W. 1842. Muscologia itineris Assamici. Calcutta J. Nat. Hist. **2:** 65–512.

Griffith, W. 1843. Muscologia itineris Assamici. Calcutta J. Nat. Hist. **3:** 56–75; 270–282.

Harvey, W.H. and Hooker, J.D. 1840. Musci Indici. Hooker's J. Bot. Kew Gard Misc. **2:** 1–21.

Hooker, W.J. and Greville, R.K. 1825a. On the genus *Hookeria* of Smith of the order Musci. Edinburgh J. Sci. 2: 221–236.

Hooker, W.J. and Greville, R.K. 1825b. On the genus *Calymperes* of Swartz and *Syrrhopodon* of Schwägrichen of the order Musci. Edinburgh J. Sci. **3:** 218–227.

Joshi, D.Y. and Biradar, N.V. 1984. Studies in the liverwort flora of Western Ghats with special reference to Maharastra, India. J. Hattori Bot. Lab. **56:** 45–52.

Kachroo, P. 1969. Hepaticae of India–A taxonomic survey and census. I. Floristic and taxonomic considerations. J. Sci., Univ. Kashmir **6:** 39–55.

Kachroo, P. 1970a. Hepaticae of India-A taxonomic survey and census. II. *Takakiaceae* through Marsupellaceae. J. Sci., Univ. Kashmir **7:** 176–200.

Kachroo, P. 1970b. Hepaticae of India–A taxonomic survey and census. IV. Lejeuneaceae. Bull. Bot. Surv. India. **12:** 226–241.

Kachroo, P. 1973. Hepaticae of India–A taxonomic survey and census. III. Plagiochilaceae through Pleuroziaceae. J. Sci., Univ. Kashmir **1:** 141–161.

Kachroo, P., Bapna, K.R. and Dhar, G.L. 1977. Hepaticae of India–A taxonomic survey and census. V (concld.). Fossombroniaceae through Anthocerotaceae. J. Indian Bot. Soc. **56:** 62–86.

Kariyappa, K.C. and Daniels, A.E.D. 2010. The African moss *Trachyphyllum dusenii* (Müll.Hal. ex Broth.) Broth. (Hypnobryales: Entodontaceae) in India. Nelumbo 52: 131–134.

Kumar, S.S. and Maniselvan, P. 1994. A report on the moss flora of Udhagamandalam in south India–I. Res. Bull. Panjab Univ., New Ser., Sci. **44:** 99–103.

Lal, J. 2005. A Checklist of Indian mosses. Bishen Singh Mahendra Pal Singh, Dehra Dun.

Linnaeus, C. 1753. Species plantarum. Vol. 2. Musci. Holmiae, Stockholm, Sweden.

Madhusoodanan, P.V. and Nair, M.C. 2004. Pleurocarpus mosses of Eravikulam National Park, Kerala-1. J. Econ. Taxon. Bot. **28:** 338–346.

Manju, C.N., Martin, K.P., Sreekumar, V.B. and Rajesh, K.P. 2012. Morphological and molecular differentiation of *Aerobryopsis eravikulamensis* Manju & Rajesh sp. nov. (Meteoriaceae) and closely related taxa of the Western Ghats of India. Bryologist **115:** 42–50.

Manju, C.N., Rajesh, K.P. and Madhusoodanan, P.V. 2008. Checklist of the bryophytes of Kerala, India. Trop. Bryol. Res. Rep. **7:** 1–24.

Manju, C.N., Rajesh, K.P. and Madhusoodanan, P.V. 2009a. Contribution to the bryophyte flora of India; the Aralam Wildlife Sanctuary in the Western Ghats. Arch. Bryol. **42:** 1–12.

Manju, C.N., Rajesh, K.P. and Madhusoodanan, P.V. 2009b. Contribution to the bryophyte flora of India: Agasthyamalai Biosphere Reserve in Western Ghats. Taiwania **54:** 57–68.

Manju, C.N., Rajesh, K.P. and Madhusoodanan, P.V. 2011. *Chiloscyphus chinnarensis* (Lophocoleaceae: Marchantiophyta): a new species from the Western Ghats of India. Acta Bot. Hung. **53:** 151–157.

Manju, C.N., Prakashkumar, P., Prajitha, B., Rajilesh, V.K. and Anoop, K.P. 2012. *Trichosteleum stigmosum* Mitt. (Sematophyllaceae) from Silent Valley National Park, a New Record for India. Taiwania **57:** 222–224.

Maniselvan, P. and Kumar, S.S. 1998. Preliminary survey of moss flora of Kodaikanal–I. Res. Bull. Panjab Univ., New Ser., Sci. **48:** 21–24.

Maniselvan, P. and Kumar, S.S. 2000a. A report on the moss flora of Udhagamandalam in south India–II. Res. Bull. Panjab Univ., New Ser., Sci. **49:** 29–34.

Maniselvan, P. and Kumar, S.S. 2000b. Cytological observations on some South Indian mosses–IV. Cryptog. Bryol. **21:** 241–245.

McNeill, J., Barrie, F.R., Buck, W.R., Demoulin, V., Greuter, W., Hawksworth, D.L., Herendeen, P.S., Knapp, S., Marhold, K., Prado, J., Prud'homme van Reine, W.F., Smith, G.F., Wiersema, J.H. and Turland, N.J. (eds.) 2012. International Code of Nomenclature for algae, fungi and plants (Vienna Code): Adopted by the eighteenth International Botanical Congress Melbourne, Australia, July 2011. Regnum Veg. **154**. Gantner.

Mitten, W. 1859. Musci Indiae Orientalis. J. Proc. Linn. Soc., Bot. **1** (Suppl.): 1–171.

Mitten, W. 1861. Hepaticae Indiae Orientalis. J. Proc. Linn. Soc., Bot. 5: 89–128.

Mohamed, M.A.H. 1981. *Rhodobryum commersonii* (Schwägr.) Paris and *R. aubertii* (Schwägr.) Thér. in South India. J. Bryol. *11*: 691–694.

Mohamed, M.A.H. 1982. *Bryum wightii* Mitt. and related species. J. Bryol. **12:** 23–29.

Mohamed, M.A.H. 1984. A synopsis of the genus *Rhodobryum* in Asia. J. Hattori Bot. Lab. **55**: 281–293.

Montagne, J.P.F.C. 1842a. Cryptogamae Nilgherienses seu plantarum cellularium in montibus Peninsulae Indicae Neel-Gherries dictis à cl. Perrottet collectarum enumeratio, Musci. Ann. Sci. Nat,. Bot. 2, **17**: 243–256.

Montagne, J.P.F.C. 1842b. Cryptogamae Nilgherienses seu plantarum cellularium in montibus Peninsulae Indicae Neel-Gherries dictis à cl. Perrottet collectarum enumeratio, Hepaticae. Ann. Sci. Nat., Bot. 2, **18**: 12–23.

Müller, C. 1853. Musci Neilgherrensis. *B*ot. Zeitung (Regensburg) **11**: 17–21; 33–40, 57–62.

Müller, C. 1854. Musci Neilgherrensis. Bot. Zeitung (Regensburg) **12**: 556–559, 569–574.

Myers, N., Mittermeier, R.A., Mittermeier, C., da Fonseca, G.A.B. and Kent, J. 2000. Biodiversity hotspots for conservation priorities. Nature **403**: 853–857.

Nair, M.C., Rajesh, K.P. and Madhusoodanan, P.V. 2005a. Bryophytes of Wayanad in Western Ghats. Malabar Nat. Hist. Soc., Kerala, India.

Nair, M.C., Rajesh, K.P. and Madhusoodanan, P.V. 2005b. Little known *Plagiochila* (Dum.) Dum. (Plagiochilaceae: Hepaticae) from India. Geophytology **35**: 39–44.

Nair, M.C., Rajesh, K.P. and Madhusoodanan, P.V. 2006. *Lejeunea exilis* (Lejeuneaceae: Hepaticae): a new record to India. Acta Bot. Hung. 48: 85–87.

Nair, N.C. and Daniel, P. 1986.The floristic diversity of the Western Ghats and its conservation: A review. Proc. Indian Acad. Sci. (Anim. Sci./Pl. Sci.) Suppl.: 127–163.

Nath, V. and Asthana, A.K. 1998. Diversity and distribution of genus *Frullania* Raddi in South India. J. Hattori Bot. Lab. 85: 63–82.

Nayar, M.P. 1996. Hot spots of endemic plants of India, Nepal and Bhutan. Trop. Bot. Gard. Res. Inst., Thiruvananthapuram, India.

Nicolson, D.H., Suresh, C.R. and Manilal, K.S. 1988. An interpretation of Rheede's Hortus Malabaricus. Regnum Veg. 119: 1–378.

Pandé, S.K. 1958. Some aspects of Indian hepaticology. J. Indian Bot. Soc. **37**: 1–26.

Parihar, N.S., Lal, B. and Katiyar, N. 1994. Hepatics and Anthocerotes of India. A new annotated checklist. Central Book Depot, Allahabad, India.

Pócs, T., Nair, M.C., Rajesh, K.P. and Madhusoodanan, P.V. 2007. Liverwort (Marchantiopsida) records from the Western Ghats (Kerala State, Peninsular India). Acta Bot. Hung. **49**: 121–129.

Potier de la Varde, R. 1922–1924. Musci Madurensis. Rev. Bryol. Lichénol. **49:** 33–44. 1922. **50:** 17–27; 72–79. 1923. **51**: 10–14. 1924.

Potier de la Varde, R. 1925. Nouvelles herborisations dans le sud de l'Inde (1er note). Rev. Bryol. Lichénol. **52**: 37–43.

Raghavan, R.S. and Wadhwa, B.M. 1968. Mosses of Agumbe-Hulical Ranges, Shimoga District. Bull. Bot. Surv. India **10:** 344–347.

Raghavan, R.S. and Wadhwa, B.M. 1970. Mosses of Agumbe (Shimoga Distr., Mysore State). II. Keys to species and new records for India. M.V.M. Patrika **5:** 28–37.

Rheede, H.A. van 1678–1703. Hortus Indicus Malabaricus. J. *van* Someren and J. *van* Dyck, Amsterdam, the Netherlands.

Schwägrichen, C.F. 1811–1842. Species muscorum frondosorum ... supplementum primum - quartum. Vols. 1–4. J. Barth, Leipzig, Germany.

Singh, D.K. 1997. Liverworts. In: Mudgal, V. and Hajra, P.K. (ed.), Floristic studies and conservation strategies in India 1: 235–300. Botanical Survey of India, Kolkata, India.

Singh, D.K. 2002. Notothylaceae of India and Nepal (A morpho-taxonomic revision). Bishen Singh Mahendra Pal Singh, Dehra Dun, India.

Singh, D.K. and Semwal, R.C. 1995. Liverworts. In: Mudgal, V. and Hajra, P.K. (ed.), Bharat Ki Vanaspathi Vividhita: 225–246. Dehra Dun, India.

Srinivasan, C. 1974. Studies in the Mosses of South India. Annamalai University, India.

Srivastava, A. and Srivastava, S.C. 2002. Indian Geocalycaceae (Hepaticae). A taxonomic study. Bishen Singh Mahendra Pal Singh, Dehra Dun, India.

Srivastava, S.C. and Sharma, D. 2000. A preliminary study on the liverwort and hornwort flora of Silent Valley (Kerala). In: Chauhan, D.K. (ed.), Recent Trends in Botanical Researches: 55–75. Department of Botany, Allahabad University, Allahabad, India.

Srivastava, S.C. and Udar, R. 1976. Indian Aneuraceae–a monographic study. Biol. Mem. **1**: 121 - 154.

Stephani, F. 1900–1924. Species Hepaticarum. Vols. 1–6. George and Cie., Génève.

Subramanian, D. 2008. Mosses of Tamil Nadu. Bishen Singh Mahendra Pal Singh, Dehra Dun, India.

Udar, R. and Srivastava, S.C. 1975. Notes on south Indian Hepaticae. I. J. Bombay Nat. Hist. Soc. **72**: 401–406.

Udar, R. and Srivastava, S.C. 1977. Notes on South Indian Hepaticae. 2. The genus *Herberta* Gray. J. Bombay. Nat. Hist. Soc. **74:** 255–263.

Vohra, J.N. and Aziz, M.N. 1997. *Mosses.* In: Mudgal, V. and P.K. Hajra (ed.), Floristic studies and conservation strategies in India 1: 301–374. Botanical Survey of India, Kolkata.

Vohra, J.N., Roychowdhury, K.N., Ghosh, R.K., Kar, B.D. and Singh, K.P. 1982. Observations on the cryptogamic flora of Silent Valley. Botanical Studies on Silent Valley (Cryptogams and Soil Study). Part I: 19–30. Botanical Survey of India, Howrah, India.

Wadhwa, B.M. 1969–1971. Checklist of the moss flora of Peninsular India including the Western Ghats and Nilgiris. Parts I and II. M.V.M. Patrika **4:** 74–93; **6:** 70–76.

Chapter 5

Vascular Flora of Thumpali Hillock

S.R. Nisha[1] *and S. Jeeva*[2]

[1]*Department of Botany, Nesamony Memorial Christian College, Marthandam, Kanyakuamri – 629 165, Tamil Nadu*

[2]*Department of Botany, Scott Christian College, Nagercoil, Kanyakuamri – 629 003, Tamil Nadu*

The current emergence of awareness and interest on biodiversity inventory and conservation can be traced to the publication of a book entitled "Tragedy of the Commons" by Barry Commoner, a British Scientist. Since then, several international organizations such as IUCN, UNESCO, UNEP, WWF, UNCED, BGCI have been playing an active role in disseminating the awareness of biodiversity conservation to individuals and groups of people in all areas of human activities. The Biological Diversity Convention held at Rio de Janeiro in 1992, roused both an interest and sense of responsibility in scientists and political leaders all over the world for preserving biological diversity and its sustainable utilization in order to make the biodiversity lasting for use by future humankind and to prevent further global warming which today threatens the very existence of humankind on earth. In view of the fast depleting floral diversity, more active interest on the part of the plant taxonomists is needed to know the endemic and vulnerable species in order to prevent their extinction by prevalent anthropogenic pressure on plant diversity (Manickam *et al.*, 2008).

Areas having an exceptional concentration of endemic species and experiencing loss of habitat have been termed biodiversity *hot-spots* (Myers *et al.*, 2000). Out of 25 hot spots of the world, two are present in India *viz.* Western Ghats and Himalayas.

The Western Ghats one of the mega endemic centres in India, which run north-south for about 1,600 km has a cradle of vascular plants. Despite the vast and varied flora in southern Western Ghats of Tamil Nadu, the biodiversity of the hillocks has not been explored sufficiently. Raj and Sukumaran (1997), Swamy *et al.* (1998), Jeeva *et al.* (2005a, 2005b), Prakash *et al.* (2006), Sukumaran and Jeeva (2008) and Sukumaran *et al.* (2008) have studied the phytodiversity of the region other than hillocks. Nayar (1959), Lawrence (1960), Henry and Swaminathan (1981), Sundarapandian and Swamy (1997), Swamy *et al.* (2000) and Brintha *et al.* (2012) explored the vegetation of Kanyakumari district, but they paid attention to forests and urban areas other than hillocks. Floristic diversity and phytosociological study of the hillocks of Kanyakumari district are lacking. Therefore, the present study has been undertaken to assess the plant biodiversity of fragmented forest hillock of Thumpali in Kanyakumari district, Tamil Nadu, India.

Study Area

The study area is about 8 km towards south of Marthandam (077°11.7022 E and 08°17.0492 N) on the way to Thengapattinam estuary, south west coast of India at an altitude of 300ft msl. It covers an area of ca. 50 ha. Topographically it has prominent natural features richly varied and crowded, majestic hills with undulating surrounding plains, paddy fields, wetlands and greenish coconut plantation, etc. gave splendid landscapes. Climate of the district is favorable agro-climatic and rainfall varies from 103 cm to 310 cm (Figure 5.1).

The hill encompasses rich repository of granites. The area was under mining till the year 2004. Partial locality of the area is degraded due to this action and hence it was banned. The pool originating from the hill adds additional scenic beauty of the area. It also serves as the water source for the people residing the vicinity of the hill.

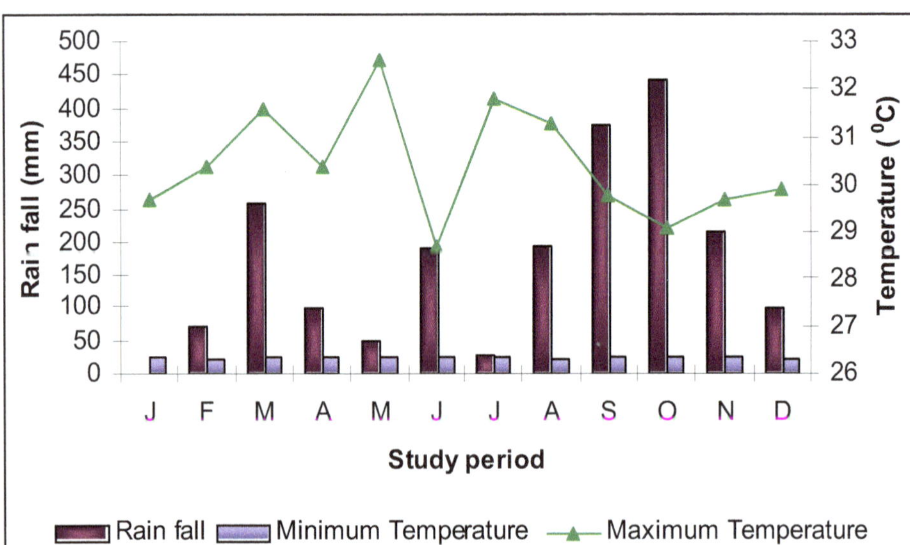

Figure 5.1: Figure Showing the Climate of the Study Area.

Floristic Inventory

An extensive floristic survey was carried out in the Thumpali hillock, Kanyakumari district, Tamil Nadu, India. Specimens of vascular plants were collected and identified taxonomically with the help of different floras (Gamble, 1915 – 1935; Nair and Nayar, 1986 – 1987; Matthew, 1983; Manickam and Irudayaraj, 1992; Karuppusamy *et al.*, 2013) and by using field keys devised by Pascal and Ramesh (1987). Economic importance of the plant species of the study area was determined by interviewing the local people of the vicinity of the hillock.

The Herbaria of Tropical Botanical Garden and Research Institute (TBGRI), Palode, Trivandrum District, Kerala, Centre for Biodiversity and Biotechnology, St. Xavier's College, Palyamkottai and Botany Department, Nesamony Memorial Christian College, Marthandam, were consulted for correct identification of plant specimens. The nomenclature of species follows the regional flora. The voucher specimens were deposited at the herbarium of Nesamony Memorial Christian College, Marthandam.

Taxonomic Diversity

Taxonomically, a total of 264 species belonging to 227 genera and 83 families including four pteriodphytic families (Adiantaceae, Hemionitidaceae, Schizaceae and Sinopteridaceae) and one gymnospermic family (Cycadaceae) was identified in the Thumpali hills of Kanyakumari district, Tamil Nadu, India (Table 5.1). Among these 48 were trees, 38 shrubs, 147 herbs and 31 climbers (Figure 5.2). The secondary invasive species were confined to the periphery and in disturbed patches of the study area showed the disturbance by cattle grazing and other socio-cultural disturbance.

Poaceae with 29 species and 22 genera was by far the largest family in the study area, Fabaceae occupied the second position having 23 species and 20 genera followed by Asteraceae (15 species), Euphorbiaceae (14 species), Acanthaceae (12 species) and Rubiaceae (10 species). Three families Malvaceae, Ascelpiadaceae and Cyperaceae having 7 species each, Verbenaceae, Amaranthaceae and Apocynaceae (6 species each), Lamiaceae, Convolvulaceae and Vitaceae (5 species each) and Rutaceae, Mimosaceeae, Moraceae and Arecaceae (4 species each). Nine families (Schropulariaceae, Oleaceae, Sapindaceae, Menispermaceae, Meliaceae, Liliaceae, Capparidaceae, Annonaceae and Anacardiaceae) had 3 species each, 11 families (Tiliaceae, Solanaceae, Sapotaceae, Myrtaceae, Melastomaceae, Guttiferae, Commelinaceae, Combretaceae, Boraginaceae, Araceae and Hemionitidaceae) were represented by 2 speceis each; whereas 42 families were monospecific (Table 5.2; Figure 5.3).

The pterdiophytic elements recorded from the study area are *Adiantum latifolium* (Adiantaceae), *Cheilanthes mysorensis* (Sinopteridaceae), *Hemionits arifolia* (Hemionitidaceae), *Lygodium flexuosum* (Schizaeaceae) and *Pityrogramma calomelanos* (Hemionitidaceae). *Cycas circinalis* commonly called *Chalai maram*, belongs to the family Cycadaceae is the only representative of gymnosperm recorded during the survey. The leaf of the species is harvested by the residents of the area for decoration purposes in marriage ceremonies.

Table 5.1: Plant Species Recorded from the Study Area.

Botanical Name	Habit	Family
Abrus precatorius L.	Climber	Fabaceae
Abutilon hirtum (Lamk.) Sweet	Herb	Malvaceae
Abutilon indicum (L) Sweet.	Shrub	Malvaceae
Acacia mangium Willd.	Tree	Fabaceae
Acanthocereus tetragonus (L.) Hummelinck	Shrub	Cactaceae
Achras sapota L.	Tree	Sapotaceae
Achyranthes aspera L.	Herb	Amaranthaceae
Actiniopteris radiata (Sw.) Link	Herb	Actiniopteridaceae
Adiantum latifolium Lam.	Herb	Adiantaceae
Aerva lanata Juss.	Herb	Amaranthaceae
Agave americana L.	Shrub	Agavaceae
Agave angustifolia Haw.	Shrub	Agavaceae
Ageratum conyzoides L.	Herb	Asteraceae
Ailanthus triphysa Roxb.	Tree	Simaroubiaceae
Albizzia lebbeck Benth.	Tree	Mimosaceae
Alloteropsis cimicina Stapf	Herb	Poaceae
Aloe vera L.	Herb	Liliaceae
Alstonia scholaris R.Br.	Tree	Apocynaceae
Alternanthera pungens Kunth	Herb	Amaranthaceae
Alternanthera sessilis R.Br.	Herb	Amaranthaceae
Alysicarpus monilifer DC.	Herb	Fabaceae
Alysicarpus vaginalis DC.	Herb	Fabaceae
Amarathus viridis L.	Herb	Amaranthaceae
Ampelocissus tomentosa Planch.	Climber	Vitaceae
Anacardium occidentale L.	Tree	Anacardiaceae
Ananas comosus Merr.	Shrub	Bromeliaceae
Andrographis echioides Nees	Herb	Acanthaceae
Andrographis paniculata Nees	Herb	Acanthaceae
Anisochilus scaber Benth.	herb	Fabaceae
Anisomeles malabarica R.Br.	Herb	Lamiaceae
Annona muricata L.	Shrub	Annonaceae
Annona squamosa L.	Shrub	Annonaceae
Apluda mutica L.	Herb	Poaceae
Argyreia osyrensis (Roth) Choisy	Climber	Convolvulaceae
Aristida adscensionis L.	Herb	Poaceae
Aristida setacea Retz.	Herb	Poaceae

Contd...

Table 5.1–*Contd...*

Botanical Name	*Habit*	*Family*
Aristolochia indica L.	Climber	Aristolochiaceae
Artocarpus heterophyllus Lam.	Tree	Moraceae
Artocarpus hirsutus Lam.	Tree	Moraceae
Asparagus racemosus Willd.	Climber	Liliaceae
Asystasia gangetica T. And.	Herb	Acanthaceae
Atalantia missionis (Wall. ex Wight) Oliv.	Shrub	Rutaceae
Atylosia scarabaeoides Benth.	Shrub	Fabaceae
Axonopus compressus (Sw.) P. Beauv	Herb	Poaceae
Azadiracta indica A. Juss.	Tree	Meliaceae
Bambusa arundinacea Willd.	Tree	Poaceae
Barleria lupulina Lindl.	Herb	Acanthaceae
Barleria mysorensis Roth	Herb	Acanthaceae
Biophytum sensitivum DC.	Herb	Oxalidaceae
Blepharis maderaspatensis (L.) Roth	Herb	Acanthaceae
Boerhavia diffusa L.	Herb	Nyctaginaceae
Borassus flabellifer L.	Tree	Arecaceae
Borreria hispida (L.) Schum.	Herb	Rubiaceae
Brachiaria setigera (Retz.) C.E. Hubb.	Herb	Poaceae
Brassica nigra (L.) Koch.	Herb	Brassicaceae
Breynia retusa (Dennst.) Alston	Shrub	Euphorbiaceae
Bryophyllum pinnatum Kurz.	Herb	Crassulaceae
Bulbostylis barbata (Rottb.) C.B.Clarke	Herb	Cyperaceae
Calophyllum inophyllum L.	Tree	Guttiferae
Calotropis gigantea (L.) R. Br.	Shrub	Asclepiadaceae
Canavalia ensiformis (L.) DC.	Climber	Fabaceae
Canthium coromandelicum (Burm.f.) Alston	Shrub	Rubiaceae
Canthium dicoccum Gaertn.	Herb	Rubiaceae
Canthium rheedii DC.	Shrub	Rubiaceae
Capparis zeylanica L.	Climber	Capparaceae
Caralluma attenuata Wight	Herb	Asclepiadaceae
Cardiospermum halicacabum L.	Climber	Sapindaceae
Carissa carandas L.	Shrub	Apocynaceae
Carissa spinarum L.	Shrub	Apocynaceae
Carmona retusa (M. Vahl) Masam.	Shrub	Boraginaceae
Caryota urens L.	Tree	Arecaceae

Contd...

Table 5.1–*Contd...*

Botanical Name	*Habit*	*Family*
Cassia occidentalis L.	Herb	Fabaceae
Cassia siamea Lam.	Tree	Fabaceae
Cassytha filiformis L.	Twiner	Lauraceae
Casuarina equisetifolia Forster & Forster f.	Tree	Casuarinaceae
Cathranthus roseus (L.) Don	Herb	Apocynaceae
Cayratia japonica (Thunb. ex Murray) Gagnep.	Climber	Vitaceae
Cayratia pedata (Lour.) A. L. Juss.	Climber	Vitaceae
Ceiba pentandra (L.) Gaertn.	Tree	Bombaceae
Centella asiatica (L.) Urban	Herb	Apiaceae
Centrosema pubescens Benth.	Climber	Fabaceae
Cheilanthus mysorensis Wall.	Herb	Sinopteridaceae
Chloris barbata Sw.	Herb	Poaceae
Cissampelos pareira L.	Climber	Menispermaceae
Cissus quadrangularis L.	Climber	Vitaceae
Cissus vitiginea L.	Climber	Vitaceae
Citrus limon Burm. f.	Shrub	Rutaceae
Cleome rutidosperma DC.	Herb	Capparaceae
Cleome viscosa L.	Herb	Capparaceae
Clerodendron inerme (L.) Gaertn.	Shrub	Verbanaceae
Clerodendron infortunatum L.	Shrub	Verbanaceae
Cocculus hirsutus (L.) Diels	Climber	Menispermaceae
Cocos nucifera L.	Tree	Arecaceae
Colocasia esculenta (L.) Schott.	Herb	Araceae
Commelina benghalensis L.	Herb	Commelinaceae
Commelina erecta L.	Herb	Commelinaceae
Crossandra infundibuliformis Salisb.	Shrub	Acanthaceae
Crotalaria juncea L.	Herb	Fabaceae
Crotalaria leschenaultii DC.	Herb	Fabaceae
Curculigo orchioides Gaertn.	Herb	Amaryllidaceae
Cuscuta reflexa Roxb.	Herb	Convolvulaceae
Cynodon dactylon Pers.	Herb	Poaceae
Cycas circinalis L.	Tree	Cycadaceae
Cymbopogon citratus L.	Herb	Poaceae
Cymbopogon flexuosus (DC.) Stapf	Herb	Poaceae
Cymbopogon travancorensis Bor	Herb	Poaceae

Contd...

Table 5.1–*Contd...*

Botanical Name	*Habit*	*Family*
Cyperus corymbosus Rotth.	Herb	Cyperaceae
Cyperus difformis L.	Herb	Cyperaceae
Cyperus esculentus Linn.	Herb	Cyperaceae
Cyperus haspan L.	Herb	Cyperaceae
Cyperus iria L.	Herb	Cyperaceae
Cyrtococcum longipes A. Cam.	Herb	Poaceae
Cyrtococcum trigonum (Retz,) A.Camus	Herb	Poaceae
Dactyloctenium aegyptium Beauv.	Herb	Poaceae
Dalbergia latifolia Roxb.	Tree	Fabaceae
Datura metel L.	Shrub	Solanaceae
Desmodium gyrans DC.	Herb	Fabaceae
Desmodium triflorum W. & A.	Herb	Fabaceae
Desmodium velutinum (Willd.) DC.	herb	Fabaceae
Dichranostachys cinera (L.) Wight & Arn.	Tree	Mimosaceae
Digitaria ciliaris (Retz.) Koeler	Herb	Poaceae
Dodonea viscosa L.	Shrub	Sapindaceae
Eclipta alba Hassk.	Herb	Asteraceae
Elephantopus scaber L.	Herb	Asteraceae
Eleusine indica Gaertn.	Herb	Poaceae
Emilia sonchifolia (L.) DC.	Herb	Asteraceae
Eragrostiella bifaria (Vahl.) Bor	Herb	Poaceae
Eragrostis gangetica (Roxb.) Steud.	Herb	Poaceae
Eragrostis unioloides (Retz.) Nees ex Steud.	Herb	Poaceae
Eupatorium odoratum L.	Herb	Asteraceae
Euphorbia hirta L.	Herb	Euphorbiaceae
Evolvulus alsinoides L.	Herb	Convolvulaceae
Evolvulus nummularis L.	Herb	Asteraceae
Ficus benghalensis L.	Tree	Moraceae
Fimbristylis dichotoma L.	Herb	Cyperaceae
Flacourtia sps.	Shrub	Flacourtiaceae
Flueggea leucopyrus Willd.	Shrub	Euphorbiaceae
Garcinia gummigatta L.	Tree	Guttiferae
Gliricidia sepium (Jacq.) Kunth ex Walp.	Tree	Fabaceae
Gloriosa superba L.	Climber	Liliaceae
Gmelina asiatica L.	Shrub	Verbanaceae
Gomphrena globosa L.	Herb	Asteraceae

Contd...

Table 5.1–*Contd...*

Botanical Name	*Habit*	*Family*
Grewia bracteata Heyne ex Roth	Herb	Tiliaceae
Gymnema sylvestre (Retz.) R. Br.	Climber	Asclepiadaceae
Gymnosporia montana (Roth) Benth.	Shrub	Celastraceaea
Heliotropium indicum L.	Herb	Boraginaceae
Hemidesmus indicus. var. indicus (L.) R. Br.	Climber	Asclepiadaceae
Hemionitis arifoilia (Burm.f.) T. Moore	Herb	Hemionitidaceae
Heteropogan contotus (L.) P.Beauv ex Roem. & Schult.	Herb	Poaceae
Hevea brasiliensis (Willd.) Muell.-Arg.	Tree	Euphorbiaceae
Hibiscus rosa-sinensis L.	Shrub	Malvaceae
Holostemma ada-kodien Schultes.	Climber	Asclepiadaceae
Hugonia mystax L.	Shrub	Linaceae
Hybanthus enneaspermus (L. f.) Muell.	Herb	Violaceae
Hydrilla verticillata (L.f) Royle	Hydrophyte	Hydrocharitaceae
Hyptis suaveolensis (L.) Poit.	Herb	Lamiaceae
Ichnocarpus frutesens (L.) R.Br.	Climber	Apocynaceae
Indigofera enneaphylla L.	Herb	Fabaceae
Indigofera tinctoria L.	Herb	Fabaceae
Ipomoea sepiaria Roxb.	Herb	Convolvulaceae
Ixora coccinea L.	Herb	Rubiaceae
Jasminum angustifolium (L.) Willd.	Herb	Oleaceae
Jasminum malabaricum Wight	Climber	Oleaceae
Jatropha glandulifera Roxb.	Shrub	Euphorbiaceae
Justicia glauca Rott.	Herb	Acanthaceae
Justicia prostrata (C.B.Clarke) Gamble	Herb	Acanthaceae
Justicia simplex D.Don	Herb	Acanthaceae
Kirganelia reticulata (Poir.) Bail.	Shrub	Euphorbiaceae
Knoxia heyneana DC.	Herb	Rubiaceae
Kyllinga squamulata Thonning ex Vahl	Herb	Cyperaceae
Lannea coromandelica (Houtt.) Merr.	Tree	Anacardiaceae
Lantana camara L.	Shrub	Verbanaceae
Leucaena leucocephala (Lam.) de Wit.	Tree	Mimosaceae
Leucas aspera (Willd.) Link	Herb	Lamiaceae
Loranthus longiflorus Desr.	Epiphyte	Loranthaceae
Lygodium flexuosum (L.) Sw.	Herb	Schizaeaceae
Maba buxifolia Pers.	Shrub	Ebenaceae
Madhuca indica J. Gmelin	Tree	Sapotaceae

Contd...

Table 5.1–*Contd...*

Botanical Name	*Habit*	*Family*
Mallotus philippensis (Lam.) Muell.	Tree	Euphorbiaceae
Mangifera indica L.	Tree	Anacardiaceae
Manihot esculenta Crantz.	Shrub	Euphorbiaceae
Maranta arundinacea L	Herb	Aracaceae
Melastoma malabathricum L.	Herb	Melastomaceae
Memecylon capitellatum L.	Shrub	Melastomaceae
Merremia emarginata (Burm. f.) Hallier f.	Herb	Convolvulaceae
Millingtonia hortensis L.f.	Tree	Bignoniaceae
Mimosa pudica L.	Herb	Mimosaceae
Mollugo pentaphylla L.	Herb	Aizoaceae
Morinda tinctoria Roxb.	Tree	Rubiaceae
Moringa pterygosperma Gaertn.	Tree	Moringaceae
Mukia maderaspatensis (L.) M. Roem.	Climber	Cucurbitaceae
Murraya koeinigii L.	Shrub	Rutaceae
Musa paradisiaca L.	Tree	Musaceae
Naregamia alata Wt. & Arn.	Herb	Meliaceae
Ocimum sanctum L.	Herb	Lamiaceae
Oldenlandia corymbosa L.	Herb	Rubiaceae
Oldenlandia umbellata L.	Herb	Rubiaceae
Olea dioica Roxb.	Tree	Oleaceae
Oplismenus burminii Beauv.	Herb	Poaceae
Orthosiphon thymiflorum (Roth) Sleensen	Herb	Lamiaceae
Panicum antidotale Retz.	Herb	Poaceae
Paspalum scrobiculatum L.	Herb	Poaceae
Passiflora foetida L.	Climber	Passifloraceae
Pennisetum polystachion (L.) Schult.	Herb	Poaceae
Perotis indica (Linn.) Kuntze	Herb	Poaceae
Phoenix pusilla Gaertn.	Shrub	Arecaceae
Phyllanthus amarus Schum. & Thonn.	Herb	Euphorbiaceae
Phyllanthus emblica L.	Herb	Euphorbiaceae
Phyllanthus simplex Retz.	Herb	Euphorbiaceae
Phylotrichum ellotii	Herb	Amaranthaceae
Piper nigrum L.	Climber	Piperaceae
Pityrogramma calomelanos (L.) Link	Herb	Hemionitidaceae
Polycarpaea corymbosa Lam.	Herb	Caryophyllaceae
Polygala javana DC.	Herb	Polygalaceae

Contd...

Table 5.1–*Contd...*

Botanical Name	*Habit*	*Family*
Pongamia pinnata Vent.	Tree	Fabaceae
Portulaca oleracea L.	Herb	Portulacaceae
Pseudarthria viscida W. & Arn.	Climber	Fabaceae
Psidium guajava L.	Tree	Myrtaceae
Rhynchelytrum repens (Willd.) C.E. Hubb.	Herb	Poaceae
Rivea hypocrateriformis Choisy	Climber	Convolvulaceae
Rivina humulis L.	Herb	Phytolocaceae
Ruellia prostrata Poir.	Herb	Acanthaceae
Ruellia tuberosa L.	Herb	Acanthaceae
Santalum album L.	Tree	Santalaceae
Schleicheria oleosa (Lour.) Oken	Tree	Sapindaceae
Scoparia dulsis L.	Herb	Scrophulariaceae
Sebastiana chamealea (L.) Muell.	Herb	Euphorbiaceae
Sida acuta Burm.	Herb	Malvaceae
Sida cordifolia L.	Herb	Malvaceae
Solanum torvum Sw.	Shrub	Solanaceae
Spermacocce hispida L.	Herb	Rubiaceae
Sporobolus indicus R. Br.	Herb	Poaceae
Stachytarpeta jamaicensis Vahl	Herb	Verbenaceae
Streblus asper Lour.	Shrub	Moraceae
Striga scottiana Jeeva *et al.*	Herb	Scropulariaceae
Stylosanthes fruticosa (Retz.) Alston	Herb	Fabaceae
Swietenia mahogoni (L.) Jacq.	Tree	Meliaceae
Synedrella nodiflora (L.) Gaertn.	Herb	Asteraceae
Syzygium cumini (L.) Skeels	Tree	Myrtaceae
Tagetes erecta L.	Herb	Asteraceae
Tamarindus indica L.	Tree	Caesalpiniaceae
Tectona grandis L.f.	Tree	Verbenaceae
Tephrosia purpurea (L.) Pers.	Shrub	Fabaceae
Terminalia catapa L.	Tree	Combretaceae
Terminalia paniculata Roth	Tree	Combretaceae
Thespesia populnea Cav.	Tree	Malvaceae
Tinospora cordifolia Roxb.	Climber	Menispermaceae
Toddalia asiatica (L.) Lam	Shrub	Rutaceae
Torenia asiatica Hk.f	Herb	Scrophulariaceae
Tridax procumbens L.	Herb	Asteraceae

Contd...

Table 5.1–*Contd...*

Botanical Name	*Habit*	*Family*
Triumfetta rhomboidea Jacq.	Herb	Tiliaceae
Tylophora indica Thw.	Climber	Asclepiadaceae
Typha angustata Bory & Chaub.	Herb	Typhaceae
Urena lobata L.	Herb	Malvaceae
Uvaria narum Wall.	Climber	Annonaceae
Vanda tessellata Hk.	Epiphyte	Orchidaceae
Vernonia cinerea Less.	Herb	Asteraceae
Vicoa indica DC.	Herb	Asteraceae
Waltheria indica L.	Herb	Streculiaceae
Wattakaka volubilis (L. f.) Stapf	Climber	Asclepiadaceae
Wedelia trilobata (L.) Hutchc.	Herb	Asteraceae
Wrightia tinctoria R. Br.	Tree	Apocynaceae
Zizyphus oenoplia Mill.	Tree	Rhamnaceae

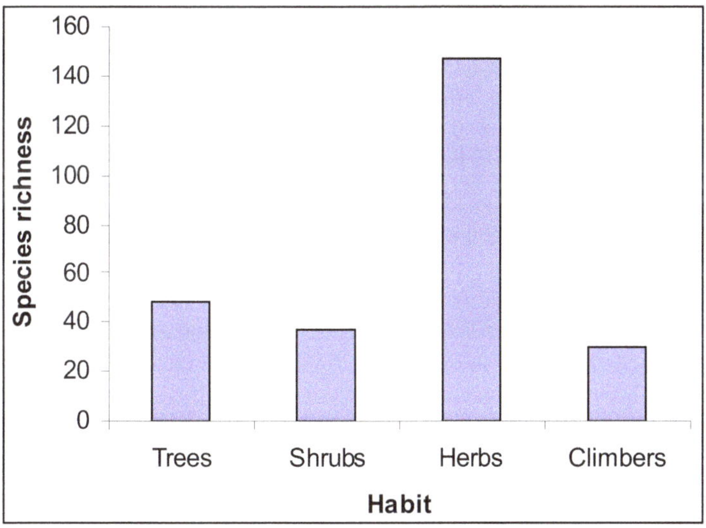

Figure 5.2: Habit-Wise Distribution of Plants in the Study Area.

The grass family Poaceae occupies the dominant position in the study area in terms of species richness. The grasses recorded in the study are *Bambusa arundinacea* Willd., *Alloteropsis cimicina* Stapf., *Apluda mutica* L., *Aristida adscensionis* L., *Aristida setacea* Retz., *Axonopus compressus* (Sw.) P. Beauv, *Brachiaria setigera* (Retz.) C.E. Hubb., *Chloris barbata* Sw, *Cynodon dactylon* Pers., *Cymbopogon citratus* L., *Cymbopogon flexuosus* (DC.) Stapf., *Cymbopogon travancorensis* Bor, *Cyrtococcum longipes* A. Cam., *Cyrtococcum trigonum* (Retz.) A.Camus, *Dactyloctenium aegyptium* Beauv., *Digitaria ciliaris* (Retz.) Koeler, *Eleusine indica* Gaertn., *Eragrostiella bifaria* (Vahl) Bor, *Eragrostis gangetica* (Roxb.)

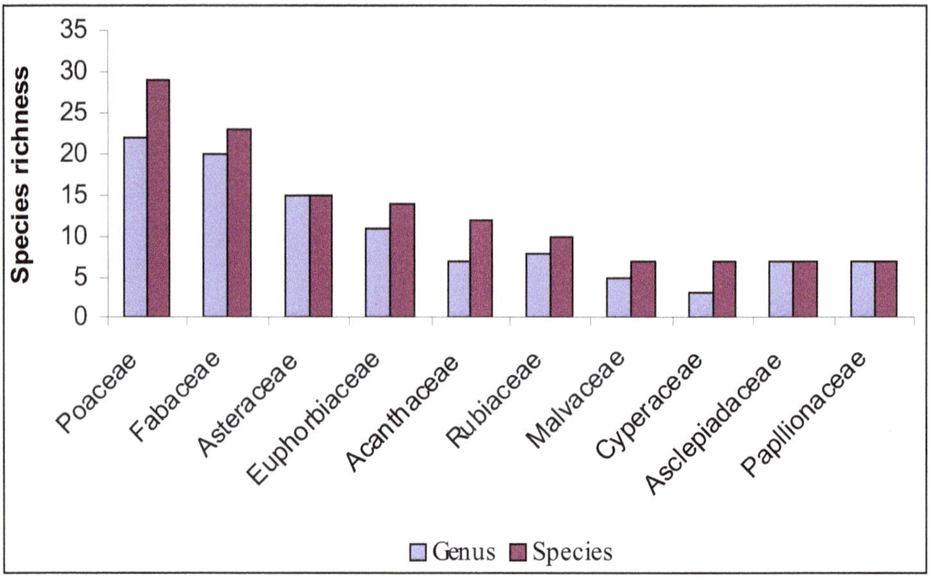

Figure 5.3: Dominant Families Recorded in the Study.

Steud., *Eragrostis unioloides* (Retz.) Nees ex Steud., *Heteropogan contotus* (L.) P.Beauv ex Roem. & Schult., *Oplismenus burminii* Beauv., *Panicum antidotale* Retz., *Paspalum scrobiculatum* L., *Pennisetum polystachyon* (L.) Schult., *Perotis indica* (L.) Kuntze, *Rhynchelytrum repens* (Willd.) C.E. Hubb. and *Sporobolus indicus* R. Br.

The upper canopy of the study area is comprised of the lofty trees like *Borassus flabellifer* (Arecaceae), *Albizia lebbeck* (Mimosaceae), *Alstonia scholaris* (Apocynaceae), *Artocarpus hirsutus, A. heterophyllus* (Moraceae), *Mangifera indica* (Anacardiaceae), *Schleicheria oleosa* (Sapindaceae), *Syzygium cuminii* (Myrtaceae), *Tamarindus indica* (Caesalpiniaceae) and *Terminalia paniculata* (Combretaceae). Trees like *Anacardium occidentale* (Anacardiaceae), *Wrightia tinctoria* (Apocynaceae), *Pongamia pinnata* (Fabaceae) and *Thespesia populnea* (Malvaceae) are well distributed in the middle canopy.

Shrubby layer of the hill is composed of *Abutilon indicum* (Malvaceae), *Breynia retusa* (Euphorbiaceae), *Calotropis gigantea* (Asclepiadaceae), *Canthium coromandelicum* (Rubiaceae), *Carrisa spinarum* (Apocynaceae), *Carmona retusa* (Boraginaceae), *Dodonaea viscosa* (Sapindaceae), *Flueggea leucopyrus* (Euphorbiaceae), *Lantana camara* (Verbenaceae), *Toddalia asiatica* (Rutaceae). The invasion of exotic weeds such as *Lantana camara* and *Eupatorium odoratum* in the periphery of the hillock depicted that the areas are under various degrees of disturbance due to mining in the past and anthropogenic activities and cattle grazing in the present.

The herby layer of the study area encompassed with some of the medicinal herbs such as *Achyranthes aspera, Aerva lanata, Ageratum conyzoides, Andrograhis panciculata, Apluda mutica, Biophytum sensitivum, Centella asiatica, Cleome rutidosperma, Cynodon dactylon, Eclipta alba, Elephantopus scaber, Emilia sochifolia, Euphorbia hirta, Justicia simplex* and *Mimosa pudica.*

Table 5.2: Family-Wise Distribution of Plant Species in the Study Area.

Family Rank	*Family*	*Genera*	*Species*
1	Poaceae	22	29
2	Fabaceae	20	23
3	Asteraceae	15	15
4	Euphorbiaceae	11	14
5	Acanthaceae	7	12
6	Rubiaceae	8	10
7	Malvaceae	5	7
8	Cyperaceae	3	7
9	Asclepiadaceae	7	7
10	Verbenaceae	5	6
11	Amaranthaceae	5	6
12	Apocynaceae	5	6
13	Lamiaceae	5	5
14	Convolvulaceae	5	5
15	Vitaceae	3	5
16	Rutaceae	4	4
17	Moraceae	3	4
18	Mimosaceae	4	4
19	Arecaceae	4	4
20	Scropulariaceae	3	3
21	Sapindaceae	3	3
22	Oleaceae	2	3
23	Menispermaceae	3	3
24	Meliaceae	3	3
25	Liliaceae	3	3
26	Capparidaceae	2	3
27	Annonaceae	2	3
28	Anacardiaceae	3	3
29	Tiliaceae	2	2
30	Solanaceae	2	2
31	Sapotaceae	2	2
32	Myrtaceae	2	2
33	Melastomaceae	2	2
34	Guttiferae	2	2
35	Commelinaceae	1	2
36	Combretaceae	1	2

Contd...

Table 5.2–*Contd...*

Family Rank	*Family*	*Genera*	*Species*
37	Boraginaceae	2	2
38	Araceae	2	2
39	Hemionitidaceae	2	2
40	Violaceae	1	1
41	Typhaceae	1	1
42	Sterculiaceae	1	1
43	Sinopteridaceae	1	1
44	Simaroubiaceae	1	1
45	Schizaeaceae	1	1
46	Santalaceae	1	1
47	Rhamnaceae	1	1
48	Portulacaceae	1	1
49	Polygalaceae	1	1
50	Piperaceae	1	1
51	Phytolocaceae	1	1
52	Passifloraceae	1	1
53	Oxalidaceae	1	1
54	Orchidaceae	1	1
55	Nyctaginaceae	1	1
56	Musaceae	1	1
57	Moringaceae	1	1
58	Loranthaceae	1	1
59	Linaceae	1	1
60	Hydrocharitaceae	1	1
61	Flacourtiaceae	1	1
62	Ebenaceae	1	1
63	Cycadaceae	1	1
64	Cucurbitaceae	1	1
65	Brassicaceae	1	1
66	Crassulaceae	1	1
67	Casuarinaceae	1	1
68	Caesalpiniaceae	1	1
69	Cactaceae	1	1
70	Bromeliaceae	1	1
71	Bombaceae	1	1
72	Bignoniaceae	1	1

Contd...

Table 5.2–*Contd...*

Family Rank	*Family*	*Genera*	*Species*
73	Aristolochiaceae	1	1
74	Apiaceae	1	1
75	Amaryllidaceae	1	1
76	Aizoaceae	1	1
77	Adiantaaceae	1	1
78	Actinopteridaceae	1	1
79	Poaceae	1	1
80	Lauraceae	1	1
81	Celastraceae	1	1
	Total	227	224

The climbers that occupied the study area are *Abrus precatorius* (Fabaceae), *Aristolochia indica* (Aristolochiaceae), *Asparagus racemosus* (Liliaceae), *Canavalia ensiformis* (Fabaceae), *Capparis zeylanica* (Capparidaceae), *Cardisopermum halicacabum* (Sapindaceae), *Cayratia japonica* (Vitaceae), *C. pedata* (Vitaceae), *Cissampelos pareira* (Menispermaceae), *Cissus quadrangularis, C. vitiginea* (Vitaceae), *Cocculus hirsutus* (Menispermaceae), *Gloriosa superba* (Liliacae), *Gymnema sylvestre* (Menispermaceae), *Hemidesmus indicus* (Asclepiadaceae), *Holostemma ada-kodien* (Asclepidaceae), *Ichnocarpus frutescens* (Apocynaceae), *Tinospora cordifolia* (Menispermaceae), *Tylophora indica* (Menispermaceae), *Uvaria narum* (Annonaceae), *Wattakaka volubilis* (Asclepiadaceae), etc.

RET Species

Presence of an IUCN red listed rare gymnospermic species *Cycas circinalis* (Cycadaceae) and the endemic and endangered keystone species *Artocarpus hirsutus* (Moraceae) denotes the conservation importance of the study area.

Economic Importance

Most of the plants recorded (170/264) from the study area are economically important. The medicinal plants ranked first, followed by edible, timber, non timber forest produce, etc (Figure 5.4). Among the edible species, *Artocarpus heterophyllus, A. hirsutus, Mangifera indica* and *Tamarindus indica* are economically valuable.

The presently studied hillock was definitely considered as repository of medicinal plants. The careful analytical studies further proved the concept and brought more facts to light. The medicinal plants which are now present may be the part of original vegetation present there or the remnant of the rich vegetation in the past or may be added from different sources in the near by grove in order to conserve them. These medicinal plants however are considered useful in folk medicines even now. Some of them are vanishing because of the habitat loss and others like lianas,

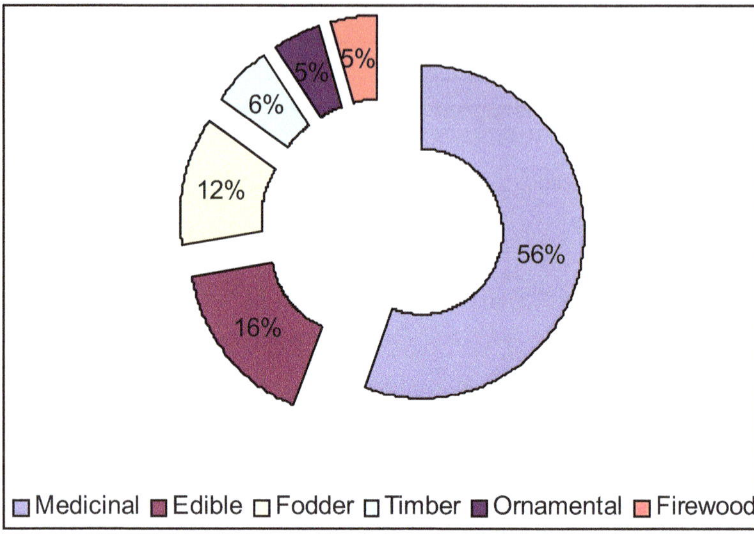

Figure 5.4: Economic Important Plants of the Study Area.

twines, and climbers due to their associate (Raj and Sukumaran, 1997). Some others are destroyed due to over exploitation or due to shift of the importance of these plants for some new medicines which are cheap and easily available.

Over exploitation of many plants like *Aloe vera, Gloriosa superba, Hemidesmus indicus, Holostemma ada-kodian,* etc. have rendered them very rare. This scarcity of medicinal plants has put the indigenous Indian medicine in a tight spot, ironically through, at a time when they are being increasingly internationally accepted as a source of alternative medicine. The present study has revealed that a single plant species is used in one type of disease, but most of them are used in different diseases. Since the efficacy of these drugs has been claimed to be very high, it is desirable to go for in-depth investigations, particularly experimental trials and clinical studies.

Anthropogenic disturbances such as unsustainable harvesting, cultivation practices and overgrazing lead to threat to such potential genetic resources. It has also lead to decline in ethnobotanical plant wealth in its natural condition. Extraction of the bark of *Anacardium occidentale* for medicinal purposes leads to attack of very popular pest called Rhinoceros beetle *(Oryctes rhinoceros*) and resulted in death of the plants in this area.

Edible Species

During the study, some species of wild edible species were reported in the sacred forests. They are *Ananas comosus, Artocarpus hirsutus, Artocarpus heterophyllous, Borassas flabellifer, Gloriosa superba, Hemidesmus indicus, Murraya koenigii, Tamarindus indicus* and *Zizyphus oenoplia*.

Children are mostly engaged in collection of wild edible and non timber forest products from the hill. They mostly prefer the fruits of *Artocarpus heterophyllus, A.*

hirsutus, Borassus flabellifer, Ixora coccinea, Carissa spinarum, Mangifera indica, Syzygium cuminii. The oil extracted from the seeds of *A. hirsutus* is used for cooking purposes. It is said to be medicinal value. The leaves of *Centella asiatica* is used as wild leafy vegetable, the stem of *Cissus quadrangularis* is used for making pickles among the inhabitants of the study area. Women are mostly engaged in collection of fuel wood in the study area. The edible and medicinal species *Caralluma* grew luxuriantly in the study area.

Anthropogenic Threats

Presently, health of the study area is deteriorating under constant anthropogenic pressure. Constant grazing, extraction of fuel-wood, collection of medicinal plants and non-wood forest products etc. are affecting the biodiversity of the area. These disturbances are also a threat for various rare, vulnerable and threatened plants of the study area. Currently the forest is utilized by the villagers and this being a community protection. The exaggerated mining practices in the past is a reason for degradation of plant diversity in the study area.

Conclusion

The overall analysis indicates that species rich communities of the presently studied hill is not only being reduced in area but they are also becoming species poor and less diverse due to rapid deforestation, mining which leads to forest fragmentation. The community organization is also changing in response to increased anthropogenic disturbance. Invasion of *Lantana camara* and *Eupatorium odoratum* is changing the forest structure, leading to species diversity loss and creation of a homogeneous, mono-specific *Lantana* invaded understorey in the forest. Serious measures are needed to preserve and conserve whatever intact forest is left in the Thumpali hills so that this alarming trend of disappearance of species rich communities is halted.

Further, educating the local people and alerting the officials towards effective implementation of the rules would be helpful in decreasing the depletion of natural forest produce, which will in turn decrease the damage caused to this fragile ecosystem due to the various human disturbances. The present study in the Thumpali hill is very preliminary, and subsequent recensus and monitoring will provide additional data on forest composition and diversity changes due to various disturbance regimes, which will be useful in forest management and conservation efforts.

Acknowledgements

The authors are thankful to the Head, Department of Botany, Nesamony Memorial Christian College, Marthandam for providing the laboratory facilities. The technical assistance provided by Mr. Z. Miller Paul and Mr. M. Mahesh, is also acknowledged.

References

Gamble, J.S. 1915-1935. Flora of the Presidency of Madras. Adlard and Sons Limited, London.

Henry, A.N. and Swaminathan, M. 1981. Observation on the vegetation of Kanyakumari district, Tamil Nadu. Bull. Bot. Surv. India, **23:** 135-139.

Jeeva, S., Kiruba, S., Mishra, B.P., Venugopal, N., Kharlukhi, L., Regini, G.S., Das, S.S.M. and Laloo, R.C. 2005a. Diversity of medicinally important plant species under coconut plantation in the coastal region of Cape Comorin. Flora Fauna, **11**: 226-230.

Jeeva, S., Kiruba, S., Mishra, B.P., Kingston, C., Venugopal, N. and Laloo, R.C. 2005b. Importance of weeds as traditional medicine in Kanyakumari district, southern Western Ghats. J. Swamy Bot. Cl. **22**: 71-76.

Karuppusamy, S., Ugraiah, A. and Pullaiah, T. 2013. *Caralluma (Sensu lato)* Antiobesity Plants. Regency Publications, New Delhi.

Lawrence, C.A. 1960. The vegetation of Kanyakumari district (Cape Comorin). J. Bombay Nat. Hist. Soc. **57:** 184-195.

Manickam, V.S. and Irudayaraj, V. 1992. Pteridophytic Flora of the Western Ghats South India. BI Publications, New Delhi.

Manickam, V.S., Murugan, C. and Jothi, G.J. 2008. Flora of Tirunelveli Hills (Southern Western Ghats), Vol. 1, Polypetalae. Bishen Singh Mahendra Pal Singh, Dehra Dun.

Matthew, K.M. 1983. Flora of the Tamil Nadu Carnatic. Rapinat Herbarium, Tiruchirapalli.

Myers, N., Mittermeier, R.A., Mittermeier, C.G., da Fonseca, D.A. and Kent, J. 2000. Biodiversity hotspots for conservation priorities. Nature, 403: 853-858.

Nair, K.K.N. and Nayar, M.P. 1986. Flora of Courtallum, Vols. 1 and 2. Botanical Survey of India, Calcutta.

Nayar, M.P. 1959. The vegetation of Kanyakumari district. Bull. Bot. Surv. India, **1:** 122-126.

Pascal, J.P. and Ramesh, B.R. 1987. A Field Key to the Trees and Lianas of the Evergreen Forests of the Western Ghats (India). Institut Francais de Pondicherry.

Prakash, J.W., Leena Suman, L., Vidhya Devi, M.S., Berin Premila, A., Asbin Anderson, N., Veni, P., Esakki, G., Amutha, M., Rajeev, R., Bensar, K., Jeeva, S., Christhudhas Williams, B., Regini, G.S. and Das, S.S.M. 2006. The medicinal plant diversity of Scott Christian College (Autonomous) Campus, Nagercoil, South Tamil Nadu, India. J. Nature Conservation, **18:** 81-89.

Raj, A.D.S. and Sukumaran, S. 1997. Observations of the sacred groves of south Tamil Nadu. In: Abstracts of National Symposium on Natural Resources Management Systems, St. Joseph College, Thiruchirapalli, Tamil Nadu.

Shynin Brintha, T.S., Edwin James, J. and Jeeva, S. Floristic spectrum of Scott Christian College Campus, Nagercoil, Tamil Nadu, India. Sciencia Acta Xaveriana, S162-166.

Sukumaran, S. and Jeeva, S. 2008. A floristic study on miniature sacred forest at Agastheeshwaram, Southern Peninsular India. EurAsian J. Biosci. **2:** 66-72.

Sukumaran, S., Jeeva, S., Raj, A.D.S. and Kannan, D. 2008. Floristic diversity, conservation status and economic value of miniature sacred groves in Kanyakumari District, Tamil Nadu, Southern Peninsular India. Turkish J. Bot. **32:** 185-199.

Sundarapandian, S.M. and Swamy, P.S. 1997. Plant biodiversity at low elevation evergreen and moist deciduous forests at Kodayar (W.Ghats, India). Intern. J. Ecol. Environ. Sci. **23:** 363-379.

Swamy, P.S., Sundarapandian, S.M. and Chandrasekharan, S. 1998. Sacred groves of Tamil Nadu. In: Conserving the Sacred for Biodiversity Management, Ramakrishnan, P.S., Saxena, K.G. and Chandrashekara, U.M. (eds), Oxford IBH Publishing Co. Pvt. Ltd., New Delhi, pp.357-364.

Swamy, P.S., Sundarapandian, S.M. and Chandrasekar, P. and Chandrasekaran, S. 2000. Plant species diversity and tree population structure of a humid tropical forest in Tamil Nadu, India. Biodiversity and Conservation, **9:** 1643-1669.

Chapter 6

Diversity, Distribution and Current Ecological Status of Botanically Less Known *Pholidota* Lindl. *ex* Hook. Orchid Species of Darjeeling Himalaya of India

Rajendra Yonzone[1] D. Lama[1], R.B. Bhujel[2], Khyanjeet Gogoi[3] and Samuel Rai[4]*

[1]*Department of Botany, St. Joseph's College, P.O. North Point, District Darjeeling – 734 104, W.B.*

[2]*Taxonomy and Ethnobiology Research Laboratory, Cluny Women's College, P.O. Kalimpong, District Darjeeling – 734 301, W.B.*

[3]*Disa Bordoloi Nagar, Talap, Tinsukia – 786 156, Assam*

[4]*Darjeeling Krishi Vigyan Kendra, Uttar Banga Krishi Viswavidyalaya P.O. Kalimpong, District Darjeeling – 734 301, W.B.*

ABSTRACT

The present paper deals botanically less known six *Pholidota* Orchid species diversity, current ecological status and distribution in Darjeeling Himalaya of India. Habitat of these species are epiphytic but frequently found in lithophytic habitat also. This attempt is the first step to correct taxonomic identification to

* Corresponding Author: E-mail: ryonzone99@gmail.com

workout currently accepted botanical names with ecological status, Voucher specimen numbers, habitat, altitudinal ranges, phenology and local and general distribution of *Pholidota* species in the regions.

Keywords: *Pholidota orchid species, Diversity, Distribution, Status, Darjeeling Himalaya.*

Introduction

Orchids are considered to be the most highly evolved in the floral specialization and diversified form among the monocotyledons. In India, Orchids form 9 per cent of our flora and are the largest and highly advanced botanical family of higher plants. It is estimated that at about 25,000-35,000 species with 800-1,000 genera are distributed throughout the world. About 1300 species with 140 genera of Orchid species are found in India with temperate Himalayas as their natural home (Yonzone and Kamran, 2008). The congenial climatic condition of the Darjeeling helps to lavish growth and development of the species but at present disturbance of natural habitat by multifarious means *viz.*, deforestation for extension of agricultural lands, timber and firewood collection, frequent landslides, over grazing, top layer soil erosion, indiscriminate collection, uncontrolled forest fire, developmental schemes, rapid urbanization and drastic environmental pollution makes all the Orchid species rare and threatened status throughout the regions.

Darjeeling Himalaya is the northernmost part of the state of West Bengal that covers the hill and mountains that includes the alpine zone to the foot hill of the plains. It is triangular in shape extending over an area of 3254.7 sq.kms. It is bordered by Sikkim in the north, Terai and Dooars in the south, Bhutan in the east and Nepal in the west. The Sub-Divisions of Darjeeling are Darjeeling, Kalimpong, Kurseong and Siliguri (Figure 6.1). The Darjeeling district has two topographical features. Darjeeling, Kalimpong and Kurseong form the hill areas whereas Siliguri is stationed at the foot hill giving way to vast stretches of the plains. The hilly region covers 2320 sq.km. and the remaining 934.75 sq.km of the area falls under Terai and plains. The altitudinal variations of the district ranges from 150m at Siliguri to 3636m at Sandakphu-Phalut with a sharp physiographic contrast between the plains and the mountainous regions. A trijunction of boundaries of Nepal, Sikkim and India is formed at the peak of Phalut (3600m). Darjeeling Himalaya is rich in Orchid diversity. In the present investigation, the current status of *Pholidota* Orchid species available in Darjeeling Himalaya has been carried out through the survey of national parks, forest area and far-flung villages of the region to find out the accurate status of the species.

Botanical Description

Plants pendent or erect, epiphytic or lithophytic. Pseudobulbs close or distant on rhizome, slender or swollen. Leaves 1 or 2, narrowly elliptic, ovate or oblong, petiolate, coriaceous or herbaceous. Inflorescence terminal from pseudobulb apex, slender, racemose, distichous, pendent, laxly to densely many-flowered; floral bracts, large, concave. Flowers small, fleshy, resupinate or non-resupinate. Dorsal sepals

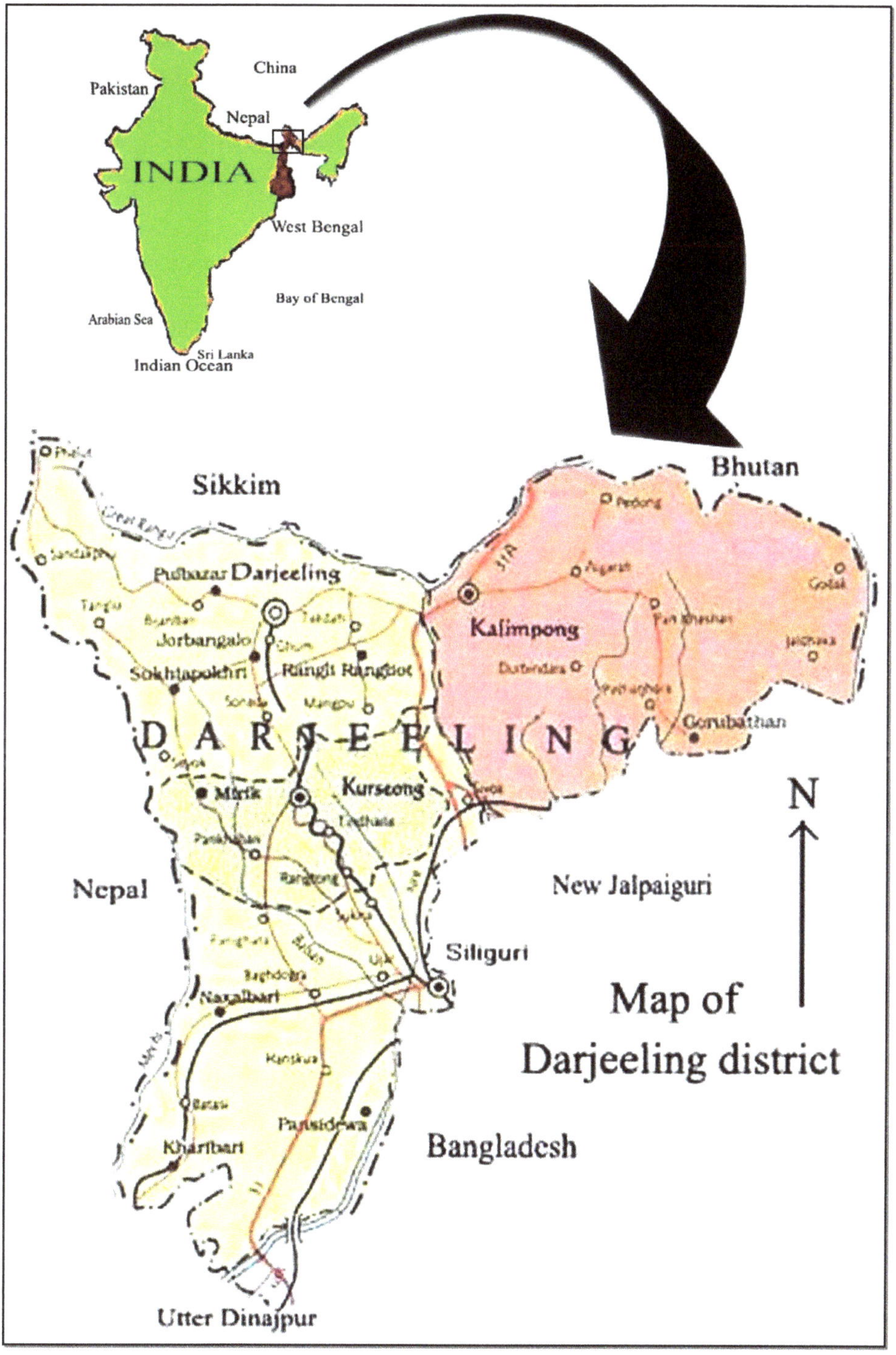

Figure 6.1: Location of Darjeeling District (Study Area) of West Bengal, India.

concave or convex, broadly ovate to elliptic; lateral sepals concave or convex, ovate to ovate-oblong, often keeled. Petals ovate to linear. Lip sessile, base saccate; epichile subentire, or 4-lobed, deflexed. Column short, apex broadly hooded; pollinia 4.

The genus *Pholidota* was established in 1825 by W.J. Hooker in his *Exotic Flora*. The genus comprises 29 species distributed in tropical S.E. Asia, the Malay Archipelago, Australia and the Pacific Islands.

Materials and Methods

The intensive field survey was conducted during the year 2007-2011 covering all the seasons of the year in the entire Darjeeling district including the forest areas, floral nurseries and farms of as low as Siliguri which is located at 150m to as high as Sandakphu-Phalut located at 3636m of entire Darjeeling district of West Bengal. All the *Pholidota* Orchid species found were recorded in the field notebook with their necessary information. The collected specimens were dissected and examined in the laboratory during flowering period. Herbarium specimens were prepared by standard methods (Jain and Rao, 1977), specimens so collected were processed, preserved and mounted on herbarium sheets and described, properly identified and authenticated with the help of the Flora of British India (Hooker, 1888 – 1890); the Orchids of the Sikkim Himalaya (King and Pantling, 1898); Indian Orchids Guide to Identification and Culture (Pradhan, 1979); Orchid Flora of Arunachal Pradesh (Chowdhery, 1998); Orchids of India (Bose and Bhattacharjee, 1999); The Flora of Bhutan (Pearce and Cribb, 2002); the Orchids of Sikkim and North East Himalaya (Lucksom, 2007) and confirmed at CAL. Finally all the Voucher specimens were deposited in the Herbarium of Department of Botany, St. Joseph's College, North Point, Darjeeling and Taxonomy and Ethnobiology Research Laboratory, Cluny Women's College, Kalimpong. All the plant specimens are arranged alphabetically as per their altitude wise distribution in the area with botanical names, date of collection, voucher specimen numbers, habitat and phenology. For ecological status, nested quadrate sampling method with plot of 10m x 10m quadrates for epiphytic species was laid down diagonally in the field. The quadrate sampling data were summed up to represent the Important Value Index (IVI) as suggested by Philips (1959).

Enumeration

Pholidota articulata Lindl., Gen. Sp. Orchid. Pl.: 38. 1830.

Plant 23-45cm long. Pseudobulbs 3.8-9cm long, slender to swollen; sheaths 0.2-3cm long, lignified. Leaves 6.5-12 x 2-3.7cm, ovate to linear-lanceolate, acute to acuminate; petiole 0.8-1.4 x 0.2-0.3cm. Inflorescence synanthous, 9 to 21-flowered; peduncle 3.2-4.5cm long; rachis 3.7-6.4cm long, straight to zigzag; internodes 1.9-5mm long; floral bracts 0.7-1.3 x 0.3-0.5cm. Flowers, 0.7-0.9cm across, resupinate, fragrant; sepals, petals and lip white-cream to greenish-pink, keels of lip buff, column whitish to purple-red; pedicellate-ovary 0.5-1.2cm long, terete. Dorsal sepal 0.6-0.8 x 0.3-0.4mm, ovate to ovate-oblong, obtuse, 5-veined; lateral sepals 0.7-0.8 x 0.3-0.4mm, ovate to ovate-lanceolate, acute to acuminate. Petals 0.4-0.5 x 0.2-0.25cm, ovate to ovate-lanceolate, obtuse to acuminate, 3 to 5-veined. Lip 5-8mm long, constricted in middle; hypochile 3-5.4 x 1.8-3.8mm, boat-shaped, inflated, tapering, 5-keeled in basal half; epichile 1.5-4 x 2-5mm, recurved, 2-lobed, lobules rounded, finely papillose. Column 1.5-2.5mm long, shortly clavate.

Voucher specimen: [Rajendra *et al.*, 0393]; *Habitat*: Epiphytic; *Altitudinal ranges*: 300-1600m; *Date of collection*: 20 May 2008; *Flowering*: April-July; *Status*: Frequent; *Local distribution within Darjeeling*: Samalbong, Kalijhora; *General distribution*: India, Nepal, Myanmar, Thailand, Cambodia, Vietnam to Sumatra, Java, Borneo and Sulawesi.

Pholidota articulata var. ***griffithii*** (Hook. f.) King & Pantling in Ann Bot. Gard. (Calcutta): 147, t.204. 1898.

Plant 22-43cm long. Pseudobulbs 4-12 x 0.4-0.5cm, cylindric, furrowed, smooth, yellowish-green, with constricted part is enclosed in short, imbricate sheaths; young pseudobulb, two leaves and the inflorescence are enclosed in 1-3cm long imbricate bracts at the base. Leaves 2, 8.5-10 x 2-3.5cm, narrowly elliptic, acute to acuminate, petiolate base. Inflorescence 1, 9-11cm long, slender, pendulous, synanthous; the young developing pseudobulb, leaves and inflorescence are enclosed in 5-6, 0.4-2cm long, overlapping, oblong-ovate, obtuse, green suffused with light brown sheathing bracts; peduncle 4-5cm long, slightly compressed, smooth, ebracteate; rachis 5-6cm long, many shortly pedicellate flowers arranged in zigzag pattern; pedicellate-ovary 3.5-4mm long; floral bracts 0.5-0.9 x 0.6-0.7cm, oblong ovate, obtuse. Flowers 8-8.5mm across, (measured form sepal to sepal), pale, yellowish-green, glabrous. Sepals subequal; dorsal sepal 5.5-6 x 2.8-3mm, elliptic-oblong, sub-acute, translucent green, suffused with light pink, its tip portion erect; lateral sepals 2.8-3 x 2.3-2.5mm, spreading, eternally keeled, concave, transulent-green. Petals 4.5-4.8x1.8-2mm, rhomboid lanceolate, sub-acute, yellowish-green, suffused with slight pink, glabrous, 3-nerved. Lip ca 5mm long, white suffused with pink, obscurely 3-lobed; the basal lobe cymbiform with 5 calli; the apical lobe broad, emarginated and folded at the base into a narrow mesochile, white suffused with yellow. Column ca 2.5mm long, slightly winged towards apex. Anther 0.7 x 0.8mm, winged; pollinia 4, oval-shaped.

Voucher specimen: [Rajendra *et al.*, 0887]; *Habitat*: Epiphytic; *Altitudinal ranges*: 400-1200m; *Date of collection*: 10 June 2009; *Flowering*: April-August; *Status*: Rare; *Local distribution within Darjeeling*: Samalbong, Yangmakum; *General distribution*: India, Myanmar, Vietnam, Java and Borneo.

Pholidota imbricata Hook., Exot. Fl. 2: t.138. 1825.

Plant 16-41cm tall, occasionally terrestrial or lithophytic; roots c.25 x 0.05-0.2cm. Pseudobulbs 5-10 x 1.6-2cm, ovoid-conical, sturdy, sheathed at base; sheaths 3-8 x 0.4-0.7cm, conspicuous, lanceolate, acute. Leaf solitary, 19-40 x 4.7-8.6cm, arises from pseudobulb apex, oblong, obovate-oblong to linear-lanceolate, acute to acuminmate, coriaceous; petiole 2.5-4cm long, grooved. Inflorescence synanthous, densely many-flowered; peduncle 11-15cm long, narrowly wiry, terete, glabrous, sheathless; rachis 17-21cm long; floral bracts 0.6-1.2 x 0.6-0.9cm, ovate to broadly ovate, acute to obtuse, puberulent to mucilaginous, membranous to papery, yellow-green to light brown with darker spots. Flowers 0.5-0.6cm across, non-resupinate, overlapping, white to cream-coloured, tinged yellowish, greenish or pinkish, lip with 1 or 2 yellow-orange spots centrally; pedicellate-ovary 4-6.3 x 1.1-1.6mm. Dorsal sepal 5-6 x 3-4mm, ovate,

P. articulata* var. *griffithii

P. articulata

P. imbricata

P. pallida

P. recurva

P. rubra

obtuse to acute, usually 5-veined, forming a hood with the petals; lateral sepals 5-7 x 2.6-3.4mm, ovate-oblong, acute, 4 or 5-veined. Petals 4.5-5 x 1.4-2.1mm, falcate, acute, usually 1-veined. Lip broadly 3-lobed, 0.6-0.9 x 0.5-0.8cm, broadly inserted, base concave; lateral lobes erect, triangular to semi-orbicular; mid-lobe deeply 2-lobed, retuse in front, lobes semi-elliptic to semi-orbicular; disc with 3 short, wing-like keels at base. Column 2.9-3.7mm long. Pollinia 4, white.

Voucher specimen: [Rajendra *et al.*, 0487]; *Habitat*: Epiphytic; *Altitudinal ranges*: 300-1200m; *Date of collection*: 14 June 2008; *Flowering*: March-August; *Status*: Frequent; *Local distribution within Darjeeling*: Samalbong-Relli, Kalijhora; *General distribution*: Nepal, India, Sri Lanka, Bangladesh, Myanmar, China, Thailand, Laos, Vietnam, Malaysia, Indonesia, the Philippines, Australia and S.W. Pacific Islands.

Pholidota pallida Lindl., in Bot. Reg. 21: Sub t.1777. 1835.

Plant 11-34cm tall. Pseudobulbs 3.5-6 x 1.3-2cm long, slender to swollen, sheathed; sheaths 5, 3-7cm long, rather inconspicuous. Leaf solitary, 14-31 x 2.5-5cm, oblong to linear-lanceolate, acute to acuminate, herbaceous. Inflorescence synanthous, densely many-flowered; peduncle 7-12cm long, narrowly wiry; rachis 8-13cm long; floral bracts 0.5-0.7 x 0.6-0.8mm, ovate-sagittate, acute to obtuse, pubescent along margins. Flowers 5-6mm across, membranous, fragrant, white or creamy-white, tinged with pink; pedicellate-ovary 1.8-3mm long. Sepals sub-equal, white; dorsal sepal 3.4-4.5 x 2.5-3.4mm, semi-elliptic to ovate, obtuse, forming a hood with the petals; lateral sepals 4-5.5 x 2-2.5mm, ovate to ovate-oblong, acute to apiculate, connate at base. Petals 3.3-4.5 x 0.9-1.2mm, weakly falcate, obtuse to acute. Lip broadly 3-lobed, 4-5mm long; lateral lobes erect, broadly rounded; mid-lobe 2-lobed, lobes irregularly quadrate to semi-orbicular; disc with 3 short, wing-like keels at base. Column 2.7-3.2 x 1.8-2.5mm, wide, wing-like, spathulate. anther 1.4 x 0.8mm dome shaped, two chambered; pollinia 4 obovate-oblong.

Voucher specimen: [Rajendra *et al.*, 1301]; *Habitat*: Epiphytic (lithophytic); *Altitudinal ranges*: 800-1500m; *Date of collection*: 7 July 2010; *Flowering*: June-October; *Status*: Sparse; *Local distribution within Darjeeling*: Samalbong, Relli, Kalijhora, Lathpanjar; *General distribution*: Nepal, India, Myanmar, China, Thailand, Laos and Vietnam.

Pholidota recurva Lindl., Gen. Sp. Orchid. Pl.: 37. 1830.

Plant 17-47cm long; aerial roots 7-14cm long. Pseudobulbs 4.3-8 x 1.2-1.4cm, swollen, slender, rugose when dry; sheaths 2.5-6.7cm long, slender. Leaves 2 arises from each node, 5-9 x 1.2-2cm, lanceolate to linear-lanceolate, acute, 3 to 5-veined; petiole 0.5-0.6cm long. Inflorescence 4.5-10cm, synanthous, 20 to 31-flowered; peduncle 1-2.5cm long; rachis 3.5-7.5cm long, curved; internodes 0.2cm long; floral bracts 6-8 x 7-8mm, distichous, transversely semi-elliptic, oblique. Flowers second, pale pinkish-brown to orange-brown; pedicellate-ovary 4-5 x 1-1.2mm long. Dorsal sepal ovate, obtuse, 3 to 5-veined, 3.6-4.2 x 1.9-2.4mm; lateral sepals 3.6-4.2 x 2.2-2.4mm, ovate to ovate-oblong, acute to apiculate. Petals 2.9-3.4 x 1-1.3mm, ovate-oblong, obtuse. Lip 3.9-4.2 x 3-4mm long; hypochile wide, concave; hypochile base with 3 low, swollen keels; epichile 0.5-0.9 x 1.8-2.2mm, faintly 2-lobed. Column 2.4-2.8mm long, clavate, hooded.

Voucher specimen: [Rajendra *et al.*, 1453]; *Habitat*: Epiphytic; *Altitudinal ranges*: 600-2000m; *Date of collection*: 31 August 2010; *Flowering*: August-September; *Status*: Threatened; *Local distribution within Darjeeling*: Todey forest; *General distribution*: Nepal, N.E. India, Myanmar, Thailand and Vietnam.

Pholidota rubra Lindl., Gen. Sp. Orchid. Pl.: 37. 1830.

Plant 18-40cm tall; rhizome scales 4, obtuse to acute, 0.5-3.5cm long. Pseudobulbs 3-9cm long, cylindric, bulging below the center, tapering to each end, 0.5-1.7cm apart; sheaths 2, 4-5cm long. Leaves 2 per pseudobulbs, linear-lanceolate, acuminate, 3 to 5-veined, 12-23 x 1.8-4cm; petiole 2-3cm long. Inflorescence proteranthous to synanthous, 30 to 60-flowered; peduncle 0.3-3cm long; rachis pendent, 11-22cm long; internodes 3-7mm long; floral bracts 7-11 x 4-5.2cm, ovate to elliptic, acute. Flowers opening widely, non-resupinate; sepals dull red, greenish-yellow to pale brownish-yellow, lip light green, yellow, pinkish to white; pedicel and ovary 3-6.5mm long. Dorsal sepal 4.3-5.5 x 2-3.6mm, ovate, acute to acuminate; lateral sepals ovate to obliquely elliptic, acute to acuminate, recurved to reflexed, 4.5-6 x 2.5-3.2mm. Petals 4.3-5.2 x 0.4-0.8mm, linear-lanceolate, obtuse to subacute, 1-veined. Lip 4-5.5m long; hypochile broadly ovate, 2.3-3.5 x 1.5-2.3mm; epichile reflexed to recurved, semi-orbicular to ovate, apex apiculate. Ecallose, 1.7-2.3x1.5-2.3mm. Column slender, hooded, 3-4mm long.

Voucher specimen: [Rajendra *et al.*, 1629]; *Habitat*: Epiphytic; *Altitudinal ranges*: 900-2200m; *Date of collection*: 6 November 2011; *Flowering*: October-December; *Status*: Sparse; *Local distribution within Darjeeling*: Algarah, Lava forest; *General distribution*: N.E. India and Vietnam.

Results and Discussion

During recent field studies in the Darjeeling Himalaya of India, six *Pholidota* Orchid species were recorded. Of them, two are sparse, two are frequent, one is rare and remaining one is threatened status in our present investigation (Figure 6.2). In general, the habitat of *Pholodota* is epiphytic but it is frequently grown in lithophytic condition also. Flowering and availability of species varies from each other. *P. articulata* generally flower in the beginning of April and persist till the month of July and available at an altitudinal range of 300 to 1600m from the mean sea level and is distribution is frequent; *P. articulata* var. *griffithii* flower during April and persist till first week of August and available in 400 to 1200m and found rare status; *P. imbricata* flower during March to August and available at 300 to 1200m and frequent in status; *P. pallida* flower

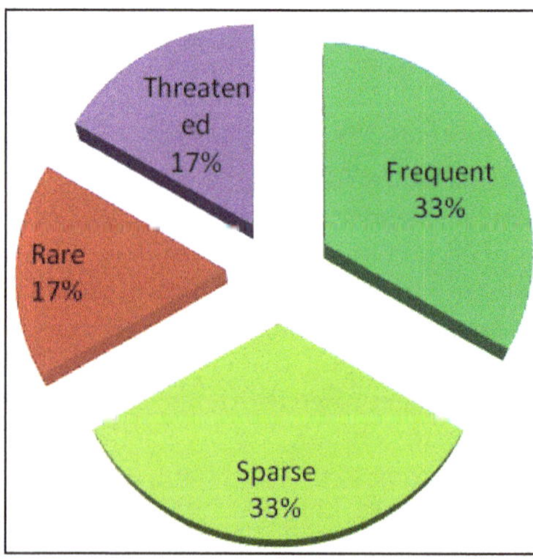

Figure 6.2: Ecological Status of *Pholidota* Orchid Species in Darjeeling Himalaya.

during June to October and available in 800 to 1500m and sparse in status; *P. recurva* flower in August to September and available in 600 to 2000m and threatened status and *P. rubra* flower in October to December and available at 900 to 2200m and found in sparse status throughout the study regions. It is observed that *P. articulata, P. articulata* var. *griffithi* and *P. imbricata* are available in low altitudinal range than *P. pallida* and *P. recurva* and finally *P. rubra* found in the highest elevation in the region.

Conclusion

It is observed that the luxuriant growth and diversity of the Orchid species in the undisturbed sites of the study area and the meager development in distressed sites clearly indicates the change or disturbance in the microclimatic conditions in habitat by anthropogenic activities. Ramdom falling of old epiphytic host trees for timber and firewood collection and indiscriminate collection by paying agents of floral nursery business makes the species threatened in the regions. An immediate step has to be falling for conservation of these species before they disappear forever especially for the species which are rare and threatened. The ever increasing population needs space which are acquired by clearing forest land. Further, frequent landslides, extension of cultivable land for cultivation, forest fire and deforestation for other developmental projects is taking a huge loss of precious Orchid resources that has to be considered seriously in order to save them.

Acknowledgements

First author is thankful to the University Grants Commission, New Delhi for awarding the Rajiv Gandhi National Fellowship for higher study leading to Ph.D in Botany.

References

Bose, T.K. and Bhattacharjee, S.K. 1999. Orchids of India. Revised Edition. Naya Prokash. Calcutta, India.

Chowdhery, H.J. 1998. Orchid Flora of Arunachal Pradesh. Bishen Singh Mahendra Pal Singh. Dehra Dun, India.

Hooker, J.D. 1888 – 1890. The Flora of British India. Vol. 5 and 6. L. Reeve and Co. London.

Jain, S.K. and Rao, R.R. 1977. Field and Herbarium Methods. Today and Tomorrow´s Printers and Publishers. New Delhi, India.

King, G. and Pantling, R. 1898. The Orchids of the Sikkim – Himalaya. In Annals of the Royal Botanic Garden, Calcutta 8, India.

Lucksom, S.Z. 2007. The Orchids of Sikkim and North East Himalaya: Development Area, Jiwan Thing Marg, Gangtok, East Sikkim, India.

Pearce, N.R. and Cribb, P.J. 2002. Flora of Bhutan. The Orchids of Bhutan. Vol. 3, part 3. Royal Botanic Garden, Edinburgh.

Philips, E. A. 1959. Methods of Vegetation study. Henry Holt and Co. Inc. New York, U.S.A. Illinios Press, Illinios, U.S.A.

Yonzone, Rajendra and Kamran, A. 2008. Ethnobotanical Uses of Orchids. Abstract in an International Seminar of XVIIIth Annual Conference of IAAT "Multidisciplinary approaches in Angiosperm Systematics" Kalyani University, West Bengal, India, October 11-13.

Chapter 7

Endemic and Threatened Plants of Indian Thar Desert, Rajasthan (India)

*Amit Kotia**

Department of Habitat Ecology, Wildlife Institute of India, Post Box # 18, Chandrabani , Dehradun – 248001, India

Introduction:

The state of Rajasthan is situated between 23°3′ and 30°12′ N latitude and 69°30 and 78°17′ E longitude. The total land area of the state is about 3, 42,239 km ^{2}, out of which about 1, 96,150 km2 is arid and rest is semi-arid, Forest covers only about 37,638 km^2 *i.e.* 11 per cent of the total area. This forest includes roughly 7 per cent of depleted and denuded forests, while the remaining area is fairly well stocked. The physical features are characterized mainly by the Aravalli and to some extent by the Vindhyan formations, and the Deccan trap, which are predominant in the south in Banswara, Jhalawar and Baran. The Aravalli hills running over a length of 680 km, divide the state into two parts, north western and south-eastern occupying 60 per cent and 40 per cent area of the state respectively. The north-western portion, with a total area of 2, 08591 km^2 has the major portion measuring about 1, 20,983 km^2 in the form of sand dunes; only 2 per cent of the total area of the region is under forest cover (Rodgers and Panwar 1988; Roy and Kumar 1995).

---

* E-mail: research4taxonomy@hotmail.com, kotia.amit@gmail.com

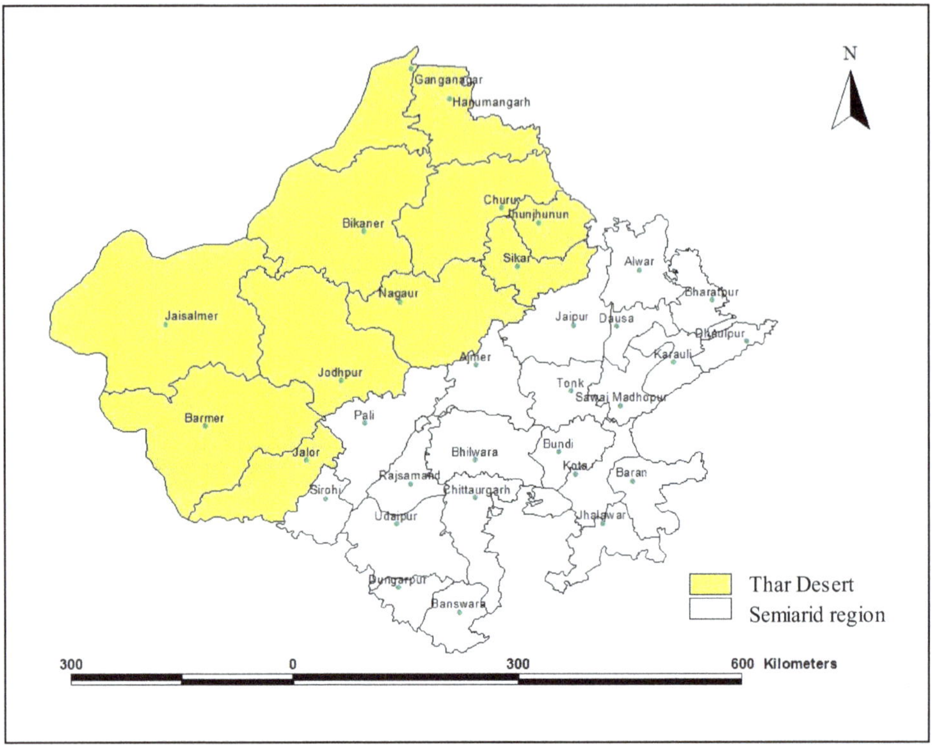

Figure 7.1: Showing Thar Desert and Semiarid Region in Rajasthan.

A major portion of western Rajasthan has desert soils and sandy plain. Sand dunes occupy a greater part of the desert. Occurrence of saline-sodic soils in Rajasthan with pH up to 9.0 is a common feature. The soils of the desert plains are loamy sand to loam in texture and lime concretion accumulation generally occur within 150 cm of soil profile (Basen and Lodha, 1971; Dhir and Jain, 1982).

The average annual rainfall in the state is 525-675 mm in the eastern region and 275 mm in the western region. The rainfall is less than 100 mm in the extreme west, in small area of Jaisalmer district. The total numbers of rainy days in the year are less than 5 at some stations in Jaisalmer district, increasing towards the Aravalli hills to 25 and thence to over 45 in Jhalawar district. About 90 per cent of the annual rainfall occurs due to south-west monsoon from June to September. The annual precipitation in different tracts of Rajasthan varies from 13 mm to 1766 mm.

The climate of the desert region also shows extremes with temperatures ranging from below freezing point in winter to as high as 52°C in summer. Dust storms in the desert tracts may reach a velocity of 130 km/h and may deposit 50-75 mm of dust on the floors. The average humidity varies from 85 per cent (in July) to 25 per cent (in April). Frosts are generally severe during winter, particularly in the sandy areas. Tree saplings up to 2 m in height, sometimes, succumb to such frosts. Most of the areas fall under the dry tropical forest and thorny scrub forest types.

General Vegetation Characteristic of Thar Desert

The present investigations were undertaken with an object to characterize Threatened plant species in reference to distribution, major threats and the present status of their population in Indian Thar Desert. The present paper represent on the basis of extensive survey report in Indian Desert mainly the Barmer, Jodhpur, Jaisalmer, Nagaur, Jalore, Hanumangarh and Bikaner area. colonising non-saline wastelands.

The Sandy Plain and Sand-Dunes

The major portion of the desert is covered by sandy plains and sand dunes they may be stabilized or stableness. Sandy plains and sand dunes are generally covered with *Calotropis procera, Leptadenia pyrotechnica, Capparis decidua,* and *Ziziphus nummularia.* The area is characteristic for its various vegetation associations in different region. Generally the stabilized sandy plains and sand dunes are covered with vegetation of shrub and tree *viz. Leptadenia pyrotechnica, Crotalaria burhia, Sericostoma pauciflorum, Calligonum polygonoides, Capparis decidua, Lycium barbatum, Ziziphus nummularia, Aerva tomentosa, Calotropis procera, Acacia jacquemontii, Acacia senegal, Prosopis cineraria, Salvadora oleoides, S. persica, Tecomella undulata* and *Maytenus emarginatus* along with herbs *viz., Argemone mexicana, Tephrosia* sp., *Farsetia hamiltonii, Convolvulus microphyllus, Boerhavia diffusa, Heliotropium* sp., *Indigofera linifolia, I. cordifolia, Tribulus* sp., *Echinops echinatus.* The *Citrullus colosynthis, C.lanatus, Cucumis* sp, are creepers which can easily seen beside this some climbers like *Cocculus pendulus, Mukia maderaspatana, Momordica dioica, Coccinia grandis* and *Pergularia daemia* are commonly grown in this area. Some grasses *viz. Aristida funiculata, A. adscensionis, Cenchrus ciliaris, C. biflorus, C. prieurii, C. setigerus, Eragrostis* sp., *Cyperus* sp., *Dactyloctenium sindicum, D. aegyptium* and *Lasiurus sindicus.* Whereas inter dune areas also appears the same vegetation but having a different abounds due to availability of moisture beside this stableness sandy area lack good vegetation only some ephemerals can be seen during the rainy seasons mostly these are included *Cleome viscosa, Polycarpaea corymbosa, Corchorus tridens, Triumfetta pentandra, Tribulus terrestris, Gisekia pharnaceoides, Mollugo cerviana, Mollugo nudicaulis, Pedalium murex, Sesamum indicum, Martynia annua, Amaranthus spinosus, Euphorbia* sp. (Roy and Kumar 1995; Bhandari, 1990, Kotia and Kumar 2001; Kotia, 2008).

Gravelly or Rocky Plain

Gravelly pediments or rocky plains are known as Magra and important area of desert and generally seen around the Pokaran- Jaisalmer, Barmer, Jodhpur and Bikaner. Common plants of this habitat are *Cleome vahliana, C. viscosa, C. gracilis, Fagonia indica, Rostellularia procumbens* and *Sericostoma pauciflora.* Besides this, some common prostrate plants growing in the gravel ground are *Tribulus terrestris, Indigofera linnaei, I. linifolia, I. cordifolia, Heliotropium rariflorum, Euphorbia prostrata, E. granulata, E. clarkeana, Mollugo cerviana* and *Mollugo nudicaulis.* The common trees and shrubs growing in gravel habitat include *Euphorbia caducifolia, Leptadenia pyrotechnica, Capparis decidua, Calotropis procera, Ziziphus nummularia, Acacia senegal, Prosopis cineraria, Salvadora oleoides* and *Maytenus emarginatus. Dactyloctenium sindicum, Blepharis sindica, Melanocenchris jaquemontii, M. abyssinica, Oropetium thomaeum* and *Tragus roxburghii* are the some common grasses for the gravel habitat. The rocky area personify different geological pattern, mainly their gravel plains and deep valley or shallow depression, particularly near the foot hills and the vegetation of the area are the mainly *Anogeissus pendula, Asparagus racemosus, Balanites aegyptiaca* and *Euphorbia caudicifolia* (Roy and Kumar 1995, Bhandari, 1990, Kotia and Kumar 2001; Kotia, 2008).

Small Hill and Rock Projection

There are many low elevation hill projection isolated of Aravalli hill such as Massuria hill, Mador, Kailana hill, Viratra hill, Neemdi and Mata ki phadi. *Capparis decidua, Grewia tenax, Euphorbia caducifolia, Rhus mysurensis, Commiphora wightii, Ziziphus nummularia, Acacia senegal, Anogeissus pendula, Boswellia serrata, Maytenus emarginatus* along with herbs *viz. Tephrosia* sp., *Convolvulus* sp., *Boerhavia* sp, *Heliotropium* sp., *Tridax procumbens, Tribulus* sp, *Anticharis senegalensis, Barleria* sp, *Rhynchosia minima, Asparagus racemosus* and *Fagonia indica.* The vegetation is comparatively in abundance near the foot hill mainly *Acacia senegal,* the tree present with combination of *Aristida adscensionis, Blepharis sindica, Tragus roxburghii, Lepidagathis trinervis* and *Melanocenchris jaquemontii* (Roy and Kumar 1995; Bhandari, 1990; Kotia and Kumar 2001; Kotia, 2008).

Saline Tracts

Saline-sodic soils is a common feature in this area with pH up to 9.0, some plants growing frequently with their hard stress tolerance ability. These plant species included *Prosopis juliflora, Tamarix indica, Trianthema portulacastrum, T. triquetra, Zaleya govidea, Chenopodium album, C. murale, Salsola baryosoma, Suaeda fruticosa, Haloxylon recurvum, Sesuvium sesuvioides,* and *Portulaca oleracea. Pulicaria rajputanae* is endemic plant for saline habitat of Rajasthan appearance with very low number (Panday *et al.*, 1983; Bhandari, 1990). Some of the grasses which grow here are *Aeluropus logopoides, Dactyloctenium aegyptium, Eleusine compressa,* and *Eragrostis ciliaris* (Roy and Kumar 1995; Bhandari, 1990; Kotia and Kumar 2001; Kotia, 2008).

Aquatic and Marshland Habitats

Some natural small fresh water ponds and salt lake and artificial water tanks support aquatic habitat in desert area and the main plant species present are *Lemna paucicostata, Naja graminea, N. welwitschii, Potamogeton crispus, P. pectinatus, Spirodela poyrhiza, Vallisneria spiralis, Nelumbo nucifera,* water hyacinth (*Eichhornia* sp.) and *Wolffia* species. According to Panday *et al.* (1983) *Ammannia desertorum* is a threatened plant for aquatic habitat (Roy and Kumar 1995; Bhandari, 1990; Kotia and Kumar 2001).

Blatter and Hallberg (1988), reported 507 flowering plant species from Indian desert, later updated by Bhandari (1990), this includes 682 species of flowering plants in Indian desert with both indigenous and naturalized plants (Shetty and Singh 1993). The migratory path and distribution of the common species shown that, floristic composition of the Indian desert is an admixture of the western (65.8 per cent taxa migrated from west to east) as well as the eastern (34.2 per cent taxa migrated from east to west), while only 9.4 per cent is endemic elements (Bhandari, 1978; Shankarnarayan 1988). The total flora comprises around 682 species and out of which following some threatened plants species are described below (Panday and Shetty, 1985; Bhandari and Shirangi, 1986; Rao *et al.*, 2003):

Enumeration of Threatened Plants Species from India Thar Desert

Family: Acanthaceae

1. *Barleria prionitis* L. var. *diacantha* Blatter & Hallberg: Plant undershrub, whitish bark, leaves up to 6cm, corolla up to 6cm long. According Bhandari (1990) this species is rare and endemic for Western Rajasthan, but it's distribution record is available in other parts of Rajasthan *viz.* Alwar, Bhilwara, Jhalawar, Pali, Tonk and Barmer. The main threat to the species is habitat loss. Fl and Fr: September-December.

Family: Molluginaceae

2. *Gisekia pharnaceoides* L. var. *pseudopaniculata* Jeffrey: A glabrous, prostrate or ascending herb, opposite sessile leaf, flowers few in lax umbels. *Gisekia pharnaceoides* L. was reported endemic by Panday *et al.* (1983), while Bhandari (1990) reported a variety *Gisekia pharnaceoides* L. *pseudopaniculata* Jeffrey. first time in India but literature surveyed cleared it was previously reported from Africa. *G. pharnacioides* L. *pseudopaniculata* is rare species for Indian Desert its distribution was recorded from Barmer and Jodhpur. The main threat to the species is habitat loss and over grazing. Fl and Fr: August-November.

Family: Asclepiadaceae

3. *Caralluma edulis* (Edgew.) Benth. & Hook.f.: Succulent herb, leaves 6-8 x 2mm, corolla broadly campanulate, follicles without soft, spine-like processes. A threatened and endemic species to Thar Desert (Pakistan and Western Rajasthan only). The young shoots are eaten and are sold in market at Jaisalmer (Mohangarh). Locally called as Pimpa. Species rarely grown for vegetable and mostly extinct in wild habitat. Fl and Fr: February-September.

4. *Odontanthera varians* (Stocks.) Mabberley (*Glossonema varians* (Stocks.) Benth. ex Hook.f.): Erect herb, leaves fleshy and pubescent, up to 6x4cm, follicles without soft, spine-like processes. Threatened species rarely found in sandy to gravelly plains. Locally called as Dodha and the follicles are known as Khirolo and eaten raw and according to past records it was sold in local market. This species distribution recorded in Barmer, Jaisalmer (Pokaran-Lavan, Kuri, Budhi Nala and Mohangarh), and Tejaraon in DNP. The main threat to the species is habitat loss and over grazing. Fl and Fr: August-December.

5. *Ceropegia bulbosa* Roxb.: A small, perennial, twining herb, tuberous root, corolla purplish-green, follicles linear, acute. Presently this species is not reported commonly in Thar Desert but during present observation only 5 locations with few individuals were recorded. This species is considered

as rare for other part of India as for its low abundances it should be considered as a rare species for Rajasthan desert. Local name-Heulo. Distributed in Jodhpur and Barmer. The main threat to the species is habitat loss with main reason of associated species cutting out. Fl and Fr: July-October.

Family: Bignoniaceae

6. *Tecomella undulata* (Smith) Seemann: A medium size tree, drooping branches, leaves simple with entire and undulating margins, corolla 5-7cm, yellow or orange, campanulate, seeds winged. A threatened tree species for Rajasthan Panday *et al.* (1983) reported as endemic but according to Bhandari (1990) distributed in Arabia, Pakistan and India (Maharastra, Gujarat and Rajasthan). The main threat to the species is over exploitation with a slow growing rate. This is a beautiful flowering tree of Rajasthan and recognized as state symbol and known as state flower for Rajasthan State. But most of the forest plantation this species is avoided. Fl and Fr: January-April.
7. *Commiphora wightii* (Arn.) Bhandari: A much branched shrub, paper like bark peelings, leaves 3-foliate, flowers solitary or in fascicles, fruit drupe, readily spitting into two parts. Panday *et al.* (1983) reported as endemic but according to Bhandari (1990) distributed in Arabia, Baluchistan, Pakistan and India (Gujarat and Rajasthan). Common in Aravali hill includes Jaipur, Alwar, Bhilwara, Udaipur, Pali and Dausa and also in desert districts Barmer, Jodhpur and Jaisalmer. The present population of Aravali is decreasing vary fast that is reason it is included as threatened species for Rajasthan and the major threat is over exploitation of the species for its medicinal value. Fl and Fr: January-May.

Family: Brassicaceae

8. *Farsetia macrantha* Blatter & Hallberg: Plant perennial, leaves linear-lanceolate, 5-22mm broad, petals 15-16mm long, white, fruit a silicula, seed winged. Local name-Motio-hiran chobbo, endangered and endemic to desert occur only in Barmer (growing on behind rocks of Mataji's temple). The plant is growing in single place in its type locality and the major threats are invasive species *Prosopis juliflora* and over grazing. Fl and Fr: August-January.

Family: Boraginaceae

9. *Heliotropium rariflorum* Stocks: Erect to semi-erect woody herb, leaves linear-lanceolate, calyx hairy, anthers not connate with their connectives, fruit 4-lobed, densely hispid with adpressed hairs. A rare species to Western Rajasthan and distributed in Jodhpur, Barmer, Jaisalmer and Bikaner. Fl and Fr: Almost throughout the year.

Family: Capparaceae

10. *Cleome gynandra* L. var. *nana* (Blatter & Hallberg) Bhandari: Small herb, stem woody at base, leaves 3-foliolate, petals 5mm long, fruit 3-4.5 x 0.3-0.35cm. Rare and endemic to Western Rajasthan and distributed in Jodhpur

(Lodi Tank, on way to Agolai), Jaisalmer (Amar Sagar). Fl and Fr: October-November.

Family: Combretaceae

11. *Anogeissus sericea* Brandis var. *nummularia* King ex Duthie: A medium size tree, young branches including leaves and inflorescences cinereo-tomentose, leaves orbicular or round. Threatened and endemic plant reported from the Rajasthan Thar Desert as well as semiarid area that includes dry habitats and distributed in Jodhpur, Jhalawar, Pali, Udaipur and Sawai Madhopur Districts. Locally called as Indoko/Dhauro. Fl and Fr: October-November.

Family: Compositae

12. *Pulicaria rajputanae* Blatter & Hallberg: A perennial herb, stem cottony tomentose, leaves not crisped, margins not recurved, pappus six times longer than achenes. Rare and endemic to Northern-Western Rajasthan and distributed in saline habitat of Jodhpur, Jaisalmer and Barmer. Local name: Dhola-ligru. Not common in its habitat hard to trace the species. Fl and Fr: October-January.

Family: Convolvulaceae

13. *Convolvulus auricomus* (A. Rich.) Bhandari var. *ferruginosus* Bhandari: Plants ferruginously silky hairy, leaves large, 7 x 3.5cm, subcordate at base. Rare and endemic to Western Rajasthan and distributed in Jodhpur (Balarwa), Jaisalmer and Barmer. Fl and Fr: September-October.
14. *Convolvulus auricomus* (A. Rich.) Bhandari var. *volubilis* (Clarke) Bhandari: Plants white silky hairy, leaves 5 x 2cm, subcordate at base with 6mm long petioles, flowers in larger heads on up to 6cm long head. Threatened species for the Rajasthan and distributed in Pakistan and Western Rajasthan (Jodhpur and Barmer), Local Name- Rota-Bel, Ratanjot, Fl and Fr: October-November.
15. *Convolvulus blatteri* Bhandari: Prostrate herb or twining herb, leaves 3.5 x 1cm, flowers on dense in axillary peduncles, corolla smaller, as long as calyx. Rare and endemic species to the Northern-Western India and distributed in Jaisalmer, Fl and Fr: October-November.
16. *Convolvulus scindicus* Stocks: Erect undershrub, plant densely hirsute with brown-grey ferruginous hairs, leaves obovate or spathulate. Rare to Northern-Western Rajasthan and distributed in Pakistan and Northern-Western Rajasthan (Barmer), Fl and Fr: November-March.
17. *Convolvulus stocksii* Boiss.: Erect undershrub, plant silky pilose or villous, leaves linear, acute, calyx glabrous. Rare in Rajasthan and distributed in Pakistan and India (Gujarat and Rajasthan), plant specimen was recorded from Barmer. Fl and Fr: August-November.
18. *Ipomoea cairica* (L.) Sweet var. *semine-glabra* (Blatter & Hallberg) Bhandari: A glabrous twiner with tuberous roots, corolla large, mauve, sepals muricate, leaves palmately 5-9-fid, with entire lobes, Rare in Rajasthan and distribution recorded in Jaisalmer (Vinjorai), Fl and Fr: August-November.

Family: Euphorbiaceae

19. *Euphorbia jodhpurensis* Blatter & Hallberg: Small herb, prostrate, with leaves less than 1.2cm and much longer than broad, seed obscurely transversely rugose. Species named after its collection of type locality. Rare and endemic to Northern-Western Rajasthan and distributed in sandy habitat of Barmer, Jodhpur and Ganganagar and locally called as Dudhli. Fl and Fr: September-November.

Family: Fabaceae

20. *Acacia jacquemontii* Benth.: A much branched small bushy shrub, stem without prickles, stipular spines present, heads yellow, on axillary peduncles, pods flat, not moniliform. This species is also facing big threat as heavily collected for fire wood purpose; this causes a drastic reduction in wild population of this species. Local name- Bu-banvali/Raio-banwal. Distribution is recorded in N W Rajasthan, and Gujarat, Punjab, Delhi, and Saurashtra. Fl and Fr: March-May.
21. *Alysicarpus monilifer* (L.) DC. var. *venosa* Blatter & Hallberg: Plant prostrate annual herb, leaves unifoliate, calyx shorter or very slightly longer than the first joint of the pods, pods moniliform, reticulately veined. Local name: Leel. Jaisalmer, N. W. Rajasthan Fl and Fr: October-November.
22. *Alysicarpus tetragonolobus* Edgew.: Diffusely branched, procumbent herb, stem marked with a line of hair, calyx deeply divided, pods furnished with a projecting ridge down the middle of each face making the pods 4-gonous, joints 2-6. Species categorized rare for Rajasthan and its distribution record in Baswara, Chhittorgarh and Churu District of Rajasthan. Fl and Fr: August-October.
23. *Psoralea odorata* Blatter & Hallberg: Herb, leaves gland dotted, flowers whitish purple, anther not dimorphous, pods indehiscent, Rare and endemic to Northern-Western Rajasthan and distributed in Ajmer, Churu, Barmer, Bikaner and Ganganar. Local name: Jhil. Fl and Fr: November-March.
24. *Rhynchosia schimperi* (Hochst.) Boiss.: Prostrate, suberect, often twining herb, pods velvety, seed orbicular, terminal leaflet obcordate. Rare for desert zone of Rajasthan and distributed in Barmer, Bikaner, Jodhpur and Jaisalmer. Fl and Fr: September March.
25. *Taverniera cuneifolia* Arn.: A much branched, erect undershurb, leaves unifoliolate, flowers yellow, pods dehiscent, anthers uniform. It is reported endemic by Panday *et al.* (1983) to Rajasthan but according to Bhandari (1990) distributed in Afghanistan, Pakistan and India (Punjab, Saurashtra and N W Rajasthan). The plant recorded in the Bhilwara, Jhalawar, Jodhpur (Sardar Samand) Districts of Rajasthan. Fl and Fr: September-November.
26. *Tephrosia falciformis* Ramasw.: Erect herb, flower pink and less than 1cm, pods highly falcate. Threatened species for Rajasthan and rarely seen, distribution recorded in Pushkar, Churu, Jaisalmer, Jodhpur and Pali, locally called as Rati Biyani. Habitat loss is major threat for the species,

plant also used in local medicine and the stem and root is used as tooth sticks. Fl and Fr: August-October.

Family: Geraniaceae

27. *Monsonia heliotropioides* (Cav.) Boiss.: A small, hoary annual, woody rootstock, leaves radical, flower umbellate, petals shorter than sepals. Rare plant for Rajasthan occurs in Jaisalmer (Khinya and Ramgarh), Bikaner (Devikund Sagar) and parts of N. W. Rajasthan, distribution records also reported from Egypt, Pakistan. Local name: Mayur Shikha. Fl and Fr: September-November.

Family: Lythraceae

28. *Ammannia desertorum* Blatter & Hallberg: An erect, scabarous, papillose plant, cymes sessile or sub-sessile, petals distinct, style distinct, capsules 2.5mm in diam, Locally known as Moto-Jal-Bhangro. Threatened for Rajasthan, distribution also recorded in Saurashtra and N W Rajasthan mainly in Jodhpur and Jaisalmer (Devikot). Fl and Fr: throughout the year (mainly September-November).

Family: Malvaceae

29. *Abutilon bidentatum* Hochst. var. *major* (Blatter & Hallberg) Bhandari: An erect perennial undershrub, leaves *ca* 20cm long, petioles 17cm long, carpels 6-8mm long. Threatened species for Indian Thar Desert, distributed in Jodhpur, Barmer, and Jaisalmer (Amar Sagar). Fl and Fr: September-December.

30. *Abutilon fruticosum* Guillemin. & Perrottet. var. *chrysocarpa* Blatter & Hallberg: An erect perennial undershrub, fruit golden–yellow, carpels truncate, scarcely beaked. Rare and endemic to Thar Desert of India as recorded from Jaisalmer (Vinjorai and Mohangarh) locally called as Imarti. Fl and Fr: November-December.

31. *Abutilon pakistanicum* Jafri & Ali: Grey tomentose undershrub, leaves 6-8cm orbicular-cordate, fruits ovoid, more than 1cm long number of mericarps around 8-10. Rare plant for Rajasthan distributed in Jodhpur and Jaisalmer (Amar Sagar). Fl and Fr: October-January.

32. *Pavonia arabica* Hochst. ex Steudel var. *massuriensis* Bhandari: An erect much branched perennial undershrub, involucral bracts 8-14, flowers white or pink, petals deeply bilobed. Rare and endemic to Jodhpur (Massuria hills). Fl and Fr: August-October.

33. *Pavonia arabica* Hochst. ex Steudel var. *glutinosa* Blatter & Hallberg: An erect much branched perennial undershrub, plant densely viscous-pubescent, involucral bracts 8-14, flower pink, Rare and endemic to Western Rajasthan and occurs in sandy and rocky habitat of Jodhpur (Kailana), Barmer and Jaisalmer. Fl and Fr: August-October.

34. *Sida tiagii* Bhandari: Much branched, perennial undershrub, densely stellately-tomentose, leaves ovate-oblong to obovate, calyx 8-9mm long, mature carpels loosely but completely enclosed by persistent calyx. Rare

plant for Rajasthan distribution is recorded in Jodhpur (Kailana, Massuria hills and Jodhapur Fort), Barmer (Mungaria), and Jaisalmer (Devikot). Fl and Fr: August-December.

Family: Rhamnaceae

35. *Calligonum polygonoides* Linn.: A much branched, perennial shrub, almost leafless shrub. Branches terete, glabrous, flower pinkish, fruiting perianth with appendages in the form of spines or wings or awns. Local name – Phog/Phogro. This species is under threat due to huge pressure of over exploitation for fire wood collection. This is the species having high calories value of fire wood. Mostly the goldsmith and coppersmith in desert focus for *Calligonum polygonoides* wood. This species distribution recorded in mostly all Western part of Rajasthan and commonly seen in Barmer, Bikaner, Jodhpur, Jaisalmer and Churu but their abundance is reduced drastically. Fl and Fr: April-May.
36. *Ziziphus truncata* Blatter & Hallberg: Much branched diffuse shrub, leaves truncate at apex and glabruous on upper surface, fruit less than 1.2cm. Rare and endemic to Northern-Western Rajasthan and distributed in Jaisalmer (Amar Sagar), Jodhpur (Balsamand) and Jalore-Siwana, locally known as Boti. Fl and Fr: October-January.

Family: Scrophulariaceae

37. *Anticharis glandulosa* Asch. var. *caerulea* Blatter & Hallberg: A densely viscid-hairy, erect, branched herb, leaves ovate, not exceeding 1.8cm in length, pedicels less than twice as long as the calyx. This is a rare and endemic to gravelly and rocky habitat of Jaisalmer (Ramgarh) in Northern-Western Rajasthan. Fl and Fr: September-November.
38. *Schweinfurthia papilionacea* Merrill: A dwarf, robust, subglabrous, herb, leaves opposite, corolla subsaccate at base. Rare plant for Rajasthan. This species is distributed in Afghanistan, Pakistan and India (Gujarat and N W Rajasthan). Locally called as sanipat (Jaisalmer-Ramgarh and Harchandari), Plant is recognized for its ethnomedicinal value; plant used for typhoid fever and stop for nose bleeding. Fl and Fr: October-December.

Family: Solanaceae

39. *Withania coagulans* (Stocks) Dunal: A undershrub, with rigid ashy-grey, leaves elliptic-lanceolate or oblanceolate, coriaceous, obtuse at apex, attenuate at base, flowers unisexual, berry with accrescent calyx. Rare plant for Rajasthan distributed in Afghanistan, Pakistan and India, it distributed in Rajasthan mainly part of Jaisalmer and Jodhpur. Plant is locally called Paneer-Bandh because the fruit is used for coagulating milk and making Paneer hence previously described as *Punneria coagulans* Stocks. Panday *et al.* (1983) reported it endemic to Rajasthan. The plant is over exploited for its market value in good medicine importance and other uses. This plant is also extinct mostly in wild habitat. Fl and Fr: November-March.

Family: Sterculiaceae

40. *Melhania futteyporensis* Munro ex Masters var. *major* (Blatter & Hallberg) Santapau: An undershrub, leaves ovate to lanceolate, 7 to 11cm long, involucral bracts ovate or acuminate, smaller than sepals, peduncles terminal. This species was recorded in both arid and semiarid region of Rajasthan. This species is categorized as rare species for Rajasthan and its distribution is recorded in Barmer, Chhittorgarh, Jodhpur, and Kota (Kishanganj). Fl and Fr: October-November.

41. *Melhania magnifolia* Blatter & Hallberg: A undershrub, leaves ovate to oblong, involucral bracts ovate or acuminate, smaller than sepals, peduncles axillary or rarely terminal. According to Bhandari (1990) this species is restricted to only in (endemic to) Northern-Western Rajasthan but distribution is also reported in Jodhpur, Kota, Jhalawar (Ameth). Fl and Fr: August-December.

Family: Zygophyllaceae

42. *Seetzenia lanata* (Willd.) Bullock: A small prostrate herb, not spinous, leaves opposite, trifoliate, flower solitary, petals absent, fruits neither spiny nor winged. This species is rare to Thar Desert, locally this species is known as Dhakari. This species is distributed in Barmer, Jaisalmer and Jodhpur in Rajasthan and also in Africa, Arabia and Pakistan (Sind). Fl. and Fr.: August-November.

43. *Tribulus rajasthanensis* Bhandari & Sharma: A prostrate, perennial herb, leaves pinnate with 3-7 pairs of leaflets, flower solitary, petals present, yellow, fruits with 2 major and 20-30 minor spines. Panday *et al.* (1983) reported as endemic but according to Bhandari (1990) it is distributed in south ward to Gujarat and west ward to Pakistan, it occurs mainly in Jodhpur (Massuria hills) and Jaisalmer of Thar Desert. Fl and Fr: August – December.

Gymnosperm

Family: Ephedraceae

44. *Ephedra foliata* Boiss. & Kotschy: A dioecious. Much branched, climbing shrub, branches slender, smooth, leaves 2-3 at each node, male flowers spikelets ovate, tetragonous, female flowers sessile, 2-3 cymes, fruit ovoid-globose. It is reported endemic by Panday *et al.* (1983) to Rajasthan. According to Bhandari (1990) distribution is reported in Afghanistan, Syria, Pakistan and India (Punjab and Rajasthan). Locally called as Lana (Macadam), Suo-Phogaro, Andho Khimp and Unth-Phog and distributed in Rajasthan mainly in Ajmer, Bikanr, Churu, Barmer, Jodhpur and Jhunjhunu. A medicinally important plant, threatened species for Rajastahn by losing it's population as fast as the habitat loss.

Discussion

Blatter and Hallberg (1988), reported 507 flowering plant species from Indian desert latter up dated by Bhandari (1990). The total flora of flowering plants in Indian desert including both indigenous and naturalized plants comprises around 682 species belonging to 87 families, 356 genera. Poaceae, Fabaceae, Asteraceae and Cyperaceae are the dominating families having about 111, 65, 41 and 36 species respectively (Bhandari, 1990). This habitat is isolated patch for Indian continent, with its characteristic topography, geology, edaphic and climatic factors has made vegetation exceptional which is not found elsewhere in India, in this condition there are 16 plants species endemic to Thar Desert (Ahmedullah and Nayar 1987; Pandey *et al.*, 1983; Singh and Pandey, 1999).

Including endemic some low abundant plants facing big threats due to their habitat loss is mainly due to Indira Gandhi Canal system which modified existing moisture regimes and adversely affected the natural desert system, for example water hyacinth, which never occurred in the region and some other *Arundo, Typha, Phragmitis* are full in waterlogged areas (Prakash, 2001). Beside the over grazing, invasive species like *Prosopis juliflora* continuously occupy the big part of the desert.

The harsh climate conditions give a unique biodiversity and nature has developed many adaptations to survival there. Anthropogenic pressure and habitat loss is a major problem which destroys biodiversity and its habitat of Indian Thar Desert. Habitat fragmentation and instrumental agriculture era is the big cause decreasing the wild population of desert species. Recent records say, around fifty plants species are categorized as threatened from Indian Thar Desert Rajasthan (Singh and Singh, 2005; Kotia, 2008; Kotia *et al.*, 2008) including the grasses species out of which 44 plant species are describe above. There are very few protected areas notified in Indian Desert, where species may exist. On the other hand the concept of Gocher/Ohran lands (Village forest/sacred groves where the agriculture and cutting is prohibited), where a few species presently survive while some are going extinct in the wild. There is no authentic record of the present status of the wild population status and proper ecology of the threatened and endemic plant of Indian Desert. There is an urgent need to attempt their conservation them in their habitat.

References

Ahmedullah, M. and Nayar, M.P. 1987. Endemic Plants of the Indian Region. Botanical Survey of India Publications, Calcutta, India.

Basen, B.L. and Lodha, P.S. 1971. Available micronutrients in sandy soil of Rajasthan. Annals of arid zone. **10:** 99-104.

Bhandari, M.M. 1978. Phytogeography of the Indian Desert. In: Tropical Botany, Larsen, K. and Holm-Nielsen, L.B. (eds.). London,143–152.

Bhandari, M.M. 1990. Flora of the Indian desert. M.P.S. Repros, Publishers, Jodhpur, Rajasthan, India.

Bhandari, M.M. and Shirangi, O.P. 1986. Conservation of threatened plants of western Rajasthan. In: Environmental degradation in western Rajasthan, Mehnot, S.M. and Bhandari, M.M. (eds.). Scientific Publishers, Jodhpur, Rajasthan, India, 129-132.

Blatter, E. and Hallberg, F. 1988. The Flora of the Indian Desert. Scientific Publishers, Jodhpur, Rajasthan, India.

Dhir, R.P and Jain, S.V. 1982. Review of soil research in India, Part II. Indian Agricultural Research Institute, New Delhi.

Kotia, A. 2008. Threatened Plants and Their Habitats in Indian Thar Desert. In: Special Habitats and Threatened Plants of India, Rawat, G.S. (ed.). ENVIS Bulletin: Wildlife and Protected Areas, Vol. 11 (1), Wildlife Institute of India, Dehradun, India, 93-99.

Kotia, A. and Kumar, A. 2001. Biodiversity characterization in arid and semiarid regions of Indian Desert. Int J Mendel, **18 (4):** 109-110.

Kotia, A., Tiwari, U. and Rawat, G.S. 2008. Semiarid Region of India: Vegetation Characteristics and Threatened Plants. In: Special Habitats and Threatened Plants of India, Rawat, G.S. (ed.). ENVIS Bulletin: Wildlife and Protected Areas, Vol. 11 (1), Wildlife Institute of India, Dehradun, India, 109-116.

Panday, R.P. and Shetty, B.V. 1985. Rare and threatened plants of Rajasthan. Geobios. 238-241.

Panday, R.P., Shetty, B.V. and Malhotra, S.K. 1983. A preliminary census of rare and threatened plants of Rajasthan. In: Proceeding an assessment of Threatened plants of India, Jain, S.K. and Rao, R.R. (eds.). Botanical Survey of India Publications, Kolkata, India, 55-69.

Prakash, I. 2001. Biological invasion and loss of endemic biodiversity in Thar Desert. Resonance, 76-85.

Rao, C.K., Geetha, B.L. and Suresh, G. 2003. Red list of Threatened Vascular Plant Species in India. ENVIS Centre for Floral Diversity, Botanical Survey of India Publications, Calcutta, India.

Rodgers, W.A. and Panwar, H.S. 1988. Planning a wildlife protection area Network in India. Wildlife Institute of India. Dehra Dun.

Roy, S. and Kumar, A. 1995. Biodiversity of Rajasthan and its energy potentials. J. Env. Poll. **2:** 105-109.

Shankarnarayan, K.A. 1988. Ecological degradation of the Thar Desert and Eco-regeneration. In: Desert Ecology (Prakash I, eds.), scientific Publishers, Jodhapur, Rajasthan, India. 1-3.

Shetty, B.V. and Singh, V. 1993. Flora of Rajasthan, Vol. I. to Vol. III. Botanical Survey of India Publications, Calcutta, India.

Singh, V. and Pandey, P.R. 1999. Rajasthan. In: Floristic Diversity and conservation Strategies in India, Mudgal, V. and Hajara, P.K. (eds.) 3: 1383-1418. BSI, Calcutta.

Singh, V. and Singh, M. 2006. Biodiversity of Desert National Park Rajasthan. Botanical Survey of India Publications, Calcutta, India.

Chapter 8

Mining and its Impacts on Aquatic Plants: A Case Study from Jaintia Hills of Meghalaya

Jasmine Therese Sawian[1], *Febreena Grace Lyndem*[1], *Roytre Christopher Laloo*[1] *and Solomon Jeeva*[1,2]

[1]*Ecology Laboratory, Centre for Advanced Studies in Botany, North-Eastern Hill University, Shillong – 793 022, Meghalaya*
[2]*Department of Botany, Scott Christian College (Autonomous), Nagercoil –629 003, Tamil Nadu*

Introduction

Plants both aquatic and terrestrial are an essential part of a healthy environment; they are a vital part of the biological diversity and an essential resource for human well-being. Plants are the basic component of agriculture, rangeland, forestry and horticulture. In many ways, the welfare of plants is consistent with that of human beings and the environment (Wang *et al.*, 1997). In recent decades, it has become evident that most of the aquatic and terrestrial ecosystems are changing under increasing pressure from anthropogenic activities (Vitousek *et al.*, 1997; Valiela and Bowen, 2002; Laloo *et al.*, 2005; Jeeva *et al.*, 2007a). The environmental status of Meghalaya has been suffering an accelerated degraded process through unscientific mining activities (Jeeva *et al.*, 2006; Mishra *et al.*, 2005a). This process began in the ca. 1960s and has been more intense in recent years, with destruction of large areas of forests creating a landscape characterized by forest fragments at different successional stages, under strong pressure from humans and with serious problems in preserving biological diversity. Thus, there is a need to understand how the vegetation and community structure is affected by mining activity, so as to develop appropriate strategy for restoration of such areas. This chapter will discuss the impacts of mining on plant diversity and community structure of aquatic ecosystem of Jaintia Hills, Meghalaya. To reach the above goal, we selected three forest sites having surface water bodies namely coal mining, limestone mining and unmined area to study the plant diversity and community structure of aquatic ecosystem of Jaintia Hills, Meghalaya.

Description of the Study Site

The present study was conducted in Sutnga Elaka of Jaintia Hills, at a distance of ca. 120 km away from Shillong, the capital of Meghalaya. Hillocks and undulating topography characterize the landscape of the area. Coal mining and limestone extraction are the major problems resulting in soil erosion, water pollution and environmental degradation of the area. To study the impact of mining on plant diversity and community structure of aquatic ecosystem, we selected three study sites (Plate 8.1: A–unmined area, B–coal mining and C–limestone mining area) in Sutnga Elaka of Jaintia Hills, Meghalaya (Figure 8.1). The unmined area in Jaintia hills is selected as the control, so as to make an assessment of the impact of mining on the area. The physiography of the study site is mentioned in Table 8.1.

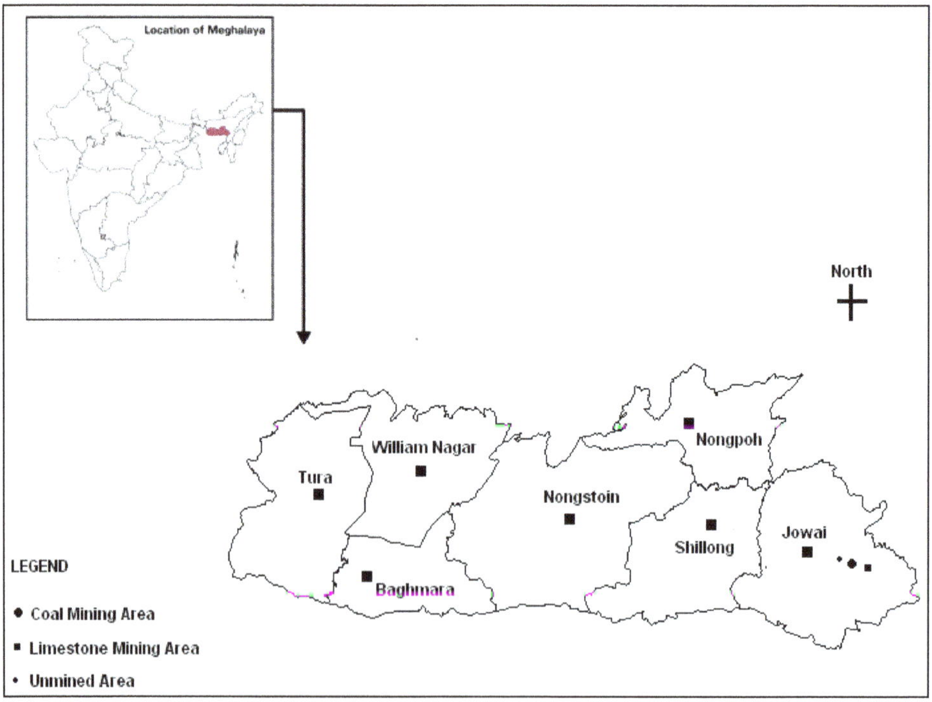

Figure 8.1: Location of Research Sites.

Plate 8.1A–C: Photographs Showing the Study Area.

A. Unmined Area

B. Coal Mining Area

C. Limestone Mining Area

Soil

The soil is mostly sandy, reddish brown to yellow brown in colour, acidic, rich in mineral content and organic matter, but deficient in phosphate and potash. Such soils are good for the cultivation of banana, potatoes, arecanut, betel vines and hill rice in hill slopes and terraces. Coal and limestone are the major deposits of the soil in the area, which are important raw material for the manufacture of Portland cement.

Table 8.1: Physiography of the Study Sites.

Study Site	*Physiography*		
	Latitude	*Longitude*	*Altitude (m)*
Unmined Area (Control)	25º20.313'N	092º28.777'E	973
Coal Mining Area	25º21.128'N	092º28.606'E	933
Limestone Mining Area	25º20.552'N	092º29.453'E	995

Climate

The climate of Jaintia Hills is subtropical monsoonic with distinct alternate wet and dry seasons. The wet season extends from April to October, followed by a dry period from November to March. During the wet season the monthly rainfall ranges from 152 to 756 mm, while in the dry period it is usually < 50 mm per month. The mean annual rainfall was 3276 mm during the study period. Relative humidity also exhibited marked seasonal variation and is closely related to precipitation. The mean

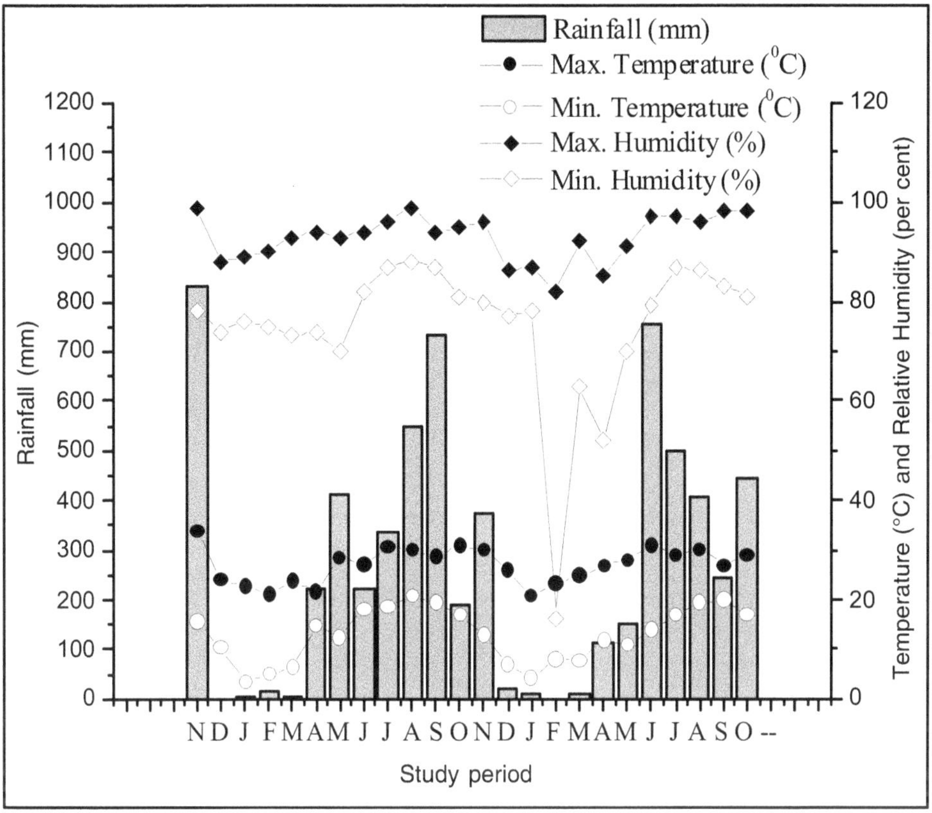

Figure 8.2: Figure Showing the Climate of the Study Area.

monthly temperature varied from a maximum of 27°C in the month of April to a minimum of 3.4°C in December (Figure 8.2).

Methodology

For studying aquatic plants, 1 × 1m quadrats were laid by using bamboo sticks in the water bodies situated in the coal mining, limestone mining and un-mined area. Importance value index (IVI) of plant species was calculated based on the methods given by Mishra *et al.* (1968). The plant species (both aquatic and wetlands) were collected and identified by using the national flora, Aquatic and Wetland Plants of India (Cook, 1996) and counterchecked with the specimens available in the Botanical Survey of India, Shillong and deposited in the herbarium of Ecology Laboratory, Centre for Advanced Studies in Botany, North–Eastern Hill University, Shillong for future reference.

Results

Altogether, 27 species belonging to 24 genera and 20 families of aquatic plants were recorded from the coal mining and unmined areas (Table 8.2). However, no aquatic plants were recorded in the limestone mining area. This may be due to the presence of high calcium content in the water of the limestone mining area. The total number of plant species found in the mining areas (13 species) was significantly less than unmined areas (20 species).

Table 8.2: Aquatic and Wetland Plants (along with their IVI) Present in the Mining and Unmined Areas of Jaintia Hills of Meghalaya.

Mining Area		*Unmined Area*	
Binomials	*IVI*	*Binomials*	*IVI*
Centella asiatica	13.45	*Alternanthera paronychioides*	6.58
Cyanotis axillaris	6.18	*Aponogeton lakhonensis*	29.84
Cyperus cuspidatus	6.49	*Centella asiatica*	29.3
Fimbristylis aestivalis	7.74	*Colocasia esculenta*	3.29
Fimbristylis dichotoma	3.17	*Cyanotis axillaris*	6.58
Juncus bufonius	10.6	*Cyperus rotundus*	8.77
Limnophila chinensis	96.53	*Eriocaulon echinulatum*	3.56
Polygonum barbatum	22.91	*Houttuynia cordata*	9.31
Polygonum lapathifolium	7.43	*Hydrocotyle sibthorpioides*	5.21
Rotala indica	9.35	*Isachne miliacea*	11.78
Rubus khasianus	6.18	*Juncus bufonius*	9.31
Sagittaria sagittifolia	6.96	*Lindernia mollis*	4.94
Spilanthes calva	3.01	*Lobelia alsinoides*	2.74
		Ludwigia prostrata	15.07
		Murdannia blumei	7.94
		Oldenlandia brachypoda	11.79
		Pandanus fasicularis	3.29
		Polygonum barbatum	11.23
		Rotala indica	16.98
		Spilanthes calva	2.47

In unmined area, Apiaceae and Commelinaceae were the dominant families having two species each. Cyperaceae and Polygonaceae with 3 and 2 species respectively, were dominant and co-dominant families in coal mining area. The number of families represented by a single species was higher in unmined area (16) than mined area (8). Generic composition had depicted that the similar trend was observed in control site, but in the mining area, the family Cyperaceae was the dominant family having two genera (Table 8.3).

In the control site, *Aponogeton lakhonensis* was the dominant (IVI 29.84) species, however, *Limnophila chinensis* was the dominant (96.40) species in mining site. *Lobelia alsinoides* (IVI 2.74) and *Spilanthes calva* (IVI 3.01) were the least dominant species in control and mining sites, respectively. The Sorensen index of similarity between two study sites calculated was very low (36.4 per cent), and only 6 species were common in both the sites.

Table 8.3: Family-Wise Distribution of Aquatic Plants in the Study Area.

Mining Area			*Unmined Area*		
Family	*Genera*	*Species*	*Family*	*Genera*	*Species*
Cyperaceae	2	3	Commelinaceae	2	2
Polygonaceae	1	2	Apiaceae	2	2
Apiaceae	1	1	Scrophulariaceae	1	1
Commelinaceae	1	1	Saururaceae	1	1
Juncaceae	1	1	Rubiaceae	1	1
Scrophulariaceae	1	1	Polygonaceae	1	1
Lythraceae	1	1	Poaceae	1	1
Rosaceae	1	1	Pandanaceae	1	1
Alismataceae	1	1	Onagraceae	1	1
Asteraceae	1	1	Lythraceae	1	1
			Juncaceae	1	1
			Eriocaulaceae	1	1
			Cyperaceae	1	1
			Campanulaceae	1	1
			Asteraceae	1	1
			Araceae	1	1
			Aponogetanceae	1	1
			Amaranthaceae	1	1

During the present study, it was noticed that in one sampling area of the mining site, agricultural wastes from nearby fields discharged into the river and got mixed with acid mine drainage resulting in eutrophication and excessive growth of aquatic macrophytye, *Limnophila chinensis*. The eutrophication process has resulted in changes in the community composition in the aquatic body.

Discussion

Aquatic plants are relatively short-lived, have rapid response times too many perturbations, are often cosmopolitan in their distributions, and are relatively easily identified (Hill, 1997). These plants are significant to the ecology of surface water bodies (Jeeva *et al.*, 2007b). Plants form the base of most aquatic food chains, produce oxygen, serve as refuge for aquatic life, and are important in nutrient cycling. Plants interact with their environment through processes that include chemical bio-concentration and excretion, shading, and organic matter production and decomposition (Carpenter and Lodge, 1986; Honnell *et al.*, 1993). As a result of these interactions, aquatic plants may significantly affect water quality (Champers and Prepas, 1994). The findings of present study revealed that mining adversely affect the plant diversity in the aquatic bodies. Since the study sites had similar edaphoclimatic condition, the differences in species composition could be attributed to the AMD discharges as a result of the mining activities.

Most aquatic plants have seasonal growth cycles, which are controlled by a variety of ecological factors such as nutrients, light and temperature. The structure and function of aquatic plant communities are influenced directly or indirectly by changes in water quality resulting from the presence of nutrients and phytotoxicants. The input of nutrients and contaminations by anthropogenic activities can inhibit or stimulate plant growth and may cause ecosystem change (Lewis and Wang, 1997).

Plants have been used frequently as biomonitors where the composition of the plant community has been related to water quality (Seddon, 1972; Sculthorpe, 1967; Franzen and McFarlane, 1980). Decline in pH in mining site is one the serious problems associated with coal mining activity. Increasing acidity strongly affects the aquatic vegetation in various ways including the availability of a large number of essential nutrients in the aquatic ecosystem (Jeeva *et al.*, 2007c). From the present study, the IVI of *Limnophila chinensis* is found to be high due to eutrophication in mining areas. Past studies reported that eutrophication plays a major role in changes of community composition in an aquatic ecosystem. In severe cases, there is a dramatic increase in growth of some species, which often results in algal blooms (Lewis and Wang, 1997). This is evident with the findings of the present study.

Conclusion

Mining operations alter ecosystem structure and function of both aquatic and terrestrial ecosystems. Environmental problems associated with mining have been felt severely because of the region's fragile ecosystems and rich biological and cultural diversity. Large scale denudation of forest cover, scarcity of water, pollution of air, water and soil and degradation of agricultural lands are some of the conspicuous environmental implications of coal mining. Besides, a vast area have become physically disfigured due to haphazard dumping of overburden, caving in of the ground, and subsidence of land. Form the present study, we found that aquatic plants are present only in the upper stream of the coal mining area and were high in the unmined area, however no aquatic plants were recorded in the limestone mining area.

The most severe environmental impact of mining activity is degradation of water quality. The water in the downstream of the coal mining area was highly acidic, on the contrary in limestone mining area it was found to be alkaline. The hardness of water, calcium and magnesium content also changed due to mining activity. Chloride content increased due to mining. The impact of AMD was more acute during rainy season. These changes make the water polluted unusable even for agricultural purposes. Low DO and high BOD were recorded especially in the down stream of the mining sites. The results indicate that mining alters the biological and physicochemical properties of water in the aquatic ecosystems of the mining areas. Based on the findings of the present study, it is revealed that coal mining and limestone extraction has adversely affected the aquatic ecosystem of Jaintia Hills, Meghalaya. There is an urgent need for reversing the trend and bringing back the area to its original state.

Filling of mine pits, channeling of seepage water for checking AMD contamination of water bodies and agricultural fields, afforestation with native and fast growing tree species, undertaking effective soil conservation and water resources management programmes are some of the measures which can mitigate the problem and restore the degraded ecology for conservation of aquatic and terrestrial ecosystems of the Jaintia Hills, Meghalaya.

References

Carpenter, S.R. and Lodge, D.M. 1986. Effect of submerged macrophytes on ecosystem processes. Aquatic Botany, **26**: 341-370.

Champers, P.A. and Prepas, E.E. 1994. Nutrient dynamics in riverbeds: the impact of sewage effluent and aquatic marcrophytes. Water Research, **29:** 453-464.

Cook, C.D.K. 1996. Aquatic and Wetland Plants of India. Oxford University Press, New York, pp: 1-385.

Franzen, W.G. and McFarlane, G.A. 1980. An analysis of the aquatic macrophytes, *Myriophyllum exalbescens*, as an indicator of metal contamination of aquatic ecosystem near a base metal smeltar. Bulletin of Environmental Contamination Toxicology, **24:** 597-605.

Hill, B.H. 1997. Aquatic plant communities for impact monitoring and assessment. In: Wang W., Gorsuch J.W. and Hughes J. S. (eds), Plants for Environmental Studies. Lewis Publishers, Boca Raton, New York, USA, pp. 277-305.

Honnell, D.R., Madsen, J.D. and Smart, R.M. 1993. Effects of selected exotic and native aquatic plant communities on water temperature and dissolved oxygen. U.S. Army Corps of Engineers Aquatic Plant Control Research Program, U.S. Army Corps of Engineers, Vicksburg, MS, Vol. A. 93-102.

Jeeva, S., Febreena G.L, Jasmine T.S. and Laloo, R.C. 2007b. Water quality and aquatic plants. In: National Conference on Adaptation Biochemistry, organized by Department of Biochemistry, North-Eastern Hill University, Shillong, Meghalaya.

Jeeva, S., Jasmine T.S., Febreena G.L. and Laloo, R.C. 2007a. Ecological studies on pollution status in coal mining areas of Jaintia Hills, Meghalaya. In: Regional Symposium on Current Thrust in Animal Sciences: Interface between end use Researchers and Stakeholders, organized by Department of Zoology, North-Eastern Hill University, Shillong–793 022, Meghalaya.

Jeeva, S., Jasmine, T. S., Febreena, G.L., Laloo, R.C. and Venugopal, N. 2007c. Acid mine drainage and its impact on water quality and aquatic plants. In: International Conference on Water Resource Management: Challenges and Opportunities in the 21st Century, organized by Department of Ecology and Environmental Sciences, Assam University, Silchar.

Jeeva S., Jasmine T. Sawian, Febreena G. Lyndem, Venugopal N. and Laloo R.C. 2006. The water quality of river Sutnga, Jaintia Hills, Meghalaya: impact of acid mine drainage. In: International Conference on Current Trends in Algal Bioresources Utilization, organized by Department of Ecology and Environmental Sciences, Assam University, Silchar, Assam.

Laloo, R.C., Jeeva, S. and Mishra, B.P. 2005. Deterioration of the quality of surface water as influenced by extraction of coal and limestone in Jaintia Hills of Meghalaya. In: International Symposium on Integrated Management of Fungal Diseases (IMF). Organised by International Society for Conservation of Natural Resources (ISCON) and Centre of Advanced Study in Botany, Banaras Hindu University, Varanasi, India, February 24-27. Abstract pp. 80-81.

Lewis, M.A. and Wang, W. 1997. Water quality and aquatic plants. In: Wang, W., Gorsuch, J.W. and Hughes, J. S. (eds), Plants for Environmental Studies. Lewis Publishers, Boca Raton, New York, USA, pp. 139-175.

Mishra, B.P., Jeeva, S. and Laloo, R.C. 2005a. Water resources and management. Souvenir of Scott Christian College, **1:** 147-159.

Mishra, B.P., Jeeva, S. and Laloo, R.C. 2005b. Effect of fragmentation on plant diversity and community characters of the sacred groves of Meghalaya, northeast India. In: National Conference on Current Trends of Research in Science and Technology, Organized by Gauhati University, Guwahati, Assam, India, Abstract No OP-L14, pp. 107.

Misra, R. 1968. Ecology Work Book. Oxford and IBH Publishing Company, Calcutta, India.

Sculthorpe, C.D. 1967. The Biology of Aquatic Vascular Plants. Edward Arnold Limited, London.

Seddon, 1972. Aquatic macrophytes as limnological indicators. Freshwater Biology, **2:** 107-130.

Valiela, I. and Bowen, J.L. 2002. Nitrogen sources to watersheds and estuaries: role of land cover mosaics and losses within watersheds. Environmental Pollution, **118:** 239-248.

Vitousek, P.M., Aber, J.D., Howarth, R.W., Likens, G.E., Matson, P.A., Schindler, D.W., Schlesinger, W.H. and Tilman, D.G. 1997. Human alteration of the global nitrogen cycle: sources and consequences. Ecological Applications, **7:** 737-750.

Wang, W., Gorsuch, J.W. and Hughes, J. S. 1997. Plants for Environmental Studies. Lewis Publishers, Boca Raton, New York, USA.

Chapter 9

Inter-Specific Variation Studies on Selected Medicinally Important *Phyllanthus* Species Using SDS-PAGE

M. Narayani, V. Kalaiarasi, M. Johnson, T. Renisheya Joy Jeba Malar and N. Janakiraman*

Centre for Plant Biotechnology, Department of Botany, St. Xavier's College (Autonomous), Palayamkottai – 627 002, Tamil Nadu

ABSTRACT

The present study was aimed to study the proteomic profiles of four medicinally important *Phyllanthus* species *viz., Phyllanthus urinaria* L., *Phyllanthus amarus* Schum & Thonn, *Phyllanthus maderaspatensis* L. and *Phyllanthus emblica* L. using SDS-PAGE electrophoresis. SDS-PAGE was carried out by the method described by Anbalagan. The similarity and variation between the bands was estimated and the cladogram was constructed using NTSsys. The selected *Phyllanthus* species expressed a total of twenty eight bands with different regions and various MW-Rf values. *P. urinaria* displayed with eight bands, *P. amarus* showed 6 bands, *P. maderaspatensis* depicted 9 bands and *P. emblica* expressed 5 bands with different MW- Rf values in the protein gel system. The results of the present study confirm the similarity and variation between the selected four

* Corresponding Author: E-mail: ptcjohnson@gmail.com

Phyllanthus species. These profiles may be used as biochemical marker for the identification in the molecular taxonomy and to distinguish the medicinally important species from its adulterant in the pharmaceutical industry.

Keywords: *Phyllanthus, Proteomics, SDS-PAGE, Taxonomy, Cladogram.*

Introduction

Plants exhibit natural variation in their form and structure. Within a single individual, parts are repeated which may differ in form and structure from other similar parts. Traditionally, genetic diversity is assessed based on morphological features such as plant height, reproductive features, day length sensitivity, local adaptation etc. A wide spectrum of simple and overlapping variations is now documented in plants (Rout *et al.*, 2010). Protein (SDS-PAGE) and Isozymes (PAGE) are simple to use in a variety of applications in plant taxonomical research. The information on polymorphism using protein and isozyme profiles in a set of genotype is useful in tagging genes of interest and genetic mapping in long run to facilitate marker assisted selection (Johnson *et al.*, 2012) where as the banding positions and regions of activity are variable based on the developments of plants (Johnson and Patric Raja, 2007). During the past 30 years, various techniques have been used for analysis of genetic variability at the molecular level in plants *viz.*, isozymes (Geetha *et al.*, 2003), RFLP (Mohapatra and Rout, 2005) and RAPD (Theerakulpisut *et al.*, 2008). Protein profiles are still powerful tools for determining genetic homology at the molecular level and for solving problems in systematic methodology.

The genus *Phyllanthus* (Euphorbiaceae) is commonly called 'carry me seed', 'stone-breaker', 'windbreaker', 'gulf leaf flower' or 'gala of wind'. All parts of the plant are employed therapeutically (Dnyaneshwar *et al.*, 2006). *Phyllanthus amarus* Schum & Thonn is an erect annual herb of not more than one and half feet tall and has small leaves and yellow flowers. In folk medicine, *P. amarus* has reportedly been used to treat jaundice, diabetes, diarrhoea, swelling, skin ulcer, gastrointestinal disturbances, weakness of male organ and blocks DNA polymerase in the case of hepatitis B virus during reproduction. *P. amarus* is also used for stomach problems and for the genitourinary system, liver, kidney and spleen (Geetha *et al.*, 2003; Oluwafemi and Debiri, 2008). *Phyllanthus emblica* L. has been used for thousands of years by Ayurvedic physicians who refer it as the 'Sustainer' and 'the fruit where the goddess of prosperity resides'. Although *P. emblica* is reputed to have the highest content of vitamin C, recent scientific interest has entered on its unique tannins and flavonoids which contain very powerful antioxidant properties and it is a potent immune stimulator since it activates natural killer cell activity (Dyneneshwar *et al.*, 2006). *Phyllanthus maderaspatensis* L. leaves are expectorant and diaphoretic, the seeds have a bad taste and are carminative, laxative, tonic to the liver, diuretic and is useful in bronchitis, griping, ophthalmia and ascites. In South India, an infusion of the leaves is given for headache relief (Praveen *et al.*, 2008). *Phyllanthus urinaria* L. is used as a traditional folk medicine for the treatment of several diseases including hepatitis B (Lam *et al.*, 2006), nephrolithiasis and some painful disorders (Chan *et al.*, 2003).

Recent *in vitro* and *in vivo* studies have shown that *P. urinaria* exhibits many biological and pharmacological functions *viz.*, anticancer, cardioprotective, hepatoprotective, antiangiogenic, antioxidant (Iqbal *et al.*, 2009), antisemicarbazide-sensitive amine oxidase and antihypertensive effects (Lin *et al.*, 2008; Srirama *et al.*, 2010).

Numerous protein profile studies have been done with various plant species such as *Capsicum* sp. (Babu *et al.*, 2009) and *Amaranthus* taxa (Babu *et al.*, 2010). However, the phenotypic or adaptive significance of molecular polymorphisms was not clear in most cases since molecular variability did not appear to be related to morphological variation inter-specifically or intra-specifically. Protein profiles strongly support the hypothesis that molecular polymorphisms are neutral in natural selection. The quantitative variations in gene product levels by electrophoretic techniques are an important basis for detection of morphological and adaptive change than classical variability (Mohapatra and Rout, 2005). Verma *et al.* (2010) fractioned the leaves and fruits of *P. emblica* using SDS-PAGE and confirmed that ascorbic acid content in fruit was found to be higher than in leaves by varying bands observation. With this knowledge, the present study was intended to study the proteomic profiles of four *Phyllanthus* species using SDS-PAGE and to record the similarity and variation between the selected *Phyllanthus* by using NTSsys software.

Materials and Methods

Plants of *Phyllanthus urinaria* L. (L1), *Phyllanthus amarus* Schum & Thonn (L2), *Phyllanthus maderaspatensis* L. (L3) and *Phyllanthus emblica* L. (L4) were collected from wild and in the green house attached to the Department of Botany, St. Xavier's College, Palayamkottai, Tirunelveli, Tamil Nadu, India. For the electrophoretic studies, 500 mg of fresh young leaves were harvested from the mother plants and washed in de-ionized water and ground using pre-chilled mortar and pestle with 0.1M Tris buffer (pH 7.0). The resultant slurry was centrifuged at 10,000 rpm for 10 min at 4°C and the supernatant were collected and separated by slab gel electrophoresis. SDS-PAGE was carried out at 25°C in the air conditioned room. After running the electrophoresis, the gel was stained with 0.1 per cent coomasie brilliant blue and destained with methanol: acetic acid: water (Anbalagan, 1999). The gel was observed using a Vilber Loubermat gel documentation system and banding profiles of protein was determined by the migration from the origin towards the anode. The similarity and variation between the bands were estimated by NTS sys software analysis and the cladogram was constructed.

Results and Discussion

Multiple regions of activity were obtained from protein electrophoretic system of different *Phyllanthus* species. The selected *Phyllanthus* species expressed 28 bands at different regions with various MW-Rf values and positions (Table 9.1). *P. urinaria* displayed 8 bands, *P. amarus* showed 6 bands, *P. maderaspatensis* depicted 9 bands and *P. emblica* expressed 5 bands in the protein gel system of *Phyllanthus*. The proteins with MW-Rf 0.551 and 0.596 were illustrated their presence only in *P. emblica.* The recorded data discussed the role of protein in similarity and variation between the selected species. Each region is occupied by different protein in the form of bands and is representative of the expression of a particular gene in the selected species.

Table 9.1: SDS-PAGE Banding Pattern of Medicinally Important *Phyllanthus* Species.

MW-Rf Values	*Regions*	*Positions*	*P. urinaria*	*P. amarus*	*P. mader-aspatensis*	*P. emblica*
0.034	1	$PP1^1$	–	+	–	–
0.045		$PP1^2$	–	–	+	–
0.089		$PP1^3$	+	–	–	–
0.179	2	$PP2^1$	+	–	–	–
0.191		$PP2^2$	–	+	+	+
0.292	3	$PP3^1$	–	–	+	–
0.337	4	$PP4^1$	+	–	–	–
0.427	5	$PP5^1$	–	–	+	–
0.528	6	$PP6^1$	–	+	–	–
0.551		$PP6^2$	–	–	–	+
0.562		$PP6^3$	+	–	–	–
0.573		$PP6^4$	–	–	+	–
0.596		$PP6^5$	–	–	–	+
0.708	7	$PP7^1$	–	–	+	–
0.730		$PP7^2$	–	+	–	–
0.742		$PP7^3$	+	–	–	–
0.831	8	$PP8^1$	+	–	–	–
0.843		$PP8^2$	–	–	+	–
0.876		$PP8^3$	+	+	–	+
0.910	9	$PP9^1$	–	+	–	–
0.933		$PP9^2$	–	–	+	+
0.988		$PP9^3$	–	–	+	–
0.977		$PP9^4$	+	–	–	–

Based on the occurrence of proteins in the gel system, the protein profiles were classified into 9 regions. Region 1 observed with three bands, $PP1^1$ (0.034) was showed by *P. amarus*, $PP1^2$ (0.034) and $PP1^3$ (0.089) were expressed only in *P. urinaria*. Region 1 showed the variation between the selected four species, the banding profile present in the Region 1 may be used as biomarker to distinguish *P. amarus* and *P. urinaria* from other selected species. Region 2 contained with two bands, $PP2^1$ (0.179) showed its presence only in *P. urinaria*, $PP2^2$ (0.191) was shared by *P. amarus, P. maderaspatensis* and *P. emblica*. Contrary to Region 1, the banding profile of region 2 showed the similarity between the selected three species and distinguished the *P. urinaria* from other selected species. Region 3 obtained only one band, $PP3^1$ (0.292) present only in *P. maderaspatensis*. This band may act as a biochemical marker to classify the medicinally important plant *P. maderaspatensis*. Region 4 illustrate with one band $PP4^1$ (0.337) and showed its expression only in *P. urinaria*. Similar to region 2, the band present in region 4 also used to characterize the *P. urinara*. The other selected

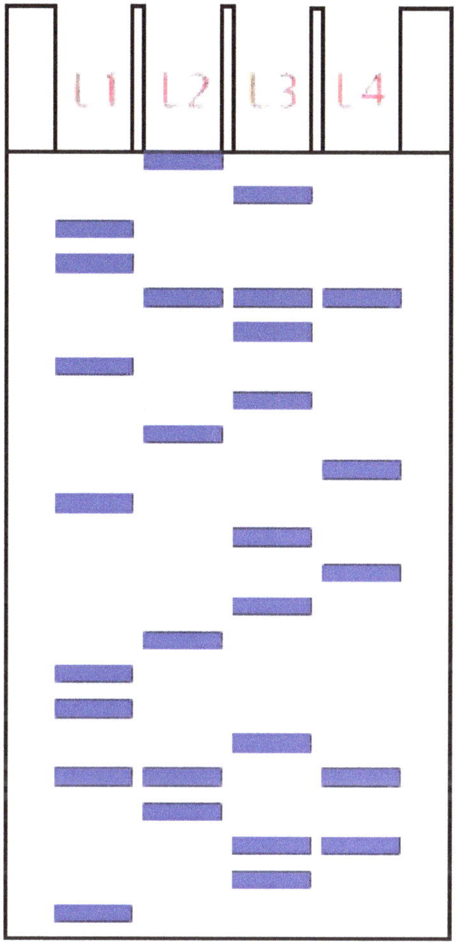

Figure 9.1: Zymogram of Selected Medcinally Important *Phyllanthus* Species. L1: *P. urinaria*; L2: *P. amarus*; L3: *P. maderaspatensis*; L4: *P. emblica.*

plants failed to express themselves in the region 3 and 4. Similar to region 3, in this region (Region 5) also we observed a single band ($PP5^1$) with the MW-Rf 0.427 in *P. maderaspatensis* only. Region 6 illustrated with 5 bands with different MW-Rf values, $PP6^1$ (0.528) and $PP6^2$ (0.551) was showed its expression only in *P. amarus* and *P. emblica* respectively. $PP6^3$ (0.562) was observed only in *P. urinaria* and $PP6^4$ (0.573) in *P. maderaspatensis.* $PP6^5$ (0.596) was found only in *P. emblica*. The region six also showed the variation in the banding profile and distinguished the selected four species with different MW-Rf values. The region 6 banding profile may be used as biochemical marker to characterize the selected four *Phyllanthus* species. Region 7 showed three bands with different MW-Rf values. $PP7^1$ (0.708), $PP7^2$ (0.730) and $PP7^3$ (0.742) was observed only in *P. maderaspatensis, P.amarus* and *P. urinaria* respectively. Similar to region 6, the banding profile of region 7 also distinguished the three species

(*P. maderaspatensis, P. amarus* and *P. urinaria*) by their unique expression. The *P. emblica* failed to express in this region. Region 8 contained three bands with varied Rf values ranged from 0.831 to 0.876. $PP8^1$ (0.831) was observed only in *P. urinaria*, $PP8^2$ (0.843) showed its unique expression only in *P. maderaspatensis*. $PP8^3$ (0.876) was shared by *P.urinaria, P. amarus*, and *P. emblica*. The banding profile in the region 8 showed the similarity and variation between the selected species equally. The protein $PP8^1$ (0.831) and $PP8^2$ (0.843) may be used to distinguish *P. urinaria* and *P. maderaspatensis*. The protein $PP8^3$ (0.876) may be used to know the similarity between *P.urinaria, P. amarus*, and *P. emblica*. Region 9 expressed with four bands, $PP9^1$ (0.910), $PP9^3$ (0.988) and $PP9^4$ (0.977) showed their presence only in *P. amarus, P. maderaspatensis* and *P. urinaria* correspondingly. $PP9^2$ (0.933) showed its presence jointly in *P. maderaspatensis* and *P. emblica*. Similar to region 8, the banding pattern of region 9 also showed similarity compared to the variation between the selected *Phyllanthus* species. The proteins with MW-Rf 0.089, 0.179, 0.337, 0.562, 0.742, 0.831 and 0.977 were present only in *P. urinaria*. The proteins with MW-Rf 0.034, 0.528, 0.730 and 0.910 showed their expression only in *P. amarus*. The proteins with MW-Rf 0.045, 0.292, 0.427, 0.573, 0.708, 0.843 and 0.988 were observed only in *P. maderaspatensis*. These banding profiles may act as biochemical marker to identify the medicinally important *Phyllanthus* species.

Based on the protein expression of selected *Phyllanthus* species, the cladogram was constructed by UPGMA method using NTSys (Figure 9.2). The cladogram showed two major clusters (Cluster 1 and Cluster 2). The cluster 1 consists of only *P. urinaria*. The cluster 2 includes the other three species *P. amarus, P. emblica* and *P. maderaspatensis*. The cluster 2 was divided into two branches $C_2N_1B_1$ and $C_2N_1B_2$. The $C_2N_1B_1$ branch includes the *P. amarus* and *P. emblica*, the $C_2N_1B_2$ branch includes only *P. maderaspatensis*. The cluster1 showed the uniqueness of *P. urinaria*, it is varied from other three selected species. The cluster 2 showed the similarity and variation between the three *Phyllanthus* species. The $C_2N_1B_1$ explained the similarity and closeness between *P. amarus* and *P. emblica*. The $C_2N_1B_2$ showed variation and individuality of *P. maderaspatensis*. Protein markers are practical, useful genetic and biochemical markers as well as good estimation of genetic variability in plant population [19]. The present study results also directly coincided and supplemented the previous observations.

Similarity and variation between the selected *Phyllanthus* was recorded by using NTSsys (Figure 9.2). *P. urinaria* and *P. amarus* expressed similar bands in $PP8^3$ (0.876) and their similarity percentage was estimated as 7.14 per cent. *P. urinaria* and *P. emblica* shows related bands in $PP8^3$ (0.876) and their similarity percentage was deliberated as 7.69 per cent. *P. amarus* and *P. maderaspatensis* uttered similar bands in $PP2^2$ (0.191) and their similarity percentage was recorded as 6.66 per cent. Analogous bands found between *P. amarus* and *P. emblica* were $PP8^3$ (0.876) and $PP2^2$ (0.191), their similar percentage was 18.18 per cent. *P. maderaspatensis* and *P. emblica* expressed parallel bands in $PP2^2$ (0.191) and $PP9^3$ (0.933), their similarity percentage was analyzed as 14.28 per cent. *P. urinaria* and *P. maderaspatensis* failed to show their similarity.

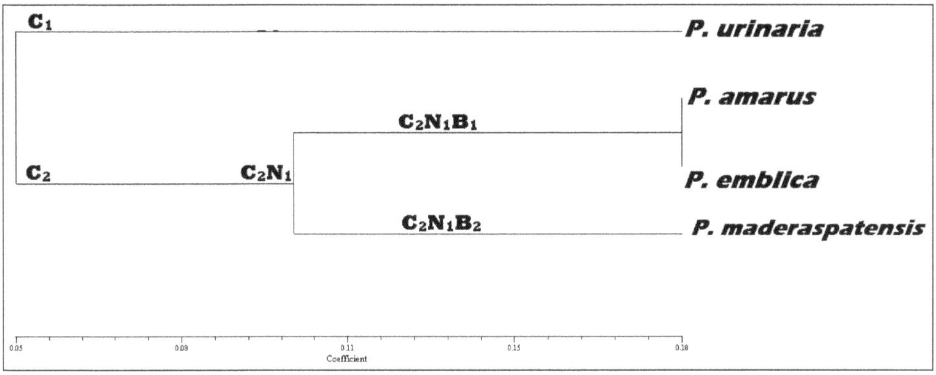

Figure 9.2: Cladogram of selected medcinally important *Phyllanthus* species

Variety is the spice of life and basic characteristic of life in its unlimited diversity. The belief that no two individuals of a sexually reproducing population are 100 per cent identical prevailed even when methods of scientific scrutiny were not available. Much of the variations in phenotype observed in natural populations of a species were earlier attributed to environmental influences (Hamrick *et al.*, 1991). Many botanists reasoned that distinct intra specific variations of plants were merely due to habitat modifications and adaptation to environment was by phenotypic fake response. A wide spectrum of simple and overlapping variations is now documented in plants. In general, all observed variations are broadly grouped into two categories, epigenetic and genetic. Genetic variations in plants are strictly heritable. Epigenetic changes in plants in general include morphological, chemical as well as physiological variations. Therefore a great deal of genetic information is necessary before the observed pattern of variation maybe interpreted. In recent years, limitations of numerical taxonomy *viz.* morphology, anatomy and cytology have been over come by biochemical and molecular markers (Johnson *et al.*, 2012). Babu *et al.* (2010) indicated that protein markers could be effectively used for genetic diversity studies among *Amaranthus* taxa. They classified the medicinally and economically important plants belonging to the genus *Amaranthus* and *Capsicum*. Similar to the previous observations, in the present study also we observed the intra-generic variation between the selected *Phyllanthus* species.

Conclusion

The present SDS-PAGE banding profile revealed the variation between four *Phyllanthus* species which may be used as a biochemical marker in the molecular taxonomy to distinguish the medicinally important species. It is necessary to study further the structural and functional roles of these reported proteins to use as a marker for the selected species.

Acknowledgement

The author Renisheya Joy Jeba Malar T, sincerely acknowledge the financial assistance provided by the Department of Science and Technology, Minsitry of Science and Technology, Govt. of India, New Delhi through the DST – INSPIRE Fellowship (Ref. No. DST/INSPIRE Fellowship/2011/IF 110640 dated on 9th December, 2011).

References

Anbalagan, K. 1999. An introduction to electrophoresis. Electrophoresis Institute, Yercaud, Tamil Nadu.

Babu, A., Johnson, M., Raja, D.P. 2009. Variability studies on selected taxa of *Capsicum* using morphological characters and SDS-PAGE profiles. Trade science inc. **3**: 1-11.

Babu, A., Raja, D.P, Johnson, M. and Sivasankari, G. 2010. Variability studies of selected taxa of *Amaranthus*. J. Basic Appl. Biol. **4 (1 and 2)**: 125-136.

Chan, H.L.Y., Sung, J.Y., Fong, W.F., Chim, A.M.L., Yung, P.P., Hui, A.Y., Fung, K.P. and Leung, P.C. 2003. Double-blinded placebo-controlled study of *Phyllanthus urinaria* for the treatment of chronic hepatitis B. J. Ethnopharmocol. **18**: 339-345.

Dnyaneshwar, W., Preeti, C., Kalpana, J. and Bhusan, P. 2006. Development and application of RAPD-SCAD marker for identification of *Phyllanthus emblica* Linn. Biol. Pharmac. Bull. **29**: 2313-2316.

Geetha, B.S., Sabu, K.K. and Seeni, S. 2003. Genetic variation in south India populations of *Phyllanthus amarus* Schum. & Thonn. (Euphorbiaceae) assessed using isozymes. New Millennium Seminar on Medicinal Plants Proceedings **15**: 196-201.

Hamrick, J.L., Godt, M.J., Murawski, D.A., Loveless, M.D., Falk, D.A. and Hosinger, K.E. 1991. Genetics and conservation of rare plants. Oxford University Press, New York, pp. 75-86.

Iqbal, M., Okazaki, Y. and Okada, S. 2009. Curcumin attenuates oxidative damage in animals treated with a renal carcinogen, Ferric Nitrilotriacetate (Fe-NTA): Implications for cancer prevention. Mol. Cell Biochem. **28**: 114-119

Johnson, M. and Patric Raja, D. 2007. Morphogenetic changes of four rare and endangered ferns of Western Ghats, South India. Biochemistry: An Indian Journal **1 (4)**: 62-64.

Johnson, M., Renola Joy Jeba Ethal, T. and Babu, A. 2012. Influence of salinity stress on proteomic profiles of *Cicer arietinum* L. J. Stress Physiol. Biochem. **8**: 5-12.

Lam, W.Y., Leung, K.T., Lee, S.M., Chan, H.L., Fung, K.P., Ooj, V.E. and Waye, M. 2006. Antiviral effects of *Phyllanthus amarus* extracts against Hepatitis B virus. Cell Biochem. **18**: 795-812.

Lin, S.Y., Wang, C.C., Lu, Y.L., Wu, W.C. and Hou, W.C. 2008. Antioxidant, anti-semicarbazide-sensitive amine oxidase, and anti-hypertensive activities of geraniin isolated from *Phyllanthus urinaria*. Food Chem. Toxicol. **46**: 2485–2492.

Mohapatra, A. and Rout, G.R. 2005. Identification and analysis of genetic variation among rose cultivars using Random Amplified Polymorphic DNA. Zeitschrift fur Naturforschung **60C**: 611-617.

Oluwafemi, F. and Debiri, F. 2008. Antimicrobial effect of *Phyllanthus amarus* and *Parquetina nigrescens* on *Salmonella typhi.* Afr. J. Biomed. Res. **11**: 215-219.

Praveen, B., Chandrasekar, M.J.N., Nanjan, M.J., Satish Kumar, M., Nataraj, N. and Suresh, B. 2008. *Phyllanthus maderaspatensis*, a dietary supplement for the amelioration of adriamycin-induced toxicity and oxidative stress in mice. J. Nat. Med. **62**: 149-154.

Rout, G.R., Sunil, K., Senapati, S. and Aparajita, S. 2010. Study of relationships among twelve *Phyllanthus* species with the use of molecular markers. Czech J. Genet. Plant breed. **46**: 135-141.

Srirama, R., Senthilkumar, U. and Sreejayan, N. 2010. Assessing species admixtures in raw drug trade of *Phyllanthus*, a hepatoprotective plant using molecular tools. J. Ethnopharmacol. **130**: 208-215.

Theerakulpisut, P., Kanawapee, N., Maensiri, D., Bunnag, S. and Chantaranothai, P. 2008. Development of species-specific SCAR markers for identification of three medicinal species of *Phyllanthus.* J. Systematic and Evolution **46 (4)**: 614-621.

Verma, K.C. and Verma, S.K. 2010. Analysis of active ingredients and SDS-PAGE profiling of Amla (*Phyllanthus emblica* L.). Agricultural science digest **30**: 29-31.

Chapter 10

Chromatographic Studies on some Selected *Selaginella* Species

A. Sivaraman and M. Johnson*

Centre for Plant Biotechnology,
Department of Plant Biology and Plant Biotechnology,
St. Xavier's College (Autonomous), Palayamkottai – 627 002, Tamil Nadu

ABSTRACT

The present study was aimed to reveal the phytoprofile of four *Selaginella* species using TLC and HPLC. The TLC phenolic and steroids profile of *Selaginella involvens* (Sw.) Spring, *Selaginella intermedia* (Bl.) Spring, *Selaginella inaequalifolia* (Hook. & Grev.) Spring and *Selaginella tenera* (Hook. & Grev.) Spring was detected using the specific mobile phase and spraying agents. The appearance of blue colour and bluish green colour spot in the TLC chromatogram indicated the presence of phenolics and steroids in the crude extracts of selected *Selaginella* species. HPLC analysis was performed on the methanolic extracts of selected *Selaginella* species and the results showed various peaks at different retention times. The developed TLC and HPLC fingerprints will help the manufacturer for quality control and standardization of herbal formulations. Such finger printing is useful in differentiating the medicinally important species from its adulterant and act as a biochemical marker for this medicinally important plant in the pharmaceutical industry and plant systematic studies.

Keywords: *Selaginella, TLC, HPLC, Phytochemistry.*

---

* Corresponding Author: E-mail: ptcjohnson@gmail.com

Introduction

Medicinal plants played a significant role to cure human diseases and disorders in the traditional medicine since time immemorial (Chandarana *et al.*, 2005). Based on the presence of one or more active constituents, the medicinal properties of the plants are authorised (Egwaikhide and Gimba, 2007). The ethnomedicinal plants have been used as driving force for development of novel drugs in the pharmaceutics (Robbers *et al.*, 1996). Due to importance and bioefficacy of herbal medicine, the growing demand for herbal product has increased in volume of plant materials traded within and across the countries (Suresh *et al.*, 2008). At present, the botanists focused their research on phytoconstituents of medicinal plants and examined secondary metabolites as a source of medicinal agents. The success story of chemotherapy lies in the continuous search of new drugs to counter the challenges possessed by resistant strains of microorganisms (Doughari *et al.*, 2008).

There are about 1200 species of Pteridophytes in India. Like flowering plants, pteridophytes are also used by tribals in India and other countries. *Selaginella,* with about 700 species, is world-wide in distribution, but most of the species are tropical and many of them grow in damp forests (Pichi-Sermolli, 1977). Alston (1945) enumerated 58 species of *Selaginella* from India and Sri Lanka. Of these, 45 species have been reported from India. According to Dixit (1984), 18 species are in South India (Tamil Nadu and Kerala) and 12 species have been reported from the Western Ghats, South India (Manickam and Irudayaraj, 1992). Considerable numbers of pteridophytes are used in Chinese medicines. The Chinese herb *Selaginella doederleinii* is used as an anticancer drug (Lee and Lin, 1988; Huang, 1992). The medicinal use of *Selaginella* in India goes back to the period of 'Ramayana' (Andel *et al.*, 2007). According to the principles of traditional Chinese medicine, *Selaginella* has sweet, spicy, bitter and cold properties, and is associated with the liver, lung and stomach meridians. Its main functions are to clear heat, reduce toxicity, and drain damp heat. More recent research has shown that *Selaginella* may be effective against both acute and chronic hepatitis. *Selaginella* may also be used externally to help stop bleeding and promote wound healing (Sah, 2008). Prepared *Selaginella* can be found in pill, powder and tablet forms. *Selaginella* should be taken with caution by patients diagnosed with cold deficiency. Taking large amounts of *Selaginella* may result in loss of appetite and abdominal discomfort. There are no known drug interactions associated with *Selaginella* (Lee *et al.*, 1999).

The morphological characters play a vital role in plant systematic studies and are used as a tool for the classification of a taxon. In recent times, in addition to morphological markers, anatomical, cytological, biochemical and molecular markers are also being used to classify the organisms. Correct identification and quality assurance of the starting material is important, therefore, an essential prerequisite to ensure reproducible quality of herbal medicine, which contributes to its safety and efficacy (Goldman, 2001). Choice of chemical markers is crucial step for the quality control of plant based medicines (Mukherjee *et al.*, 2006). Chemical fingerprinting has been demonstrated to be a powerful technique for the quality control of herbal medicines (Bhutani, 2000). According to regulatory guidelines and pharmacopoeias macroscopic and microscopic evaluation and chemical profiling of the botanical materials is used for quality control and standardization (IHP, 2002; BHP, 1996). Thin layer chromatography (TLC), High Performance Liquid Chromatography and Spectrophotometric methods are frequently used for quality control and standardization of plant based drugs (WHO, 1998). The chromatogram derived using HPLC, HPTLC and GC analysis, Rf values and Retention time obtained through the chromatographic analysis are used as a phytochemical marker to characterize the medicinal plants and identify the chemical constituents present in the medicinal plants. Thin layer chromatography (TLC) is very important technique for qualitative study in both small and large scale and therefore widely-employed laboratory technique. A number of studies were reported to characterize the plants using TLC and HPLC profiles (Janakiraman *et al.*, 2012; Johnson *et al.*, 2012). Only few reports are available for pteridophytes (Talukdar *et al.*, 2010; Irudayaraj and Johnson, 2011). But there is no report on the chemical characterization of *Selaginella* using TLC and HPLC profile. With this knowledge, the present study was aimed to reveal the phytoprofile of four *Selaginella* species using TLC and HPLC.

Materials and Methods

Plant Collection

Healthy, disease free fern allies of *Selaginella involvens* (Sw.) Spring, *Selaginella intermedia* (Bl.) Spring, *Selaginella inaequalifolia* (Hook. & Grev.) Spring and *Selaginella tenera* (Hook. & Grev.) Spring were collected from Kakachi stream, Tirunelveli Hills (1000 m) in the Western Ghats. The fern allies were identified by Dr. V. Irudayaraj, St. Xavier's College, Palayamkottai, Tamil Nadu, India.

Plant Extraction

The fresh materials were washed in running tap water for 5 min and dried using blotting papers. The washed plant materials were air and shade dried for two weeks and pulverized to powder using tissue blender. The powdered materials (25 g) were successively extracted using Soxhlet extractor with 150 mL of various solvents *viz.*, petroleum ether, chloroform, acetone and methanol for 8 h at a temperature not exceeding the boiling point of the solvent. The extracts were filtered using Whattman filter paper (No.1) and then concentrated in vacuum at 40°C using rotary evaporator. The residues obtained were stored in a freezer -70°C until further tests.

TLC Analysis

The phenolic and steroids profile of *S. involvens, S. intermedia, S. inaequalifolia* and *S. tenera* was detected by TLC. For phenolics, chloroform and methanol at 9: 1 ratio was used as mobile phase and benzene and methanol at 9: 1 ratio for steroids. The folin-ciocalteau reagent and 5 per cent alcoholic sulphuric acid was used as spraying agent for phenolics and steroids respectively. After spraying with specific spraying agents, the appearance of blue colour and bluish green colour spot in the TLC chromatogram indicated the presence of phenolics and steroids in the crude extracts of *S. involvens, S. intermedia, S. inaequalifolia* and *S. tenera*.

HPLC Analysis

Based on the preliminary results, methanolic extracts of *S. involvens, S. intermedia, S. inaequalifolia* and *S. tenera* were taken for the HPLC analysis. The extracts were centrifuged at 3000 rpm for 10 min and then filtered through Whatmann No.1 filter paper using high pressure vacuum pump. The sample was diluted to 1: 10 with the same solvents. HPLC method was performed on a Shimadzu LC-10AT VP HPLC system, equipped with a model LC-10AT pump, UV-Vis detector SPD-10AT, Rheodyne injector fitted with a 20 µL loop and auto injector SIL-10AT. A Hypersil BDS C-18 column (4.6 × 250 mm, 5 µm size) with a C-18 guard column was used. The elution was carried out with gradient solvent systems with a flow rate of 1 mL min-1 at ambient temperature (25-28°C). The mobile phase was consisted of 0.1 per cent v / v methanol (solvent A) and water (solvent B). The mobile phase was prepared daily, filtered through a 0.45 µm and sonicated before use. Total running time was 15 min. The sample injection volume was 20 µL while the wavelength of the UV-Vis detector was set at 254 nm.

Results

Phenolics profile present in the various crude extracts of *S. involvens, S. intermedia, S. inaequalifolia* and *S. tenera* was revealed by the TLC and the Rf values of the phenolics profile of various extracts was tabulated in Table 10.1. Of which, acetone extracts of *S. involvens* showed 8 phenolic spots, chloroform extracts of *S. intermedia* and *S. inaequalifolia* illustrated 7 and 12 phenolic bands respectively. 8 phenolic spots were detected in petroleum ether extracts of *S. tenera.* The steroids profile present in the various extracts of different *Selaginella* species was tabulated in Table 10.2. Among these, chloroform extracts of *S. involvens* showed 8 steroids spots, chloroform extracts of *S. intermedia* showed 8 steroids spots, and *S. inaequalifolia* illustrated 9 steroids spots in acetone extracts. 10 steroids spots were detected in petroleum ether extracts of *S. tenera.*

The qualitative HPLC fingerprint profile of methanolic extracts of *S. involvens, S. intermedia, S. inaequalifolia* and *S. tenera* were selected based on the sharpness of the peaks and proper baseline. Methanolic extract prepared by hot extraction was subjected to HPLC for the separation and identification of constituents present in various *Selaginella* species. In *S. involvens,* seven peaks were observed and the profile displayed one prominent peak at a retention time of 2.933 min and some moderated peaks at different retention time 1.637, 2.553, 3.433, 3.793, 4.043 and 4.673 min respectively (Figure 10.1). Methanolic extracts of *S. intermedia* showed the presence of three peaks with one prominent peak at a retention time of 3.043 min and two moderate peaks were also observed at a retention time 3.437 and 4.030 min respectively (Figure 10.2). In *S. inaequalifolia* also, three peaks were obtained, of which, one prominent peak at a retention time of 3.053 min and two moderate peaks at different retention time (3.467 and 4.043 min) was observed (Figure 10.3). *S. tenera* illustrated the presence of three peaks with one prominent peak at a retention time of 3.190 min and two moderate peaks were also observed at a retention time 3.540 and 3.677 min respectively (Figure 10.4).

Table 10.1: Phenolics TLC Profile of different *Selaginella* sps.

Rf	*S. involvens*				*S. intermedia*				*S. inaequalifolia*				*S. tenera*			
	P	*C*	*A*	*M*	*P*	*C*	*A*	*M*	*P*	*C*	*A*	*M*	*P*	*C*	*A*	*M*
0.06	–	–	–	–	–	–	–	–	–	–	–	–	–	–	–	+
0.09	–	+	–	–	–	–	–	–	–	–	–	–	–	–	–	–
0.10	–	–	–	–	–	–	–	–	–	–	+	–	–	–	–	–
0.12	–	–	+	–	–	–	–	–	–	–	–	–	–	–	–	–
0.14	–	–	–	–	–	+	+	–	–	+	–	–	–	–	–	–
0.19	–	–	–	–	–	–	–	–	–	–	+	+	–	–	–	–
0.20	–	–	+	–	–	–	–	–	–	–	–	–	–	–	–	–
0.21	–	–	–	–	–	–	–	–	–	–	–	–	–	–	–	+
0.25	–	–	–	–	–	+	–	–	–	–	+	–	–	–	–	–
0.29	–	–	–	–	–	–	–	–	–	+	–	–	–	–	–	–
0.30	–	–	+	–	–	–	–	–	–	–	–	–	–	–	+	–
0.32	–	–	–	–	–	+	–	–	–	–	–	+	–	–	–	–
0.36	–	–	–	–	–	+	–	–	–	–	+	–	–	–	–	–
0.38	–	–	–	–	–	–	–	–	–	+	–	–	–	–	+	–
0.39	+	–	–	–	–	–	–	–	–	+	–	–	–	–	–	–
0.45	–	–	–	–	–	–	+	+	+	–	–	–	+	+	–	–
0.52	–	–	+	–	–	–	–	–	–	+	+	–	+	–	–	–
0.55	–	–	–	–	–	–	–	–	+	+	+	–	–	–	+	–
0.57	–	–	–	–	+	–	–	–	+	+	–	–	+	–	–	–
0.59	–	–	–	–	–	–	–	–	–	–	–	–	–	+	–	+
0.60	–	+	+	+	–	–	–	–	–	–	–	+	–	–	–	–
0.62	–	–	–	–	+	–	–	–	–	–	–	–	–	–	–	–
0.64	–	–	–	–	–	+	–	–	–	+	–	–	–	–	–	–
0.66	+	+	–	–	–	–	–	–	–	–	+	–	–	–	–	–
0.67	–	–	–	–	–	–	–	–	+	–	+	–	+	–	–	–
0.70	+	–	–	–	–	–	–	+	–	+	–	–	–	–	–	–
0.72	–	–	–	–	–	–	–	–	–	–	–	–	–	+	–	–
0.75	–	–	–	–	+	+	–	–	–	–	–	–	+	–	–	–
0.76	–	–	+	–	–	–	–	–	–	–	–	–	–	–	–	–
0.77	–	+	–	–	+	–	–	–	–	–	–	–	–	–	–	–
0.79	–	–	–	–	–	–	–	–	–	+	–	–	–	–	–	–
0.80	+	–	–	–	–	–	–	–	–	–	+	+	+	–	–	–
0.82	–	–	–	–	–	–	–	–	–	+	–	–	–	–	–	–
0.84	–	–	+	–	–	–	–	–	–	–	–	–	–	–	–	–
0.86	–	–	–	–	–	–	–	–	–	–	+	–	+	–	–	–

Contd...

Table 10.1–*Contd...*

Rf	*S. involvens*				*S. intermedia*				*S. inaequalifolia*				*S. tenera*			
	P	*C*	*A*	*M*	*P*	*C*	*A*	*M*	*P*	*C*	*A*	*M*	*P*	*C*	*A*	*M*
0.88	–	–	–	–	–	–	–	–	–	+	–	–	–	–	–	–
0.90	+	–	+	–	–	+	–	–	–	–	–	–	+	+	–	–
0.93	–	–	–	–	–	–	–	–	–	–	–	–	–	–	–	+
0.95	+	–	–	–	–	–	+	–	–	–	+	–	–	–	+	–
0.97	–	–	–	–	–	–	–	+	–	–	–	–	–	–	–	–

P: Petroleum ether; C: Chloroform; A: Acetone; M: Methanol.

Table 10.2: Steroids TLC Profile of different *Selaginella* sps.

Rf	*S. involvens*				*S. intermedia*				*S. inaequalifolia*				*S. tenera*			
	P	*C*	*A*	*M*	*P*	*C*	*A*	*M*	*P*	*C*	*A*	*M*	*P*	*C*	*A*	*M*
0.02	–	–	–	–	–	–	–	–	–	–	+	–	–	–	+	–
0.06	–	–	+	–	–	–	–	–	–	–	–	–	–	+	–	–
0.08	+	–	–	–	–	–	–	–	–	–	–	–	–	–	–	–
0.10	–	+	–	–	–	–	–	–	+	–	–	–	–	–	–	–
0.11	–	–	–	–	–	+	+	–	–	–	–	–	–	–	–	–
0.12	–	–	–	–	–	–	–	–	–	–	–	–	+	–	–	–
0.14	–	+	–	–	–	–	–	–	–	–	–	–	+	–	–	–
0.15	–	–	–	–	–	+	–	–	–	–	–	–	–	–	–	–
0.16	+	–	–	–	–	–	–	–	–	–	–	–	–	+	–	+
0.18	–	–	–	–	+	–	–	–	–	–	–	–	–	–	–	–
0.20	–	–	–	–	–	+	–	–	+	–	+	+	+	–	–	–
0.21	–	+	–	–	–	–	–	–	–	–	–	–	–	–	–	–
0.22	–	–	–	–	–	–	–	–	–	–	+	–	–	–	–	–
0.25	+	–	–	–	+	–	–	–	–	–	–	–	–	–	–	–
0.26	–	–	–	–	–	–	–	–	–	–	+	–	–	–	–	–
0.29	–	+	–	–	–	–	–	–	–	–	–	–	–	–	–	–
0.30	–	–	+	–	–	–	–	–	–	–	–	–	–	–	–	–
0.33	–	–	–	–	–	+	–	–	–	–	+	–	–	–	–	–
0.34	–	–	+	–	+	–	–	–	–	–	–	–	–	–	–	–
0.36	–	–	–	+	–	–	–	–	–	–	–	–	–	–	–	–
0.37	–	–	–	–	–	+	–	–	–	–	–	–	–	–	–	–
0.40	–	–	+	–	–	+	–	+	+	–	+	–	–	+	–	–
0.41	+	–	–	–	–	–	–	–	–	–	–	–	–	–	–	–
0.46	–	–	–	–	–	–	–	–	–	+	+	–	–	+	–	–

Contd...

Table 10.2–*Contd...*

Rf	*S. involvens*				*S. intermedia*				*S. inaequalifolia*				*S. tenera*			
	P	*C*	*A*	*M*	*P*	*C*	*A*	*M*	*P*	*C*	*A*	*M*	*P*	*C*	*A*	*M*
0.50	–	+	+	–	+	–	–	–	–	–	–	–	+	–	–	–
0.54	–	–	–	–	–	–	–	–	–	–	–	–	+	–	–	–
0.55	–	–	–	–	–	+	–	–	–	–	+	–	–	–	–	–
0.56	–	–	–	–	–	–	–	+	+	–	–	–	–	–	–	–
0.58	–	+	+	–	+	–	+	–	–	+	–	–	+	+	–	–
0.60	–	–	–	–	–	–	–	–	+	–	–	–	+	+	–	–
0.66	+	–	–	–	–	–	–	–	+	–	–	–	–	–	–	–
0.69	–	–	–	–	–	–	+	–	–	–	–	–	–	–	–	–
0.70	–	–	–	–	+	–	–	–	–	–	–	–	+	–	–	–
0.72	–	+	–	–	–	–	–	–	–	–	–	–	–	+	–	–
0.74	–	–	+	–	–	–	–	–	–	–	–	–	–	–	–	–
0.77	–	–	–	–	–	+	–	–	–	–	+	–	–	–	–	–
0.80	–	–	–	–	+	–	–	–	+	–	–	–	+	+	–	–
0.83	–	+	–	–	–	–	–	–	–	–	–	–	–	–	–	–
0.88	+	–	–	+	–	–	–	–	–	–	–	–	–	–	–	–
0.90	–	–	–	–	–	–	–	–	+	–	–	–	+	+	–	
0.92	–	–	–	–	–	–	–	–	–	–	–	–	–	–	+	+
0.97	–	–	–	–	–	–	–	–	–	+	–	–	–	–	–	–

P: Petroleum ether; C: Chloroform; A: Acetone; M: Methanol.

Discussion

TLC studies employed the Rf values to distinguish the plants from other species and adulterant. In the present study also, we developed the phenolics and steroids TLC profile for various extracts of *S. involvens, S. intermedia, S. inaequalifolia* and *S. tenera*. This provides valuable information in authenticating the genuine mother plants along with the nature of phytoconstituents present in it. HPLC identification test are required to confirm the presence of the active constituents and potential adulterant in ayurvedic drugs. In the present study, HPLC profile for methanolic extracts of *S. involvens, S. intermedia, S. inaequalifolia* and *S. tenera* were evolved. Standardization is an important aspect for establishing the quality and efficacy of ayurvedic formulations or any multiple ingredient herbal formulations. Generally, two approaches being used for standardization are fingerprint analysis by HPLC/HPTLC and quantification of individual chemical markers. It ensures reproducible pharmaceutical quality of herbal products. Preliminary analysis of the methanolic extracts of various *Selaginella* species showed varying patterns in HPLC chromatogram suggesting that there might be differences in the process of manufacturing and therefore a proper scientific validation (chromatographic fingerprinting of major constituents) is needed for quality control purposes. These studies exhibited novel markers in

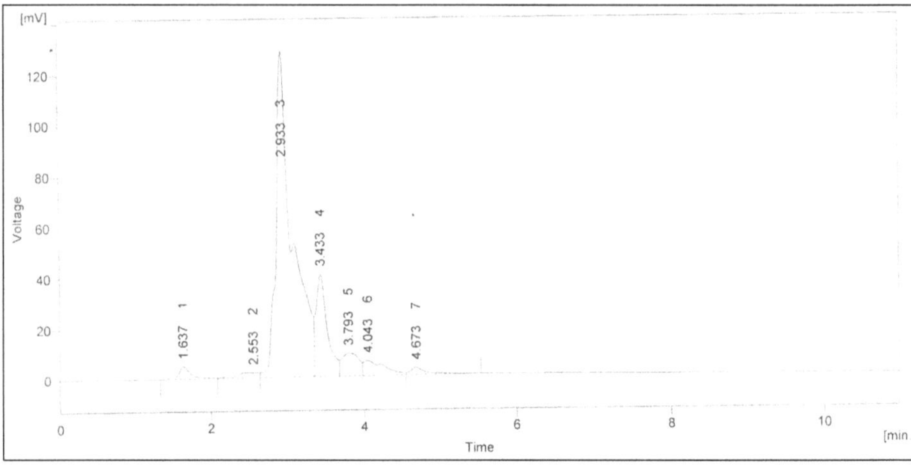

Figure 10.1: HPLC Chromatogram of *Selaginella involvens* (Sw.) Spring.

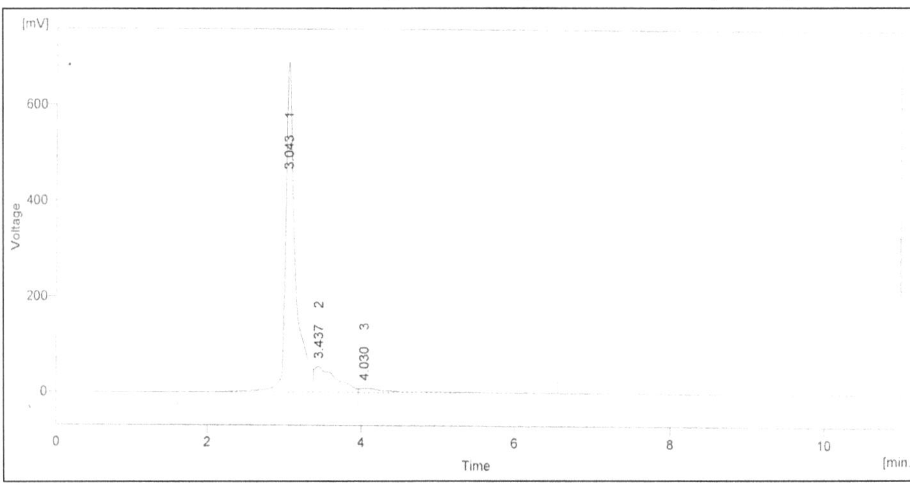

Figure 10.2: HPLC Chromatogram of *Selaginella intermedia* (Bl.) Spring.

standardization as useful analytical tools to check not only the quality of the powder but also the presence of adulterants in ayurvedic drugs.

Chemosystematics is the attempt to classify and identify organisms, according to demonstrable differences and similarities in their biochemical compositions. To date, the coupling of morphological characters with other biochemical/phytochemical data has yielded valuable insight into underlying biochemical processes and has contributed to numerous advances in the area of chemotaxonomy. Many modern techniques have been employed to characterize and identify metabolites present in plants and also distinguish the medicinal plants from its adulterants. Furthermore, it can hardly provide any information for unknown samples. All these reasons have prompted the hyphenation of high performance liquid chromatography (HPLC).

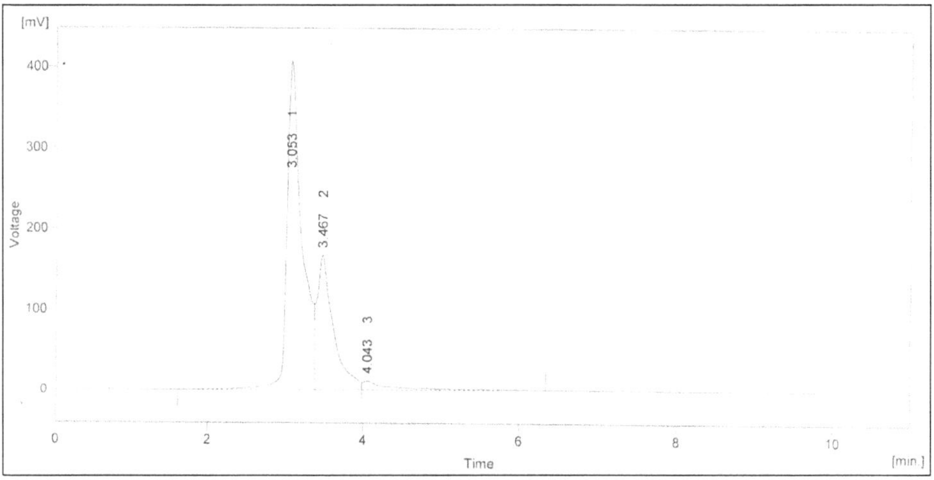

Figure 10.3: HPLC Chromatogram of *Selaginella inaequalifolia* (Hook. & Grev.) Spring.

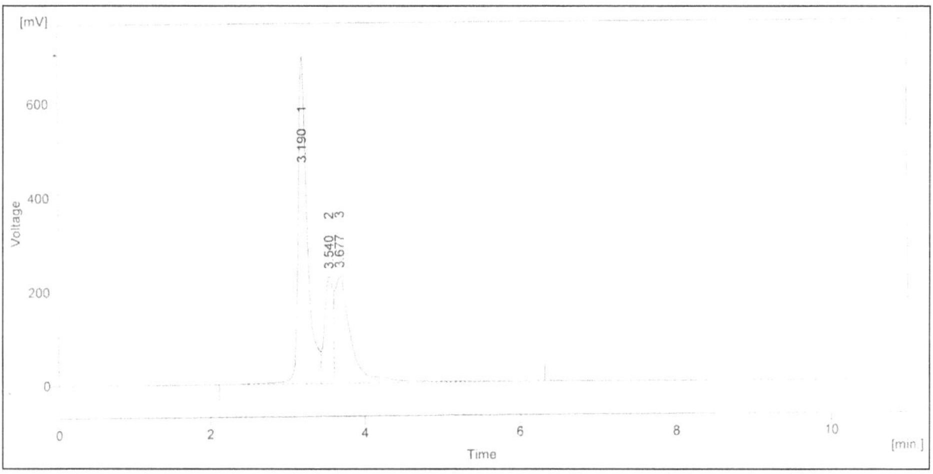

Figure 10.4: HPLC Chromatogram of *Selaginella tenera* (Hook. & Grev.) Spring.

Conclusion

The developed TLC and HPLC fingerprints will help the manufacturer for quality control and standardization of herbal formulations. Such finger printing is useful in differentiating the medicinally important species from its adulterant and act as a biochemical marker for this medicinally important plant in the pharmaceutical industry and plant systematic studies.

Acknowledgement

The authors are thankful to St.Xavier's College management for providing infrastructure, constant support and encouragements.

Financial Support

The authors are sincerely acknowledging the financial assistance provided by the University Grants Commission through College with Potential for Excellence (Ref. No. 16-44/2004/2010 (NS/PE) dated 22. 12. 2010.

References

Alston, A.H.G. 1945. An enumeration of the Indian species of *Selaginella*. Proc. Natl. Inst. Sci. India **11B:** 211-235.

Andel, T.R. van, Behari-Ramdas, J., Havinga, R.M. and Groenendijk, S. 2007. The medicinal plant Trade in Suriname. Ethnobotany Research and Applications. **5:** 351-372.

Bhutani, K.K. 2000. Finger-Printing of Ayurvedic Drugs. The Eastern Pharmacist **2:** 21-26.

British Herbal Pharmacopoeia. 1996. British Herbal Medicine Association.

Chandarana, H., Baluja, S. and Chanda, S.V. 2005. Comparison of antibacterial activities of selected species of Zingiberaceae family and some synthetic compounds. Turk. J. Biol. **29:** 83-97.

Dixit, R.D. 1984. A census of Indian Pteridophytes. Howrah: Botanical Survey of India, Dept. Environment, pp. 1-177.

Doughari, J.H., El-mahmood A.M and Tyoyina, I. 2008. Antimicrobial activity of leaf extracts of *Senna obtusifolia* (L). African J. Pharm. Pharmacol. **2 (1):** 007-013.

Egwaikhide, P.A. and Gimba, C.E. 2007. Analysis of the phytochemical content and anti-microbial activity of *Plectranthus glandulosis* whole plant. Middle East J. Scientific Res. **2 (3-4):** 135-138.

Goldman, P. 2001. Herbal Medicines Today and the Roots of Modern Pharmacology. Annals of Internal Medicine **148**: 594-600.

Huang, K.C. 1992. The Pharmacology of Chinese herbs. CRC publishers, pp. 388.

Indian Herbal Pharmacopoeia. 2002. Indian Drug Manufacturers' Association, Mumbai.

Irudayaraj, V. and Johnson, M. 2011. Pharmacognostical studies on three *Asplenium* species. J Phytol. **3 (10):** 1-9.

Janakiraman, N., Jasmin Jansi, J., Johnson, M., Jeeva, S. and Babu, A. 2012. Phytochemical Investigation of *Abrus precatorius* L. using TLC, GCMS and FTIR. Asian Pacific Journal of Tropical Biomedicine (In press); 1-9.

Johnson, M., Babu, A., Janakiraman, N. and Renisheya Joy Jeba Malar, T. 2012. Phytochemical studies on *Laurencia obtusa* (Hudson) Lamourux. Intern. J. Biomed. Advance Res. **3 (4):** 225-232.

Lee, H. and Lin, J.Y. 1988. Antimutagenic activity of extracts from anticancer drugs in Chinese medicine. Mutat. Res. **204 (2):** 229-234.

Lee, I., Nishikawa, S., Furukawa, A., Kasahara, F.K. and Kim S.U. 1999. Effects of *Selaginella tamariscina* on *in vitro* tumor cell growth, p53 expression, G_1 arrest and *in vivo* gastric cell proliferation. Cancer Lett. **44 (1):** 93-99.

Manickam, V.S. and Irudayaraj, V. 1992. Pteridophyte Flora of the Western Ghats, South India. B.I. Publications, Pvt. Ltd. New Delhi.

Mukherjee, P.K., Wahile, A., Kumar, V., Rai, S., Mukherjee, K.and Saha, B.P. 2006. Marker profiling of botanicals used for hepatoprotection in Indian system of medicine. Drug Information J. **40:** 131-139.

Pichi-Sermolli, R.*E.G.* 1977. Tentamen Pteridophytorum genera in Taxonomicum ordinem reoigendi. Webbia **31:** 313-512.

Robbers, J., Speedie, M. and Tylor, V. 1996. Pharmacognosy and Pharmacobiotechnology. Williams and Wilkins, Baltimore, pp. 1-14.

Sah, P. 2008. Does the Magical Himalayan Herb "Sanjeevani Booti" really exist in Nature? J. Amer. Sci. **4 (3):** 65-67.

Suresh, K., Deepa, P., Harisaranraj, R. and Vaira Achudhan, V. 2008. Antimicrobial and phytochemical investigation of the leaves of *Carica papaya* L., *Cynodon dactylon* (L.) Pers., *Euphorbia hirta* L., *Melia azedarach* L. and *Psidium guajava* L. Ethnobotanical Leaflets **12:** 1184-91.

Talukdar, A.D., Choudhury, M.D., Chakraborty, M. and Dutta, B.K. 2010. Phytochemical screening and TLC profiling of plant extracts of *Cyathea gigantea* (Wall. ex Hook.) Holtt. and *Cyathea brunoniana* (Wall. ex Hook.) Cl. & Bak. Assam University Journal of Science and Technology: Biological and Environmental Sciences **5 (1):** 70-74.

WHO. 1998. Quality Control Methods for Medicinal Plant Materials, Geneva.

Chapter 11

Ecology and Distribution of the Genus *Lepisorus* (Polypodiaceae: Pteridophyta) from South India with the Incidence of Apogamous Populations from the Nilgiris

***P. Murugan*[1], *V. Irudayaraj*[1] *and S. Jeeva*[2]**

[1]*Department of Botany, St. Xavier's College (Autonomous), Palayamkottai – 627 002, Tamil Nadu*

[2]*Department of Botany, Scott Christian College (Autonomous), Nagercoil – 629 003, Tamil Nadu*

ABSTRACT

Ecology and distribution of the species of the polypodioid genus *Lepisorus* from South India were studied, based on the herbarium specimens housed in XCH (St. Xavier's College Herbarium, Palaymkottai). Two species of *Lepisorus* {*L. amaurolepidus* (Sledge) Bir & Trikha, *L. nudus* (Hook) Ching} are commonly distributed as epiphytes or lithophytes inside the evergreen forests at high altitude on the Eastern Ghats and on the Western Ghats. Both the species are small plants with creeping rhizome, simple monomorphic fronds with exindusiate sori. Both the species are differentiated by the difference in rhizome scale. Based on spore count, two populations from the Nilgiris are proved to be of apogamous.

Keywords: *Ecology, Distribution, Lepisorus, South India, Apogamous.*

Introduction

Pteridophytes are primitive tracheophytes and non-flowering vascular cryptogams with conducting tissues xylem and phloem. These plants produce definite embryos during the life cycle, but they are not developed within definite seeds. Hence, Pteridophytes are regarded as seedless vascular plants or seedless embryophytes which include club-mosses, spike-mosses, horse-tails and ferns. There are about 12,000 living species of ferns and 1000 species of fern allies in the world.

India's biological diversity is one of the most significant in the world, since India has only 2 per cent of the total land mass of the world containing about 6 per cent of the worlds and with ca. 10 per cent of world's pteridophytes. The Pteridophytes constitute an important component of Indian flora and they are rich in high altitude mountains like Himalayas, Western Ghats and Eastern Ghats. About 1,100 species of Pteridophytes have been reported from India. Most of them inhabit a variety of niches in the mountainous regions particularly the Himalayas and Western Ghats. The Western Ghats of South India is one of the richest centres of the fern floras of the world. It is observed that most of the pteridophytes prefer moist tropical, sub-tropical, temperate, sub-alpine types of climatic conditions to grow; some also prefer dry tropical, alpine type of climate. The Pteridophytes are very important component of ground flora of the temperate forest. They are most conspicuous among ground flora and epiphytes and definite associations with Phanerogamic flora in the altitudinal zone between 1500-2700m.

Polypodiaceae is an advanced family of Pteridophytes which are dominant in evergreen forest as epiphytic plants along with orchids. Various studies revealed that the vascular epiphytes communities are dominated by the division of Pteridophyta especially the two families of ferns: Aspleniaceae and Polypodiaceae. They were the commonest species that can be found in the three different categories of host trees. *Asplenium* species have been noted for their abundance in tropical forest throughout the world (Ellwood and Foster, 2004). Polypodiaceae was also the most diversified family in the tropical rain forest (Zapfack *et al.*, 1996). Ferns, especially *Pyrrosia lingua* (Thunb.) Farw. (Polypodiaceae), are most abundant (40 per cent) in the outer canopy of subtropical rain forest in north- eastern Taiwan (Hsu *et al.*, 2002). Out of 15 vascular epiphytes, eleven are pteridophytes at Pusat Sejadi, Kawang Forest Reserve, Sabah Malaysia with the abundance of Aspleniaceaeous and Polypodiaceous pteridophytes (Mojiol *et al.*, 2009). According to Dixit (1984), the different studies carried out on the diversity of pteridophytes throughout the country reveal that on all India basis, maximum number of 150 species and seven varieties occur in the family Polypodiaceae.

After Beddome (1864), Manickam and Irudayaraj (1992) explored the South Indian Hill areas and published the book *"Pteridophyte Flora of the Western Ghats, South India"* with 20 species of polypodioid ferns under 12 genera. Although it is of a comprehensive account on both ferns and fern allies of South India with recent nomenclature, they have given only the distributional pattern of each species in South India without citing any voucher specimen. In 1998 Nampy and Madhusoodanan published the book on *"Fern Flora of South India: Taxonomic revision of Polypodioid Ferns"* with 25 species under 11 genera with voucher specimens mainly from Calicut University Herbarium and Madras Herbarium. Although enormous number of specimens for each South Indian pteridophyte is housed in XCH (St. Xavier's College Herbarium, Palayamkottai) they were not consulted/observed/ included by Nampy and Madhusoodanan (1998). Manickam and Irudayaraj (2003) reported 15 polypodioid ferns from the Nilgiris in their book *"Pteridophyte Flora of the Nilgiris, South India"* by citing original vouchers for each species. So far, complete account on materials of South Indian pteridophytes is not available for the scientific community even for a single taxonomic family.

Although the polypodioid genus *Lepisorus* is with only two species (*L. nudus* and *L. amaurolepidus*), it is one of the dominant epiphytic fern genus in the evergreen forests of the Western Ghats, South India. Plants of *Lepisorus* species are usually small growing lithophytically or epiphytically at high altitude inside the forests. Long or short creeping rhizome with lateral branches bears simple, monomorphic fronds with articulated stipe. Veins anastomose with included veinlets in most areoles. Sori are round, on both sides of the costa, exindusiate. Soral paraphyses are peltate, conspicuous in young sori.

The taxonomic revision of *Lepisorus* (J. Sm.) Ching sect. *Lepisorus* in China based on herbarium specimen examinations, field observations, and microscopic study of rhizome scales, soral paraphyses, leaf epidermis, and spores revealed the presence of nine species including *Lepisorus nudus* (Hook.)Ching. According to them the rhizome scales and paraphyses are the most useful characters for species delimitation as well as for infrageneric classification. Characteristics of the leaf epidermis and spore ornamentation are usually stable and thus of great significance in understanding the relationships among groups within the genus (Qi and Zhang, 2009).

Investigation on phylogenetic relationships within the paleotropical genus *Lepisorus* (Polypodiaceae) using plastid DNA sequences from four regions: *rbcL*, *rps4* and *rps4-trnS* IGS, *trnL* intron plus *trnL-F* IGS, *rbcL-atpB* IGS by sequencing over 4000 nucleotides for 77 specimens belonging to 54 species revealed the paraphyly of *Lepisorus* with the monotypic *Drymotaenium miyoshianum* and of the paleotropical genus *Belvisia* nested within the *Lepisorus* clade. The phylogenetic results provided new evidence for the sectional classification of *Lepisorus*. The evolution of three morphological characters, clathrateness of rhizome scales, margin of rhizome scales and defoliated leaves, and the evolution of the karyotype, has been reconstructed to identify lineage specific phenotypic character states or combination of characters. Unique character combinations, rather than synapomorphies, were found to be of systematic value in sectional delimitation. It has also been mentioned that the variation of chromosome numbers is largely due to a single aneuploidy event instead of a stepwise reduction during the evolutionary history of this genus (Wang *et al.*, 2010).

According to Kholia *et al.* (2012), the species of the genus *Lepisorus* (J. Sm.) Ching Kumaon Himalaya can be grouped into two categories, based on the studies on ecology: **i.** species with thin, membranaceous seasonal lamina which abscise before winter months, **ii.** in the others, the lamina is thick, coriaceous and remain attached on the rhizome for more than one year.

In the present study, it has been aimed to enumerate the specimens of the common South Indian polypodioid species of the genus *Lepisorus* by citing all the vouchers housed in XCH (St. Xavier's College Herbarium). The literature survey shows that *Lepisorus nudus* is either diploid or tetraploid sexual species. In order to know the presence of any apogamous population in South India, spore count was made on selected specimens.

Materials and Methods

The present study is mainly based on the herbarium specimens housed in St. Xavier's College Herbarium (XCH), Palayamkottai, India, along with field studies on few localities of the Western Ghats, South India. For spore count, the undehisced sporangia were dissected out from the mature sori of the selected specimens. For each specimen the details about the collector, date, locality and habitat-ecology were enumerated. They were mounted in glycerin. Number of spores within a sporangium was counted by observing under microscope after giving a gentle press on the cover glass with a needle to release the spores from the sporangium.

Results and Discussion

Lepisorus amaurolepidus (Sledge) Bir & Trikha

It is a small epiphytic or lithophytic plant with short creeping scaly rhizome bearing simple fronds of about 3-5 inches. Fronds are elliptic-lanceolate, simple with two rows of exindusiate sori on either side of the midrib (Plate 11.1).

Thirty eight specimens were observed from St. Xavier's College Herbarium (Table 11.1). It is very common on Tirunelveli Hills, occasional on Palni Hills, Kerala Ghats and rare on Anamalais. Nampy and Madhusoodanan (1998) have cited 10 specimens from Kerala and four specimens from Tamil Nadu.

Lepisorus nudus (Hook.) Ching

It is also a small plant morphologically similar to *L. amaurolepidus*, except the difference in rhizome scale. Ninety five specimens were observed in St. Xavier's College Herbarium. It is more common than *L. amaurolepidus.* It is common throughout South India (Kerala, Tamil Nadu, Karnataka) both on Eastern Ghats and Western Ghats (Plate 11.1, Table 11.2). It has also been collected from Kodaikanal, Palni Hills in the present study. It is common in Kodaikanal.

Cytological reports on this species from India shows that only diploid sexual populations are present in Eastern and Western Himalayas and in Eastern Ghats (Bir and Verma, 2010). *Lepisorus amaurolepidus* and *L. nudus* differ in rhizome scales, apex and texture of the leaves. Rhizome scales are ovate, entire and uniformly thickened in *L. nudus*, while the rhizome scales are cordate, toothed and bicoloured (dark at the centre pale brown at the periphery. In Taiwan, out of eleven species of *Lepisorus*, five are with concolorous rhizome scales and six are with discolorous rhizome scales (Shieh *et al.*, 1994).

In order to check the reproductive nature (sexual or apogamous) of this species, spore count was made on South Indian populations of *Lepisorus nudus*, from Palni

Table 11.1: *Lepisorus amaurolepidus*–Herbarium Specimens Observed.

Voucher Number	*Localities and Altitude*	*Date*	*Collector*	*Notes*
RHT31623	TN/Kodaikanal/Stream banks before Thandikudi1200m.	01-05-1981	VSM	Compact cluster on stone walls and shady places.
RHT31064	TN/Kanyakumari/Vilavancodu/Kothaiyar Hills, Nalumookku. 1320m.	08-06-1984	VSM	Lithophytes; Occasional.
RHT 31189	TN/Kodaikanal, Perumal malai, First Stream on Palni Road. 1700m.	02-07-1984	VSM	Rhizome short-creeping.
RHT31249	TN/Tirunelveli/Nanguneri/Kalakad Hills, Sengaltheri-Naterikkal path. 1200m	18-07-1984	VSM	Epiphyte, Rhizome long-creeping.
RHT31452	Kerala/Iddukki/Devikulam/Munnar Hills, Kallar coupe. 1300m.	20-08-1984	VSM	Epiphyte in dry place; common; Rhizome short-creeping.
RHT31704	TN/Tirunelveli/Ambai/Kannikatty Hills, Stream near lodge. 750m.	14-11-1984	VSM and KMM	Lithophyte by streams; Rhizome short-creeping.
RHT31834	TN/Madurai (Anna)/Kodaikanal/Palni Hills1300m.	16-01-1985	VSM	Rhizome short-creeping.
RHT32025	TN/Kanyakumari/Kalkulam/Kothayar Hills, Stream on stones. 1100m.	16-02-1985	VSM	Rhizome short-creeping. Lamina dark green.
RHT 32107	TN/Kanyakumari/Kalkulam/Kothayar Hills/Kakachi-Sengaltheri path . up to kusavan kuli Ar. 1100m	17-02-1985	VSM	Forest floor on stone.
RHT32408	TN/Tirunelveli/Kannikkatty Hills, Sangumuthirai (Just below Agasthiar hill top). 1500m.	09-04-1985	VSM	Lithophyte; evergreen shola; Occasional.
RHT32447	TN/Ambai/Agasthiar Hills, Peak shola 1500m	10-04-1985	VSM	Lithophyte.
RHT32601	TN/Madurai (Anna)/Kodaikanal/Palni Hills, Gundar Shola. 2100m.	02-05-1985	VSM and KMM	
RHT32914	TN/Coimbatore/Valparai/Anamalai Hills, Attakatti forest. 1200m.	19-06-1985	VSM and KMM	Epiphyte on trees. Rhizome short-creeping. Lamina dark green.
RHT33174	TN/Salem/Namakkal/Kolli Hills Solakkadu pongal koil Shola1200m.	04-07-1985	VSM and KMM	On shaded rocks in shola.
RHT33748	Kerala/Trivandrum/Nedumangad/Ponmudi Hills, Ghat Road. 400m.	14-09-1985	VSM and KMM	Roadside, on trees, in shade, lamina dark green.

Contd...

Table 11.1–*Contd...*

Voucher Number	*Localities and Altitude*	*Date*	*Collector*	*Notes*
RHT 33571	Kerala/Kottayam/Meenachil/Kurusumalai1000m.	31-08-1985	VSM and KMM	Forest interior.
RHT33814	Kerala/Trivandrum/Nedumangad/Ponmudi Hills, Road east of R.C. church. 800m.	15-09-1985	VSM and KMM	On trees.
RHT33896	TN/Kanyakumari/Thuvalai/Mahendragiri Hills shola at the top. 800m.	01-10-1985	VSM and KMM	Lithophytes along stream banks.
RHT34042	TN/Tirunelveli/Nanguneri/Sengaltheri Hills, Sengaltheri-Neterical path. 1000m.	18-12-1985	VSM and KMM	Epiphyte in shade.
RHT34275	TN/Kanyakumari/Vilavancodu/Kothayar Hills/Muthukuzhivayal, palmore estate1200m.	21-01-1986	VSM and KMM	Lithophyte.
RHT34781	TN/Kanyakumari Hills/Kothayar-Muthukuzhivayal Road.	13-10-1986	VSM and KMM	Way side.
RHT34796	TN/Kanyakumari/Vilavancodu/Kothayar Hills, IInd Stream crossing the Road from Kakachi.	17-11-1986	VSM and KMM	Lithophytes.
RHT34974	TN/Madurai/Palni Hills, Tiger shola-Kodai1700m.	13-12-1986	VSM and KMM	Lithophyte.
RHT34992	TN/Madurai/Palni Hills, Pannaikadu Ist Stream. 1300m.	14-12-1986	VSM and KMM	Lithophyte.
RHT35253	TN/Tirunelveli/Kothayar/Kothaiyar1200m.	08-07-1990	VSM and KMM	Lithophyte, Occasionally found in the stream banks.
RHT35292	TN/Kothaiyar/Kakachi. 1700m	14-09-1990	VSM and KMM	Lithophyte; Occasional on dry places. Rhizome short-creeping.
XCH 001	TN/Ooty/Nilgiris/Dodabetta nalu Road. 2500m.	05-09-1991	VSM	Epiphytic on fully shaded trees in evergreen forest; Occasional.
XCH 1096	TN/Nilgeri/Coonoor/Pachaikadu stream (Coonoor Alakaral Road). 1700m	05-01-1992	VSM	Rare, on epiphyte, lithophytes on Stream banks.
XCH 1164	TN/Nilgiris/Coonoor/Droog R.F (Coonoor) 2000m.	07-01-1992	VSM	Rare; On the forest interior. Epilithic; Sori mature.

Contd...

Table 11.1–*Contd...*

Voucher Number	*Localities and Altitude*	*Date*	*Collector*	*Notes*
XCH 1274	TN/Nilgiris/Coonoor/Guliar/R.F.urlikal R.F. 1800m.	22-01-1992	VSM	Epilithic in the shola sori mature.
XCH 1307	TN/Nilgiris/Forest little after Kodumalai. 1700m.	24-01-1992	VSM	
XCH 2269	TN/Coimbatore/Valparai/Solayar dam-Chalakudi Road. 1050m.	04-06-1992	VSM	Rare on fully shaded streams banks in the evergreen forest. Rhizome short-creeping sori mature.
XCH 2348	TN/Tirunelveli/Nanguneri/Sengaltheri; Stream around the village. 1000m.	27-06-1992	VSM and KMM	Rare, on partially exposed places in the vallicolic forest. Sori mature.
XCH 2704	Karnataka/Coorg/Medikeri/Bhagamadala forest. 1200m.	19-08-1992	VSM	Rare on fully shaded Stream banks; Rhizome long-creepings. Sori mature.
XCH 2996	TN/Kamarajar/Sriviliputhur/Deviar-Nagariar Estate. 1100m.	29-12-1992	VSM	On partially exposed the dry deciduous forest, Epiphyte, rare, Sori mature.
XCH 3008	TN/Kamarajar/Sriviliputhur, Kamarajar Estate. 1100m.	29-12-1992	VSM	Epiphyte, on dry deciduous forest, Rare, on Sori mature.
XCH 3303	TN/Anna/Kodaikanal/Ghat Road1800m.	21-08-1993	VSM	Rare; Sori mature, stone wall.
XCH 3443	TN/Coimbatore/Valparai/Akkamalai-forest near gate. 1800m.	26-05-1993	VSM	On forest interior of evergreen Shola; Epiphyte rare. Only one plant, once one leaf fertile.

VSM: V.S. Manickam; KMM: K.M. Matthew.

Table 11.2: *Lepisorus nudus*–Herbarium Specimens Observed.

Voucher Number	*Localities and Altitude*	*Date*	*Collector*	*Notes*
RHT31623	TN/Kodaikanal/Stream banks before Thandikudi1200m.	01-05-1981	VSM	Compact cluster on stone walls and shady places.
RHT 17653	TN/Kodaikanal/Palni Hills, Bear Shola. 2400m.	02-05-1981	VSM	Common on rocks and tree trunks large colonies on rocks, Big clumps on rocks, tree trunks etc.
RHT 17733	TN/Kodaikanal/Palni Hills, Berijam western end, a little after the bank. 2400m.	05-05-1981	VSM	One thick clump on exposed stones.
RHT 17919	TN/Kodakanal/palni Hills/Berijam, Kodai. 2300m.	10-05-1982	VSM	Epiphyte, Forest east of Berijam lake view. Occasional.
RHT 18043	TN/Kodaikanal/Palni Hills, shembaganur-Periakulam Coolie path. 1600m.	15-05-1982	VSM	Shaded, small colonies, frequent on mud banks; Small patches; in shade.
RHT 31362	Kerala/Trivandrum/Nedumangad/Ponmudi Hills, Shola north of Upper Sanitorium, 900m.	28-07-1984	VSM	Epiphyte, Rhizome long-Creeping.
RHT 31377	Kerala/Idukki/Devikulam/Munnar Hills, Naimakad. 1700m	17-08-1984	VSM	Epiphyte, Rhizome Long-Creeping.
RHT 31492	TN/Madurai (Anna)/Uthamapalayam top station Hills. 2000m.	13-09-1984	KMM	Lithophyte in shade, Rhizome long-creeping; 0.2 cm across. Sori orange in colour.
RHT 31514	TN/Madurai (Anna)/Uthamapalayam/Top station Hills. 2000m.	14-09-1984	VSM	Epiphyte, Rhizome creeping.
RHT 31526	TN/Madurai (Anna)/Kodaikanal/Palni Hills, Shembaganur below village along path to Periakulam. 1800m.	26-09-1984	VSM	Epiphyte; among debris with scant soil.
RHT 31563	TN/Valparai/Anaimalai hill, Valparai 5km towards sholayar dam. 1000m.	29-09-1984	VSM and KMM	Epiphyte; Common.

Contd...

Table 11.2–*Contd...*

Voucher Number	*Localities and Altitude*	*Date*	*Collector*	*Notes*
RHT 31655	TN/Valparai/Anamalai Hills, Jeep Road from Akkamalai to Grass hill 1. 600m.	01-10-1984	KMM	Epiphyte on shola trees. Common long-creeping rhizome, sori mature.
RHT 31831	TN/Kodaikanal/Palni Hills, way to vellakavi. 1300m.	16-01-1985	VSM	Rhizome long-creeping.
RHT 31911	TN/Madurai Kodaikanal/Palni Hills, Shola by perumal peak.	18-01-1985 1700m	VSM	Shola floor on stones. Rhizome long-creeping.
RHT 32875	TN/Palni Hills, Top station-vattavadai Road. 1900m.	13-03-1985	VSM and KMM	Unusual for terrestrial habitat and larger size.
RHT 32592	TN/Kodaikanal/Palni Hills, Gundar shola. 2100m.	02-05-1985	VSM and KMM	Epiphyte/Lithophyte.
RHT 32663	TN/Kodaikanal/Palni Hills, Bear shola. 2100m	04-05-1985	VSM	Lithophytes; Common.
RHT 32728	TN/Kodaikanal/Palni Hills, Monioa forest Roadside1650m	06-05-1985	VSM and KMM	Rhizome short-Creeping.
RHT 32765	TN/Kodaikanal/Palni Hills, Berijam-Mediketten shola. 2200m.	08-051985	VSM and KMM	Epiphyte, common in the shola forest.
RHT 32816	TN/Kodaikanal/Palni Hills, Kodaikanal-Berijam Road shola opposite to Thoppi thukkiparai. 1150m.	11-05-1985	VSM and KMM	Shola trees; locally abundant.
RHT 32842	TN/Kodaikanal/Palni Hills, Berijam, N shola near tea stall. 2050m.	11-05-1985	VSM and KMM	On shola trees
RHT 32852	TN/Kodaikanal/Palni Hills, Marion shola. 2250m.	12-05-1985	VSM and KMM	On Shola trees.
RHT 32930	TN/Coimbatore/Valparai/Anamalais Hills, Attakatti forest 1,200m.	10-06-1985	VSM and KMM	Epiphyte
RHT 33076	TN/Coimbatore/Valparai/Anamalai Hills, Neeyar Road. 1000m.	13-06-1985	VSM and KMM	On trees.
RHT 33111	TN/Coimbatore/Valparai/Anamalai Hills, Jeep Road from Akkamalai to Grass hill1800m.	14-06-1985	VSM and KMM	On branches of shola trees; Common; Rhizome long-Creeping.
RHT 33152	TN/Salem/Namakkal/Kolli Hills, Sothuparai Stream. 1100m	04-07-1985	VSM and KMM	Lithophytes
RHT 33228	TN/Salem/Namakkal/Kolli Hills, Arioor-Panniparai Shola. 1200m.	06-07-1985	VSM and KMM	On forest trees; locally abundant.

Contd...

Table 11.2–*Contd...*

Voucher Number	*Localities and Altitude*	*Date*	*Collector*	*Notes*
RHT 33255	TN/Salem/Namakkal/Kolli Hills, Arioor-Shola. 1200m.	06-07-1985	VSM and KMM	Forest trees.
RHT 33299	TN/Salem/Yercad/Servarayan Hills, Servarayan Temple hill1. 500m.	08-07-1985	VSM and KMM	Epiphyte
RHT 33311	TN/Salem/Namakal/Servarayan hill, Kaka Shola. 1400m.	08-07-1985	VSM and KMM	Epiphyte.
RHT 33319	TN/Salem/Yarcad/Servarayan hill, Yarcad loop Road-Manjakuttaiar. 1400m.	09-07-1985	VSM and KMM	Epiphyte.
RHT 33367	Kerala/Idukki/Peermedu/Thekkadi Hills, Mangaladevi. 1250m.	24-08-1985	VSM and KMM	On forest trees.
RHT 33792	Kerala/Trivandrum/Nedumangad/Ponmudi Hills, Shola north of Upper Sanitorium. 950m.	15-09-1985	VSM and KMM	Lithophytes by streams.
RHT 33816	Kerala/Ponmudi Hills, Road east of R.C Church. 800m.	15-09-1985	VSM and KMM	Lithophytes; Occasionally seen.
RHT 34456	Kerala/Idukki/Devikulam/Vagvarai Hills, Vaguvarai-Eravikulam path. 2000m.	06-03-1986	VSM and KMM	On forest trees.
RHT RHT 34495	Kerala/Idukki/Devikulam/Vagavarai Hills, Vagavarai-Eravikulam Camp. 2000m.	06-03-1986	VSM and KMM	On forest trees.
RHT 34831	TN/Palni Hills, Shembaganur, SHC, Tannery stream. 1700m.	05-12-1986	VSM and KMM	Lithophyte by streams.
RHT 34855	TN/Palni Hills/Kodai pambar. 2000m.	05-12-1986	VSM and KMM	Lithophyte along Stream.
RHT 34866	TN/Palni Hills. 2300m.	06-12-1986	VSM and KMM	Epiphyte along stream banks
RHT 34898	TN/Palni Hills. 2100m.	07-12-1986	VSM and KMM	Epiphyte on forest floor.
RHT 34908	TN/Palni Hills 2100m.	07-12-1986	VSM and KMM	On forest trees.
RHT 34923	TN/Kodaikanal/Palni Hills, Kodai-Bear shola. 2100m.	12-12-1986	VSM and KMM	Lithophytes by stream banks.
RHT 34964	TN/Madurai/Palni Hills, Kodai-Tiger Shola.1700m.	13-12-1986	VSM and KMM	Lithophyte by streams.
RHT 35038	TN/Kodai-Munnar Road IInd stream from Kerala border. 2400m.	09-01-1987	VSM	Lithophyte by stream

Contd...

Table 11.2–*Contd...*

Voucher Number	*Localities and Altitude*	*Date*	*Collector*	*Notes*
XCH 008	TN/Nilgiris/Ooty/Dodabetta, Nalu Road. 2500m.	06-09-1991	VSM	Very common as epiphyte on tree in forests.
XCH 021	TN/Nilgiris/Eill-Hill. 2500m.	08-09-1991	VSM	Abundant as lithophytes on clearings.
XCH 030	TN/Nilgiris/stream near Avalanche forest rest house. 2000m.	09-09-1991	VSM	Very common epiphyte on Stream banks in fully forest trees.
XCH 047	TN/Nilgiris/Muthurai. 2000m.	09-09-1991	VSM	Very common and dominant in streams along with *Pyrrosia porosa* and *Pleopeltis macrocarpa*.
XCH 060	TN/Kanyakumari/Vilavancode/Forest on the left of Road1500 m.	17-09-1991	VSM	Common epiphyte or lithophytes in the forest interior or partially exposed spot.
XCH 072	TN/Kanyakumari/Vilavancode/Forest on the left of Road before lodge. 1500 m.	17-09-1991	VSM	Rare epiphyte in the forest interior.
XCH 161	TN/Kanyakumari/Kalkulam/Upper kothyar; First stream next to power house. 1500 m.	19-09-1991	VSM	Occasional, on fully exposed rocks on vallicolic open forest.
XCH 257	TN/Nilgiris/Ooty-Naduvattum Roadside. 2300 m.	13-10-1991	VSM	Common and locally abundant on fully shaded Vallicolic closed forest.
XCH 299	TN/Nilgiris/Shola before diversion parson valley2300 m.	14-10-1991	VSM	Epiphyte as fully shaded parts of shola abundant in all forest at high plateau.
XCH 308	TN/Nilgiris/Shola near parson valley diversion. 2400 m.	14-10-1991	VSM	Common epiphyte in the interior of shola.

Contd...

Table 11.2–*Contd...*

Voucher Number	*Localities and Altitude*	*Date*	*Collector*	*Notes*
XCH 326	TN/Shola near parson valley diversion. 2400 m.	14-10-1991	VSM	Common epiphyte in the shola.
XCH 381	TN/Nilgiris/Shola after Naduvattum2250 m	15-10-1991	VSM	A lithophyte in shola.
XCH 408	TN/Nilgiris/Dodabetta Road. 2650 m.	24-10-1991	VSM	Occasional epiphyte in the interior of sholas.
XCH 471	TN/Nilgiris/Terace Estate, Forest Bangalow; 2200m.	25-10-1991	VSM	Locally abundant in the Streams Epiphytes.
XCH 501	TN/Nilgiris/Valley after Belluve terrace Estate Road. 2200m.	26-10-1991	VSM	Common on tree trunks in vallicolic closed forest.
XCH 554	TN/Nilgiris/Sholas between T.R. Bazaar and Belluve2300m.	27-10-1991	VSM	Very common epiphyte in the partially exposed places in the vallicolic forest.
XCH 590	TN/Nilgiris/Shola short cut from the T.R. Bazaar to Naduvattum. 2300m.	28-10-1991	VSM	Very common epiphyte in sori mature.
XCH 663	TN/Nilgiris/T.R.Bazaar Stream. 2200m	18-11-1991	VSM	Rare on partially exposed stream banks.
XCH 698	TN/Nilgiris/Anger board Shola2200 m.	19-11-1991	VSM	Very common on partially exposed stream banks.
XCH 711	TN/Nilgiris/stream on the way Governor shola. 2200 m.	19-11-1991	VSM	Very common epiphyte on fully shaded stream bank in the shola.
XCH 750	TN/Nilgiris/Ellakandy (Avalanchi Road). 2200 m.	20-11-1991	VSM	Very common epiphyte on fully shaded places.

Contd...

Table 11.2–*Contd...*

Voucher Number	*Localities and Altitude*	*Date*	*Collector*	*Notes*
XCH 758	TN/Nilgiris/Ellkandy forest. 2100m.	20-11-1991	VSM	Very common epiphyte on fully shaded vallicolic closed forest.
XCH 767	TN/Nilgiris/Ellkandy forest. 2100m.	20-11-1991	VSM	Very common on fully shaded stream banks in the vallicolic closed forest.
XCH 772	TN/Nilgiris/stream near avalanche forest rest house. 2100m.	20-11-1991	VSM	Very common epiphyte on fully shaded Stream banks.
XCH 777	TN/Nilgiris/stream along Ooty-Kothagiri Road. 2000m.	21-11-1991	VSM	Epiphyte, common on Stream banks in the vallicolic open forest.
XCH 853	TN/Nilgiris/Avalanchi forest. 2100m.	06-12-1991	VSM	Epiphyte, on common and locally abundant.
XCH 869	TN/Nilgiris/Avalanchi forest. 2100m.	07-12-1991	VSM	Very common and locally abundant epiphyte in the shola interior.
XCH 908	TN/Nilgiris/Avalanchi forest. 2100m.	07-12-1991	VSM	Very common epiphyte in the shola.
XCH 937	TN/Nilgiris/Round Road on the plateau via uppar Bhavani (Manjoor). 2300m.	08-12-1991	VSM	Occasional on fully shaded places in the shola.
XCH 1030	TN/Nilgiris/Coonoor/Lamb's rock shola from coonoor. 1900m.	12-12-1991	VSM	Rare; epiphyte in shola.
XCH 1193	TN/Nilgiris/Coonoor/Pakkasuramalai R.F (Coonoor). 1900m.	08-01-1992	VSM	Occasional on partially exposed stream banks; sori mature.

Contd...

Table 11.2–*Contd...*

Voucher Number	*Localities and Altitude*	*Date*	*Collector*	*Notes*
XCH 1273	TN/Nilgiris/Coonoor/Urlikal R.F. 1700m.	22-01-1992	VSM	Rare; on epilithic in the Sholas.
XCH 1172	TN/Nilgiris/Coonoor/Droog R.F. (Coonoor). 2000m.	07-02-1992	VSM	Rare, on fully exposed places in forest interior; sori mature.
XCH 1681	TN/Anna (Madurai)/Kodaikanal/First stream on the perumal malai Road. 1600m.	30-03-1992	VSM	Epilithic on fully shaded Stream banks, sori mature.
XCH 1779	TN/Coimbatore/Valparai/Calparai-Nirarar dam Road. 1300m.	02-05-1992	VSM	Occasional on partially exposed Roadside; Sori mature.
XCH 1918	TN/Coimbatore/Valparai/Akkamalai forest. 1700m.	05-05-1992	VSM	Common on fully exposed places and Roadside. Rhizome long-creeping sori mature.
XCH 2097	TN/Coimbatore/Valparai/Akkamalai forest east of the factory. 1700m.	22-05-1992	VSM	On fully shaded Stream banks in the vallicolic closed forest. Sori mature.
XCH 2138	TN/Anamalys Hills/Akkamalai grass hill path. 1700m.	22-05-1992	VSM	Epiphyte, occasion on Stream banks in the vallicolic closed forest.
XCH 2243	TN/Coimbatore/Valparai/Shekalmudi forest. 1200m.	02-06-1992	VSM	Occasional epiphyte on a tree trunk at the edge of evergreen forest; Sori mature.
XCH 2422	TN/Nellai/Nanguneri/Sengaltheri-Naterikal Road1250 m.	29-06-1992	VSM	Epiphyte, on partially exposed the stream banks; Occasional.

Contd...

Table 11.2–*Contd...*

Voucher Number	*Localities and Altitude*	*Date*	*Collector*	*Notes*
XCH 2438	TN/Nellai/Nanguneri/Forest east of lodge1200 m.	30-06-1992	VSM	Epiphyte, on fully shaded the stream banks in evergreen forest; Rare, Rhizome short-creeping. Sori mature.
XCH 2613	Karnataka/Coorg/Medikeri/Medikeri-Siddapur Roadside. 1100 m.	16-08-1992	VSM	Occasional on partially exposed Stream banks, Rhizome short-creeping. Sori mature.
XCH 2640	Karnataka/Coorg/Medikeri/Thalacauvary Forest/ 1200 m	17-08-1992	VSM	Occasional Epiphyte on Semi-exposed stream banks. Sori mature.
XCH 2727	Karnataka/Coorg/Medikeri/Bhagamandala–Panathur Road. 1100 m.	20-08-1992	VSM	Rare, on fully shaded forest edge. Rhizome long-creeping. Sori mature.
XCH 2772	Karnataka/Coorg/Medikeri/Bhagamandala-Napoklu Road. 900 m.	21-08-1992	VSM	Epiphyte, on partially exposed Roadside; Occasional. Sori mature.
XCH 2803	Karnataka/Coorg/Medikeri/Bhagamandala–Medikeri Road. 1000 m.	22-08-1992	VSM	Epiphyte, on partially exposed the Roadside; Occasional, sori mature.
XCH 2825	Karnataka/Coorg/Sampaje/Sampaje Road. 700 and 500 m.	23-08-1992	VSM	Epiphyte, on partially exposed the Roadside; Occasional, Sori mature.
XCH 2892	Karnataka/Coorg/Sampaje/Sampaje-Homerpet Road. 1100 m.	24-08-1992	VSM	Epiphyte, on Roadside; Occasional rhizome long-creeping. Sori mature.

Contd...

Table 11.2–*Contd...*

Voucher Number	*Localities and Altitude*	*Date*	*Collector*	*Notes*
XCH 3133	Kerala/Waynad/Waynad/Tirvnelvely way to Pachibath-alam. 900 m.	19-04-1993	VSM	Rare, on Roadside; sterile.
XCH 3195	Kerala/Mannerghat/Silent valley/Silent Valley forest. 900 m.	04-05-1993	VSM	Epiphyte, on stream banks in forest interior rare, Sori mature.
XCH 3249	Kerala/Silent Valley/Kummanthodu. 850 m.	05-05-1993	VSM	Epiphyte, Rarely found.
XCH 3442	TN/Coimbatore/Valparai/Akkamalai forest near gate. 1800m.	26-06-1993	VSM	Epiphyte, on forest interior of evergreen Shola sori mature.

Table 11.3: Spore Count in different Populations of *Lepisorus nudus* from South India to know Whether Sexual or Apogamous.

Voucher Number	*Localities and Altitude*	*Spore Number per Sporangium*	*Sexual/ Apogamous*
RHT 17653	TN/Kodaikanal/Palni hills, Bear Shola. 2400m.	Above 60	Sexual
RHT 31563	TN/Coimbatore/Valparai/Anaimalai hill, Valparai 5km towards sholayar dam. 1000m.	Above 60	Sexual
RHT 33299	TN/Salem/Yercad/Servarayan hills, Servarayan Temple hill1. 500m.	Ca. 64	Sexual
XCH 590	TN/Nilgiris/Shola short cut from the T.R. Bazaar to Naduvattum. 2300m.	Above 60	Sexual
XCH 698	TN/Nilgiris/Anger board Shola. 2200 m.	Above 60	Sexual
XCH 758	TN/Nilgiris/Ellkandy forest. 2100m.	64	Sexual
XCH 767	TN/Nilgiris/Ellkandy forest. 2100m.	64	Sexual
XCH 853	TN/Nilgiris/Avalanchi forest2100m.	Above 60	Sexual
XCH 908	TN/Nilgiris/Avalanchi forest. 2100m.	Above 60	Sexual
XCH 1030	TN/Nilgiris/Coonoor/Lamb's rock shola from coonoor. 1900m.	Ca. 64	Sexual
XCH 1193	TN/Nilgiris/Coonoor/Pakkasuramalai R.F (Coonoor). 1900m.	Above 60	Sexual
XCH 1681	TN/Anna (Madurai)/Kosaikonal/First stream on the Perumal malai road. 1600m	Ca. 64	Sexual
XCH 2422	TN/Nellai/Nanguneri/Sengaltheri-Naterikal road1250 m	Above 60	Sexual
XCH 2772	Karnataka/Coorg/Medikeri/Bhagamandala-Napoklu road 900 m.	Above 60	Sexual
XCH 021	TN/Nilgiris/Eill-Hill. 2500m.	32	Apogamous
XCH 030	TN/Nilgiris/Stream near avalanche forest rest house. 2000m.	32	Apogamous

Plate 11.1: A-B: *Lepisorus amaurolepidus* (A: Habit; B: Sporophyll); C-H: *L. nudus* (from Nilgiris); C-F: Habit; G-H: Sporophyll; I: Herbarium specimen.

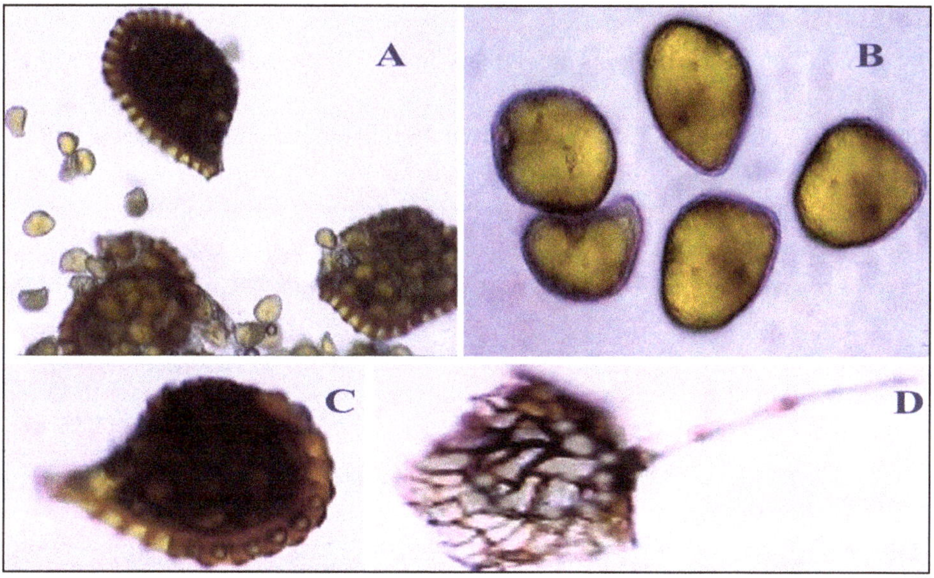

Plate 11.2: *Lepisorus nudus*.

A and C: Sporangia with Spores; B: Spores, D: Paraphysis of sorus.

Hills, Tirunelveli Hills, Anamalais, Nilgiris (Western Ghats) and Shevaroy Hills (Eastern Ghats). Totally 16 populations were checked. All the populations, except two from the Nilgiris, showed the presence 60-64 normal spores in each sporangium, indicating the sexual nature. Only two populations from the Nilgiris showed the presence of 32 spores per sporangium, indicating the apogamous nature of the populations. This is the first report for the presence of apogamy in *Lepisorus nudus* from India. Out of eight populations from the Nilgiris, six are sexual and two are apogamous (Table 11.3, Plate 11.2).

In the present study 38 specimens of *Lepisorus amaurolepidus* and ninety five specimens of *L. nudus* were observed in St. Xavier's College Herbarium, Palayamkottai. Out of 16 specimens of *L. nudus* from various localities of South India, fourteen specimens are with 60-64 normal spores per sporangium with the indication for sexual reproduction and two specimens of Nilgiris are with 32 spores per sporangium with the indication for apogamous reproduction. This is the first report for the presence of apogamy in *Lepisorus nudus*. It has to be further confirmed by cytological study.

Acknowledgement

Authors are thankful to Rev. Dr. John Peter Arulanandam S J, the Director of Centre for Biodiversity and Biotechnology (CBB), St. Xavier's College, Palayamkottai for his kind permission to observe the herbarium specimens of XCH (St. Xavier's College Herbarium). Technical helps rendered by Mr. Ganthikumar, CBB, are also acknowledged.

References

Beddome, R. H. 1864. The Ferns of Southern India. Today and Tomorrow's Printers and Publishers, New Delhi.

Bir, S. S. and Verma, S. C. 2010. Chromosome Atlas of the Indian Pteridophytes (1951-2009). Bishen Singh Mahendra Pal Singh. Dehra Dun, India.

Dixit, R. D. 1984. A Census of the Indian Pteridophytes. Bot. Surv. India, Howrah.

Hsu, C. C. Horng, F. W. and Kuo, C. M. 2002. Epiphyte biomass and nutrient capital of a moist subtropical forest in north-eastern Taiwan. J. Tropical Ecology, **18**: 659-670.

Kholia, B. S., Bhakuni, K., Punetha, R. and Bankoti, N. S. 2012. Taxonomic studies on central Himalayan species of fern genus *Lepisorus* (Polypodiaceae) with a note on thickness of rhizome and deciduousness of the lamina. **NeBIO 3 (3) 29-41.**

Manickam, V. S and Irudayaraj, V. 1992. Pteridophyte Flora of the Western Ghats, South India. BI-Publications. New Delhi.

Manickam, V. S. and Irudayaraj, V. 2003. Pteridophyte Flora of the Nilgiris, South India. Bishen Singh and Mahendra Pal Singh, Dehra Dun.

Mojiol, A. R., Jitinu, A. M. A., Adella, A., Ganang, G. M. and Nasly, N. 2009. Vascular epiphytes diversity at Pusat Sejadi, Kawang Forest Reserve, Sabah Malaysia. J. Sust. Development, **2** (1): 121-127.

Nampy, S. and Madhusoodanan, P. V. 1998. Fern Flora of South India: Taxonomic Revision of Polypodioid Ferns. Daya Publishing House. Delhi.

Qi, X. P. and Zhang, X. C. 2009. Taxonomic revision of *Lepisorus* (J. Sm.) Ching sect. *Lepisorus* (Polypodiaceae) from China. J. Syst. Evol. **47** (6): 581-598.

Shieh, W.-C., DeVol, C. E. and Kuo, C. M. 1994. Polypodiaceae. In: Huang, T.-C. *et al.* (eds.), *Flora of Taiwan*, 2nd ed. 1: 469-519. Editorial Committee, Dept. Bot., NTU, Taipei, Taiwan.

Wang, L., Qi, X. P., Xiang, Q. P., Heinrichs, J., Schneider, H. and Zhang, X. C. 2010. Phylogeny of the paleotropical fern genus *Lepisorus* (Polypodiaceae, Polypodiopsida) inferred from four chloroplast DNA regions. Molecular Phylogenetics and Evolution, **54**: 211-225.

Zapfack, L., Nkongmeneck, A. B., Villiers, J. F., and Lowman, M. 1996. The importance of Pteridophytes in the epiphytic Flora of some phorophytes of the Cameroonian semi-deciduous rain forest. Selbyana, **17:** 76 – 81.

Chapter 12

A Study on Non Timber Forest Products Diversity and their Utility Modes in Dry Deciduous Forests of Kadapa Division, Andhra Pradesh

M. Sridhar Reddy*[1] and B. Sadasivaiah[2]

[1]*Department of Environmental Sciences, Yogi Vemana University, Kadapa, Andhra Pradesh*

[2]*Department of Botany, Govt. Degree and P.G. College, Wanaparthy – 509 103, Mahabubnagar District, Andhra Pradesh*

Introduction

Natural Forests harbor rich biodiversity that provides both direct and indirect services to human kind. Non Timber Forest Products (NTFP) are one such direct service that forms an important source of livelihood and economy for several million people living in and around forests especially in developing countries (Bhat *et al.*, 2003). It is estimated that 275 million poor rural people in India–27 percent of the total population– depend on NTFPs for at least part of their subsistence and cash livelihoods and thus resulting in crucial livelihood linkage to these forest dwellers

* Corresponding Author: E-mail: sridharmullangi@yahoomail.com

(Malhotra and Bhattacharya, 2010). Around 3000 NTFP species belonging to 328 flowering plant families are found to have one or many potential NTFP resource and about 384 wild plant species products are found to be edible of which nearly 30 per cent is of tribal diet (Bhuyan *et al.*, 2010). Around 900 plant species are found to be involved in medicinal plant trade and about 85 per cent are collected from the wild (Ved *et al.*, 2001). There are around 9,500 licensed herbal industries and many unregistered herbal units that depend upon continuous supply of medicinal plants for manufacture of herbal formulations (Ved and Goraya, 2007). Atleast 2,400 plant species are used by the codified Indian system of medicine and western medicine (Johnson *et al.*, 2013). Further, India exports NTFPs worth of Rs. 446 crores, they contribute to a national trade amounting about Rs. 1000 crores, for National Indian System of Medicine market of 2500 crores, for Global Market US$ 60 billion (Bhuyan *et al.*, 2010).

Across all Indian forests types the strikingly common feature is the NTFP collection that form a wide range of products and utilization patterns. NTFP collection includes forest resources that are marketable as well as for household use like gathering of gums, resins, edible fruits, flowers, tubers, roots, medicinal plants and leaves, stem barks for food and medicines; fire wood for cooking, fencing, fodder for livestock. These activities form an important employment generation source, food security and generate 40-60 per cent of their total income (Chopra, 1993; Kumar, 2003). In accordance to a forest type, certain NTFPs may be pronouncedly extracted and marketed, like *Asparagus* tubers in Gujarat forests, tooki leaves in central India, gums in dry forests of southern Eastern Ghats, medicinal plants in Western Ghats (TFRI, 2011). NTFPs constitute certain characteristics like wide variety of products and of species, the shorter frequency of harvest cycles and the typically smaller yields per unit area (Shahabuddhin and Soumya, 2004). Their collection provides varied kind of challenges to forest conservation as it may often lead to over exploitation of resources and can also enhance the chances of forest conservation by community effort by making them direct beneficiaries. Policy initiatives like Joint Forest Management (JFM), organization of Vana Samrakshana Samithis (VSS) have rightly send the message that–as the long-term financial returns from sustainable NTFP harvest could far outweigh the net economic benefits of timber production or conversion of the same area of land to agricultural fields (Nepstad and Schwartzman, 1992) to certain extent. In addition, a large number of NTFP still do not enter the markets and are locally consumed by producers/collectors and therefore has no market price and were sold at very low rate (Chopra, 1997). Moreover, due to commercialization of few NTFPs, bulk and quick harvests are becoming common that leads to destructive harvesting (Giridhar *et al.*, 2008).

In India, the tropical dry forests occupy 38 per cent of the total forest area and these forest support large number of forest dwelling people and are getting degraded at alarming rate owing to deforestation for extraction of timber and forest produce (Raghubanshi and Tripathi, 2009). Thus in these kind of rapid declining climax forests, great significance should be provided to NTFPS that support locally than timber. Compiled information of NTFP in a particular dry forest area certainly helps in designing the marketing and conservation measures with active involvement of the local people.

Study Area

Eastern Ghats along the East coast of India constitutes hill ranges that support a diverse array of tropical dry forests. Field explorations have reported 3000 plant species and about 100 species that are endemic to this region (Pullaiah and Muralidhara Rao, 2002). Kadapa forests are part of hill ranges namely Thurupu kondas, Lankamalas and Palakondas with an altitude range of 99-902 m of southern Eastern Ghats that support dry deciduous and scrub forests. These forests are subjected to varied rainfall, extreme conditions of temperature in summer, low relative humidity for about 7-9 months and frequent occurrence of droughts that affect the natural regeneration potential. In addition to these factors, anthropogenic disturbances like grazing, repeated fires, tree felling have reduced the forests to thorn and inferior dry mixed deciduous forests in reserved forests (Rawat, 1997). The total forest area in Kadapa division includes 1,70,180.6 ha of which only 6 per cent includes the dense forest area and the rest is occupied by open dry forest area and scrub land of 47 per cent each. The Kadapa division includes a certain portion of two wild life sanctuaries namely Lankamalla and Pensula Narasimha swamy, thus the effective area meant for NTFP collection is 1,32,609ha. A total of 118 VSS were actively involved in managing the 46,500 ha of degraded forests (Anon, 2000). VSS members are entitled to collect NTFP in these managed forests. Yanadis and Boyas are the tribal people that reside in these forests, collection and selling of NTFPs forms one of their main occupations. Further large number of people residing near the forest tracts depends on forest products.

Methodology

The present study was carried out in the Kadapa forest division of YSR Kadapa district. It includes both primary data acquired through field enumeration studies and interactive studies with VSS members in the division, while the secondary information was gathered from Girijan Cooperative Corporation and forest department working plans.

Results and Discussion

The hilly topography of Eastern Ghats provides a variety of microhabitats which support rich biodiversity and resources. The basic feature of these forests is low canopy and high diversity of trees (Rawat, 1997). Understanding the species diversity and distribution patterns helps in evaluating the resources of tropical forests to certain extent (Kumar *et al.*, 2010). Field explorations have reported 3000 plant species and about 100 species that are endemic to Eastern Ghats eco region. A total of 689 plant species, that constitute 183 trees, 162 shrubs, 90 climbers, 254 herbs was observed in wild among Kadapa forests of southern Eastern Ghats ((Pullaiah and Muralidhara Rao, 2002). Field survey, interviews with the VSS members and forest dwelling people reveal that, 112 plants are found to have certain NTFP value either as medicinal, household or for market purpose. Trees form the major component with 71 species (63 per cent) followed by 15 shrubs (13 per cent), 13 climbers (12 per cent) and 13 herbs (12 per cent). It indicates that trees in these dry deciduous form not only a major component in forest structure that generates ecosystem goods and services but also provides major NTFPs as also observed in dry forests of Adilabad (Omkar *et al.*, 2012).

Sustenance and Livelihood Sources

Local people use NTFPs either for personal use or as procurement to GCC and as well to sell them in the local market. The recorded 112 NTFP generating plants are categorized into three groups – those used for sustenance, NTFPs that can generate additional livelihood income and those meant for both sustenance and as well can generate income. The analysis revealed that the local forest dwellers use a total of 53 species (47 per cent) for their sustenance to meet their primary health care needs, food, house hold items and 37 species products (33 per cent) are being traded in the local market or procured for GCC, while 22 species (20 per cent) are being used both household and market purpose. Among the plants that are being used for sustenance, considerable plant products (51 per cent) are used for medicinal purpose, notable among them are *Balanites roxburghii, Lannea coromandelica, Phyllanthus amarus, Schleichera oleosa, Tribulus terrestris.* This reflects the importance of NTFPs in providing the primary health care needs of the local people. Products like fibre from *Bauhinia racemosa, Hardwickia binata* and thatching grass from *Cymbopogon* spp., *Saccharum spontaneum* and branch wood from *Chloroxylon swietenia, Dodonaea viscosa, Givotia moluccana* and *Gyrocarpus americanus* to make house hold wood implements and toys play a significant role in improving the quality of life of the locals. They were of the opinion that, when they venture into forests to collect NTFPS, fruits from seven species like *Cordia dichotoma, Carissa spinarum* form an important supplementary diet and leaves of *Polyalthia cerasoides, Ziziphus xylopyrus* are of fodder value to their domestic animals. These features indicate that NTFPs play an important role in meeting their local needs for day to day sustenance. Around 37 plants yield NTFPs like gums, medicinally important tubers, leaves, fruits and barks, beedi leaves, plate making leaves and both edible and non edible fruits that can be traded in the local market and thus can generate an additional income that will improve the quality of local people life. This category includes important medicinal plants like *Asparagus racemosus, Cassia fistula, Gymnema sylvestre, Hemidesmus indicus, Litsea glutinosa, Strychnos nux-vomica Tinospora cordifolia, Decalepis hamiltonii*; 11 plants produce gums and of them *Sterculia urens, Cochlospermum religiosum, Boswellia serrata* and *Anogeissus latifolia* and fruits from *Buchanania lanzan, Terminalia chebula* and *T. bellirica* and bark from *Terminalia arjuna,* fibre from *Thespesia lampas* and broom sticks made of *Apluda mutica* and *Heteropogon contortus* grasses are of most sought after in the local market. Leaves of *Diospyros melanoxylon* used for beedi making and plate leaves from leaves of *Butea monosperma* and *Bauhinia vahlii* generates considerable employment generation. The group of NTFPs that are mainly helpful in sustenance and also can generate additional income in certain seasons include 22 species. Edible fruits of *Annona squamosa, Limonia acidissima, Mangifera indica, Gardenia gummifera, Borassus flabellifer, Phoenix sylvestris, Tamarindus indicus, Phyllanthus emblica, Syzygium cumini* and non edible fruits like *Sapindus emarginatus, Maduca indica* are of most important. About six medicinal plants like *Azadirachta indica, Holarrhena pubescens* and five species like *Murraya koenigii, Dendrocalamus strictus* play major role in providing additional income and sustenance sources for day to day sustenance (Table 12.1). The chi square test between the

Table 12.1: List of Plants that Yield Non-Timber Forest Products, their Usage Pattern and their Role in Sustenance and Livelihood.

Sl.No.	Name of the Species and Family	Common Name	Part Used	Habit	Purpose	Sustenance (S) and Livelihood (L)
1.	*Abrus precatorius*–Papilionaceae	Gurivinda	Seeds	C	Medicinal, Ornamental	S
2.	*Andrographis paniculata*–Acanthaceae	Nela vemu	Whole plant	H	Medicinal	S
3.	*Acacia chundra*–Mimosaceae	Sundra	Bark	T	Bark paste is applied on skin diseases	S
4.	*Acacia leucophloea*–Mimosaceae	Tella tumma	Bark	T	Gum is Medicinal, bark yields fibre	L
5.	*Acacia nilotica*–Mimosaceae	Nalla thumma	Gum	T	Adhesive	L
6.	*Achyranthes aspera*–Amaranthaceae	Uttareni	Stems, roots	H	Young leaves edible	S
7.	*Aegle marmelos*–Rutaceae	Maredu	Bark, leaves, Fruits	T	Medicinal, Edible	S and L
8.	*Ailanthus excelsa*–Simaroubaceae	Peddamanu	Bark	T	Bark powder is used to cure fever symptoms	S
9.	*Alangium salvifolium*–Alangiaceae	Ooduga	Bark and fruits	T	Barks, fruits possess laxative property	S
10.	*Albizia odoratissima*–Mimosaceae	Chinduga	Gum	T	Adhesive	L
11.	*Albizia procera*–Mimosaceae	Tella chinduga	Bark	T	Medicinal	S
12.	*Alternanthera sessilis*–Amaranthaceae	Ponnaganti	Leaves	H	Young leaves edible	S
13.	*Annona squamosa*–Annonaceae	Seethaphalam	Fruits	S	Fruit edible, can be supplied to local market in the season	S and L
14.	*Anogeissus latifolia*–Combretaceae	Thiruman	Gum	T	Medicinal	L
15.	*Apluda mutica*–Poaceae		Whole plant	H	Household thatching purpose	S
16.	*Aristida setacea*–Poaceae	Nakkusa	Whole plant	H	Brooms	S
17.	*Aristolochia indica*–Aristolochiaceae	Eswari	Roots	C	Extract from roots are considered as antidote for snake bites	S

Contd...

Table 12.1–*Contd...*

Sl.No.	*Name of the Species and Family*	*Common Name*	*Part Used*	*Habit*	*Purpose*	*Sustenance (S) and Livelihood (L)*
18.	*Asparagus racemosus*–Asparagaceae	Pilli gadda	Roots	C	Root powder is considered as aphrodisiac, Medicinal, market	L
19.	*Atalantia monophylla*–Rutaceae	Adivi Nimma	Fruits	T	Household	S
20.	*Azadirachta indica*–Meliaceae	Vepa	Fruits, leaves	T	Oil from seeds is used in medicines, leaves are used to cure chicken pox	S and L
21.	*Balanites roxburghii*–Balanitaceae	Gara	Fruits	T	Fruits paste is considered as anti-diabetic	S
22.	*Bauhinia racemosa*–Caesalpiniaceae	Are	Bark	T	Fibre, medicinal	S
23.	*Bauhinia vahlii*–Caesalpiniaceae	Addaku	Leaf	L	Plate leaves	L
24.	*Boerhavia diffusa*–Nyctaginaceae	Atukamamidi	Leaf	H	Medicinal, Jaundice	S
25.	*Bombax ceiba*–Bombacaceae	Buruga	Gum	T	Medicinal	S and L
26.	*Borassus flabellifer*–Arecaceae	Thadi	Leaves, fruits, toddy	T	Household, market, edible, thatching	S and L
27.	*Boswellia serrata*–Burseraceae	Guggilam	Gum	T	Medicinal	L
28.	*Bridelia retusa*–Euphorbiaceae	Verri karaka	Fruits, Bark	T	Edible, tannins	S
29.	*Buchanania axillaris*–Anacardiaceae	Sara	Fruits	T	Kernel in the seed is the edible part, used in Sweets preparation	L
30.	*Buchanania lanzan*–Anacardiaceae	Sara	Fruits	T	Kernel in the seed is the edible part, used in Sweets preparation	L
31.	*Butea monosperma*–Papilionaceae	Moduga	Flowers, leaves, seeds	T	Dye, leaf plate	L
32.	*Caesalpinia bonduc*–Caesalpiniaceae	Gatchakaya	Seed	C	Medicinal	S
33.	*Capparis grandis*–Capparaceae		Bark	T	Bark paste is usually applied on wounds	S
34.	*Capparis zeylanica*–Capparaceae	Adonda	Fruits	S	Household	S

Contd...

Table 12.1–*Contd...*

Sl.No.	*Name of the Species and Family*	*Common Name*	*Part Used*	*Habit*	*Purpose*	*Sustenance (S) and Livelihood (L)*
35.	*Carissa spinarum*–Apocynaceae	Kalivi	Fruits	S	Fruits are edible	S
36.	*Cassia auriculata*–Caesalpiniaceae	Tangedu	Bark	S	Tannins can be obtained	L
37.	*Cassia fistula*–Caesalpiniaceae	Rella	Bark	T	Medicinal	L
38.	*Cassine glauca*–Celastracea		Leaves, bark	T	Leaves and bark powder is given to fever patients	S
39.	*Chloroxylon swietenia*–Flindersiaceae	Billudu	Stem branches	T	Household, to make toys	S
40.	*Cissampelos pareira*–Menispermaceae	Adavi banka teega	Roots	C	Medicinal	S
41.	*Cochlospermum religiosum*–Cochlospermaceae	Kondagogu	Gum	T	Medicinal	L
42.	*Cordia dichotoma*–Cordiaceae	Iriki	Fruits	T	Edible, household	S
43.	*Crataeva magna*–Capparaceae	Mogalinga	Leaves	T	Medicinal	S
44.	*Curculigo orchioides*–Hypoxidaceae	Nela thadi	Roots	H	Medicinal	S
45.	*Cymbopogon flexuosus*–Poaceae	Nimma gaddi	Oil	H	Medicinal, Market	L
46.	*Cyperus rotundus*–Cyperaceae	Thunga	Tubers	H	Medicinal	S
47.	*Decalepis hamiltonii*–Asclepiadaceae	Maredu	Roots	C	Cooling agent	S and L
48.	*Dendrocalamus strictus*–Poaceae	Veduru	Stems	S	Making baskets, Market	S and L
49.	*Dioscorea bulbifera*–Dioscoriaceae	Chinna gadda	Tubers	C	Edible, Household	S
50.	*Dioscorea pentaphylla*–Dioscoriaceae	Duka pendalam	Tubers	C	Edible, Household	S
51.	*Diospyros chloroxylon*–Ebenaceae	Ullinda	Fruits	T	Edible, household	S
52.	*Diospyros melanoxylon*–Ebenacceae	Tooniki	Fruit, leaves	T	Edible, Beedis preparation	L
53.	*Dodonaea viscosa*–Sapindaceae	Bandedu	Stem, branches	S	Household wood implements	S

Contd...

Table 12.1–*Contd...*

Sl.No.	*Name of the Species and Family*	*Common Name*	*Part Used*	*Habit*	*Purpose*	*Sustenance (S) and Livelihood (L)*
54.	*Dolichandrone atrovirens*–Bignoniaceae	Chittu Neerudhi	Bark	T	Household, fish poison	S
55.	*Dolichandrone falcata*–Bignoniaceae	Neerudhi	Bark	T	Fibre	S
56.	*Eclipta prostrata*–Asteraceae	Gunta Galagara	Leaves	H	Household, Market, Medicinal	S and L
57.	*Erythroxylum monogynum*–Erythroxylaceae	Devadaru	Leaves	T	Medicinal	S
58.	*Evolvulus alsinoides*–Convolvulaceae	Vishnu kranthi	Aerial parts	H	Medicinal	S
59.	*Gardenia gummifera*–Rubiaceae	Bikki	Fruits	T	Edible, Market	S and L
60.	*Gardenia latifolia*–Rubiaceae	Pedda bikki	Gum	T	Market	L
61.	*Gardenia resinifera*–Rubiaceae	Yerri bikki	Gum, Fruits	T	Shoot tips contain gums, medicinal Market	L
62.	*Garuga pinnata*–Burseraceae	Garuga	Gum	T	Medicinal	L
63.	*Givotia moluccana*–Euphorbiaceae	Tella poliki	Stem, branches	T	Household, to make toys	S
64.	*Gloriosa superba*–Liliaceae	Agnisikhi, Nabhi	Roots	C	Medicinal, Market	L
65.	*Gmelina arborea*–Verbenaceae	Adavi Gummadi	Gum	T	Bark paste is used against snake bites	S
66.	*Grewia abutilifolia*–Tiliaceae		Bark	S	Fibre	S
67.	*Grewia tiliifolia*–Tiliaceae	Jana	Fruits	S	Edible	S
68.	*Gymnema sylvestre*–Asclepiadaceae	Poda patri	Leaves	C	Leaf powder is considered as anti-diabetic	L
69.	*Gyrocarpus americanus*–Hernandiaceae	Poliki	Stem, branches	T	Household, to make toys	S
70.	*Hardwickia binata*–Caeslapiniaceae	Yepi	Bark	T	Fibre	S
71.	*Helicteres isora*–Sterculiaceae	Gooba thada	Leaves, Bark	S	Fodder, fibre	S
72.	*Hemidesmus indicus*–Asclepiadaceae	Sugandhapala	Roots	C	Dried root powder is considered as anti diarrhoea	L
73.	*Heteropogon contortus*	Yerragoddi	Whole plant	H	Brooms, Household, Market	S

Contd...

Table 12.1–*Contd...*

Sl.No.	*Name of the Species and Family*	*Common Name*	*Part Used*	*Habit*	*Purpose*	*Sustenance (S) and Livelihood (L)*
74.	*Holarrhena pubescens*–Apocynaceae	Kolamukku	Stem and branches	T	Bark is used as anti dysentery, stems are used to make toys	S and L
75.	*Holoptelia integrifolia*–Ulmaceae	Tapasi	Leaves	T	Fodder	S
76.	*Lannea coromandelica*–Anacardiaceae	Gumpena	Gum, Bark	T	Bark paste is applied on wounds	S
77.	*Limonia acidissima*–Rutaceae	Velaga	Bark, Fruits	T	Gum, edible	S and L
78.	*Litsea glutinosa*–Lauracceae	Naramamidi	Bark	T	Bark paste used to set bones	L
79.	*Madhuca indica*–Sapotaceae	Ippa	Flowers, seeds	T	Oil can be extracted from seeds, fleshy corolla is edible, country liquor (ippa sara) is prepared, Market	S and L
80.	*Manilkara hexandra*–Sapotaceae	Pala	Fruits	T	Edible, Market	S and L
81.	*Maytenus emarginatus*–Celastraceae	Danti	Leaves	S	Leaf paste is applied on wounds	S
82.	*Mitragyna parviflora*–Rubiaceae	Rudraganupu	Bark	T	Bark powder is used to cure fever Medicinal; Fibre	S
83.	*Murraya koenigii*–Rutaceae	Karivepaku	Leaves	S	Used in house hold in curries preparation, Market	S and L
84.	*Phoenix sylvestris*–Arecaceae	Eetha	Leaves, Fruits	T	Leaves are used for making basket, brooms, mat. Fruits Edible	S and L
85.	*Phyllanthus amarus*–Euphorbiaceae	Nela Usiri	Leaves	H	Medicinal	S
86.	*Phyllanthus emblica*–Euphorbiaceae	Usiri	Fruits	T	Household, Market	S and L
87.	*Polyalthia cerasoides*–Annonaceae	Gutti	Leaves	T	Fodder for goats	S
88.	*Pongamia pinnata*–Papilionaceae	Kanuga	Seeds	T	Bodiesel	L
89.	*Pterocarpus marsupium*–Papilionaceae	Peddegi	Gum	T	Medicinal	L

Contd...

Table 12.1–*Contd...*

Sl.No.	*Name of the Species and Family*	*Common Name*	*Part Used*	*Habit*	*Purpose*	*Sustenance (S) and Livelihood (L)*
90.	*Sapindus emarginatus*–Sapindaceae	Kunkudu	Fruits	T	Used in shampoo preparation, Market	S and L
91.	*Schleichera oleosa*–Sapindaceae	Busi	Bark	T	Medicinal	S
92.	*Semecarpus anacardium*–Anacardiaceae	Jeedi	Fruits	T	Marking nut, seed coat resin	L
93.	*Soymida febrifuga*–Melaiceae	Somi	Gum	T	Medicinal	L
94.	*Sterculia urens*–Sterculiaceae	Yerra poliki	Gum	T	Gum is used in ice cream, pharmaceutical products	L
95.	*Streblus asper*–Moraceae	Barrenka	Branches, Leaves	S	Wood implements, leaves are rough enough to polish wood implements	S
96.	*Strychnos nux-vomica*–Loganiaceae	Visha -mushti	Fruits	T	Strychnine is extracted, Medicinal	L
97.	*Strychnos potatorum*–Loganiaceae	Chilla	Seeds	T	Seeds are useful for cleaning water, Household, market	L
98.	*Syzygium alternifolium*–Myrtaceae	Mogi	Fruits	T	Household, market	S and L
99.	*Syzygium cumini*–Myrtaceae	Neredu	Fruits	T	Household, market	S and L
100.	*Tamarindus indica*–Caesalpiniaceae	Chinta	Leaves, fruits	T	Young leaves, fruits are edible Household, Market	S and L
101.	*Terminalia alata*–Combretaceae	Nalla maddi	Bark	T	Tannins, laxative	L
102.	*Terminalia arjuna*–Combretaceae	Tella maddi	Bark	T	Medicinal	L
103.	*Terminalia bellirica*–Combretaceae	Tanikaya	Fruits	T	Myrobalan, fruits are used to prepare triphala	L
104.	*Terminalia chebula*–Combretaceae	Karakkaya	Fruits	T	Myrobalan, fruits are used to prepare triphala	L
105.	*Thespesia lampas*–Malvaceae	Konda patthi	Branches	S	Fibre	L
106.	*Tinospora cordifolia*–Menispermaceae	Tippateega	Aerial Roots	C	Aerial roots paste is usually applied on scorpion bites	L

Contd...

Table 12.1–*Contd...*

Sl.No.	*Name of the Species and Family*	*Common Name*	*Part Used*	*Habit*	*Purpose*	*Sustenance (S) and Livelihood (L)*
107.	*Tribulus terrestris*–Zygophyllacee	Palleru	Leaves	H	Medicinal	S
108.	*Vitex negundo*–Verbenaceae	Vavili	Leaves	S	Leaves powder is used for curing fever	S
109.	*Wrightia tinctoria*–Apocynaceae	Reppala	Stems and branches	T	To make toys	S
110.	*Ziziphus mauritiana*–Rhamnaceae	Regu	Fruits	T	Edible	L
111.	*Ziziphus oenoplia*–Rhamnaceae	Parika	Fruits	L	Edible, Household, market	S and L
112.	*Ziziphus xylopyrus*–Rhamnaceae	Gotti	Leaves	T	Fodder	S

frequency of medicinal and non medicinal plants that are used for either sustenance or for livelihood income generation revealed no significant difference $X^2_{(1)0.05} = 0.144$). It suggests that natural forests meet their primary health care needs as well as their livelihood needs and nothing else can supplement it.

Management of NTFPS

GCC plays a major role in collection and marketing the NTFPs throughout the state, in Kadapa division it procures 24 products and for about 13 products the corporation has fixed prices (Table 12.2). In regard to beedi leaf *(Diospyros melanoxylon)* which forms the major income source for state and employment for locals, the forest department auctions the units in the division to the private contractor for a fixed period of three years. The analysis of the procured 24 NTFPs in the decade (1992-2002) indicated that Gum Karaya *(Sterculia urens),* Maredu gaddalu *(Decalepis hamiltonii*), Chillaginjalu *(Strychnos potatorum),* Mushti *(Strychnos nux-vomica*), Myrobalan *(Terminalia bellirica*), Kondagogu gum *(Cochlospermum religiosum)* and Honey were collected steadily throughout the decade. Seven items namely Thirumanu gum *(Anogeissus latifolia*), Guggilam gum *(Boswellia serrata),* Pillitheegalu *(Asparagus racemosus*), Naramamidi bark *(Litsea glutinosa),* Mohua flower and seed *(Madhuca indica),* Kunkudu *(Sapindus emarginatus*) and Bee wax were procured in 5-10 years while the rest were procured in less than five years. This infrequent collection rates and the coefficient of variation values of the income generated in the order of 27 per cent to 168 per cent revealed a highly skewed demand and fluctuation of prices in the NTFPs in this area. This situation leads to unsustainable harvesting of few products, which was observed for *Decalepis hamiltonii* tuberous roots and under value for certain products as observed for broom sticks, Sarapappu (*Buchanania axillaris*) etc. The tuberous roots of *Decalepis hamiltonii* are usually harvested in monsoon period and cut into pieces and dried ones will be sold in the market. Due to huge demand for these tubers the roots are being extracted totally throughout the year affecting its growth and regeneration. The lack of support for Rella (*Cassia fistula*) bark in the local market led to severe logging of this tree for timber leading to its decline in population. It was observed that in recent times the demand for medicinal plant products of *Phyllanthus maderaspatanus, Eclipta prostrata, Cyperus rotundus, Tribulus terrestris, Boerhaavia diffusa* had increased in the local market and they are being harvested heavily. It suggests that value addition for certain products and sustainable harvest practices are necessary for increase in income for local people and also for forest conservation.

Conclusions

Globally, it is certain that NTFPs play an important role in the socio-economic life of the forest dwellers and the study strengthens this view point. The NTFP diversity and usage patterns revealed a significant contribution of medicinal plants to the primary health needs of the local people and as well for the sustenance of rural livelihoods; strengthening the view point that tropical dry deciduous forests provide socio-ecosystem services worth in comparison to the timber value. But these tropical dry forests are under immense pressure owing to the high population density and demand for fuel wood and timber especially of *Pterocarpus santalinus*. The results

Table 12.2: List of Non-Timber Forest Products and the Price Index Provided by GCC and the Procurement Details of Kadapa Division Forests.

Sl.No	*Non Timber Forest Product*	*Species and Family Name*	*Part Used*	*Price Index Provided by GCC (Rs)*	*Coefficient of Variation (SD/Mean X 100) of the Traded Amount (1992-2002)*
1.	Thirumanu	*Anogeissus latifolia*–Combretaceae	Gum	Variable, local market	38.9
2.	Pillitheegalu	*Asparagus racemosus*–Asparagaceae	Roots	12/kg	104
3.	Guggilam gum	*Boswellia serrata*–Burseraceae	Gum	18/kg	65.87
4.	Sarapappu	*Buchanania lanzan*–Anacardiaceae	Fruit	300/kg	
5.	Tangedu bark	*Cassia auriculata*–Caesalpiniaceae	bark	Variable, local market	Procured only twice
6.	Rella bark	*Cassia fistula*–Caesalpiniaceae	bark	Variable, local market	Procured only once
7.	Kondagogu gum	*Cochlospermum religiosum*–Cochlospermaceae	Gum	67/kg	49.8
8.	Maredu gaddalu	*Decalepis hamiltonii*–Asclepiadaceae	Tubers	15/kg, Heavily extracted and used in this area	127
9.	Beedi leaves	*Diospyros melanoxylon*–Ebenaceae	Leaf	As wage laborers under the contractor and now 50 per cent to VSS members if collected in the VSS area	Contract for 3 yrs
10.	Podaparti leaf	*Gymnema sylvestre*–Asclepiadaceae	Leaves	Variable, local market	Procured only thrice
11.	Chinna sugandhapala	*Hemidesmus indicus*–Asclepiadacceae	Root	Variable, local market	Procured only twice
12.	Naramamidi bark	*Litsea deccanensis*–Lauraceae		25/kg	70
13.	Mohua flower and seed	*Madhuca indica*–Sapotaceae	Corolla, Seed	11/kg	97
14.	Usiri	*Phyllanthus emblica*–Euphorbiaceae	Fruit	Variable, local market	Dry amla is procured twice
15.	Pongum seed	*Pongamia pinnata*–Papilionaceae	Seed	10/kg	Procured only twice
16.	Kunkudukayalu	*Sapindus emarginatus*–Sapindaceae	Fruit	Variable, local market	92.5

Contd...

Table 12.2–*Contd...*

Sl.No	*Non Timber Forest Product*	*Species and Family Name*	*Part Used*	*Price Index Provided by GCC (Rs)*	*Coefficient of Variation (SD/Mean X 100) of the Traded Amount (1992-2002)*
17.	Gum Karaya	*Sterculia urens*–Sterculiaceae	Gum	95/kg	7.1
18.	Mushti	*Strychnos nux-vomica*–Loganiaceae	Fruits	—-	168.4
19.	Cleaning nuts	*Strychnos potatorum*–Loganiaceae	Seed	6/kg	95
20.	Chintapandu	*Tamarindus indica*–Caesalpiniaceae	Fruit	25/kg	
21.	Myrobalan	*Terminalia bellerica*–Combretaceae	Fruit	Variable, local market	100.1
22.	Karakkayalu	*Terminalia chebula*–Combretaceae	Fruit	Variable, local market	Procured only twice
23.	Honey			80/Kg	42
24.	Beewax			—	27

indicate rich NTFP diversity in this area, and the NTFP based activities indicate that locals gain employment for major part of the year. Harvest of *Decalepis hamiltonii* tubers formed the major source of income for the local people and collection of gum karaya for GCC provided additional income and gathering of Tooki leaves formed the major agent for employment generation. On the other hand the procurement data revealed a highly skewed collected quantities and prices through years suggesting a demand based unsustainable harvest practices for few NTFPs. As the NTFPs involved large number of people and has a potential to generate employment throughout the year, policy initiatives in the lines of Mahatma Gandhi National Rural Employment Guarantee Scheme (MGNRES) that can provide decent amount of money for work can be initiated.

References

Anonymous. 2000. Working plan for the forests of Kadapa division. Forest Department, Government of Andhra Pradesh.

Bhat, P.R., Murali, K.S., Hegde, G.T., Shastri, C.M., Bhat, D.M., Murthy, I.K. and Ravindranath, N.H. 2003. Annual variation in non-timber forest product yield in the Western Ghats, Karnataka, India. Curr. Sci. **85 (9):** 1350-1355.

Bhuyan, L.R., Rao, A.N. and Chaudhry, P. 2010. Non timber forest products of Arunachal Pradesh. Scope, Prospects and problems. Bulletin of Arunachal Forest Research. **27 (1):** 47-53.

Chopra, K. 1993. The value of non-timber forest products: An estimation for tropical deciduous forests in India. Economic Botany. **47:** 251-57.

Chopra, K. 1997. The valuation and pricing of Non-Timber Forest Products: Conceptual issues and a case study from India. In F. Smith (ed.) Environmental Sustainability: Practical Global Implications, pp. 107.40. Florida: St Lucie Press.

Giridhar, A.K., Rao, R. J. and Lawrence, A. 2008. Sustainable harvesting of medicinal plants/non-timber forest products: development of methodology- concepts and practices. In Giridhar, A.K. and Rao, R. J. (eds.) Adaptive management of medicinal plants and Non timber forest products. Strategies, implications and policy. Bishen Singh Mahendra Pal Singh Publishers, Dehra Dun, India. pp. 18-37.

Johnson, T.S., Agarwal, R.K., and Agarwal, A. 2013. Non-timber forest products are source of livelihood option for forest dwellers: role of society, herbal industry and government agencies. Curr Sci. **104 (4):** 440-443

Kumar J.I.N., Rita N. Kumar, Rohit Kumar Bhoi and Sa-jish, P.R. 2010. Tree species diversity and soil nutrient status in three sites of Tropical Dry Deciduous Forest of Western India. Trop. Ecol. **51 (2)**: 273-279.

Kumar, N. 2003. 'Sustainable management of NWFPs through their sustainable utilization: Madhya Pradesh experience', pp. 61-67, in Proceedings of the National Workshop on Non-Wood Forest Products: New Mechanism and Strategies. Organized by the Ministry of Environment and Forests, Government of India and the Madhya Pradesh State Minor Forest Produce Co-operative Federation Ltd., 29-30 May 2003, Vigyan Bhavan, New Delhi.

Malhotra, K.C. and Bhattacharya, P. 2010. Forest and Livelihood. Published by CESS, Hyderabad. pp.246

Nepstad, D.C. and Schwartzman, S. 1992. Non-Timber products from Tropical forests: Evaluation of a conservation and development Strategy. New York: New York Botanical Garden.

Omkar, K., Sateesh, S., Samata, A., Ragan, A. and Raju, V.S. 2012. Diversity of NTFPs and their utilization in Adilabad District of Andhra Pradesh, India. J. Plant studies. **1:** 33-46.

Pullaiah, T. and Muralidhara Rao, D. 2002. Flora of Eastern Ghats Hill Ranges of Southeast India. Vol.1 (Ranunculaceae-Moringaceae). Regency Publications New Delhi.

Raghubanshi, A.S. and Tripathi, A. 2009. Effect of disturbance, habitat fragmentation and a line invasive plants on floral diversity in dry deciduous tropical forest of Vindhyan highlands – A review. Tropical Ecol. **50:** 57-69.

Rawat, G.S. 1997. Conservation status of forests and wildlife in the Eastern Ghats, India. Enviromental Conserv. **24**: 307-315.

Shahabuddhin,G. and Soumya, P. 2004. Assessing Ecological sustainability of Non Timber Forest produce extraction: The Indian scenario. Conservation and Society **2:** 235-250.

TFRI 2011. National Workshop on Non-timber Forest Products Marketing: Issues and Strategies (Background paper). Tropical Forest Research Institute, Jabalpur.

Ved, D.K. and Goraya, G.S. 2007. Executive summary. In Demand and supply of medicinal plants in India. NMPB, N.Delhi and Foundation for Revitilisation of Local Health Traditions, Bangalore.

Ved, D.K., Prathima, C.L., Morton, N. and Shankar, D. 2001. Conservation of India's medicinal plant diversity through a novel approach of establishing a network of *in situ* gene banks. In: Shankar, U., Ganeshaiah, K.N. and Bawa, K.S. (eds.). Forest Genetic Resources: Status, Threats and Conservation Strategies. Oxford and IBH Publishing Company. New Delhi, India. pp. 183-195.

Chapter 13

Biodiversity of Medicinal Plants in Ponmalai Hills, Kalkulam Taluk in Kanyakumari District, Tamil Nadu, India

A. Eruthaya Angel and B. Parthipan *

P.G. and Research Centre, Department of Botany,
South Travancore Hindu College, Nagercoil – 629 002, Tamil Nadu

Introduction

India is recognized as a country rich in all aspects of Biodiversity *viz.* ecosystem, species and genetic diversity mainly due to its tropical location, disparate physical features and climatic types. This is enhanced by confluence of the three major bio-geographic realms *viz.*, Indo-Malayan, the European and Afro-tropical. The floral wealth of India is around 40,000species (Jain, 1987). India is also estimated to have around 30 per cent of all flowering plants and 18 per cent all plants considered to be endemic (Kothari, 1994). This continent has well-documented traditional knowledge, long standing practice of traditional medicine and the potential for social and economic development of medicinal and aromatic plants constitute the basis of primary health care for the majority of the population in Asia and are critical source of income for rural population. They earn their living by selling wild or cultivated plant materials. Approximately 90 per cent of the plants are still collected from forests.

---

* Corresponding Author: E-mail: parthipillai64@ gmail.com

The global demand for herbal medicines are not only large, it has a growing demand. An estimate of the EXIM Bank puts the international market of medicinal plants related trade at US $ 60 billion per year growing at the rate of 7 per cent only. Though India has a rich biodiversity, the growing demand is putting heavy strain on the existing resources (Ministry of Health, 1991 and LSPSS, 1993). Approximately 166 companies import crude extract of medicinal and aromatic plants from India (Samydurai, *et al.*, 2012). Nearly 90 per cent of plant species used in industries is collected from wild habitat (Srivastava, 2000). In India it is estimated that traditional healers use 2500 plant species of plants serve as regular sources of medicine (Pie, 2001). Tamil Nadu is endowed with 3,000 species of plants of which 123 species are considered endangered by the Botanical Survey of India (Anon, 1990). The conservation of a wide variety of species in natural forests depends on maintaining essential components of the relevant ecosystem (Gilbert, 1980 and Whittmore, 1990).

A number of published data was available regarding the flora of various hill ranges of Kanyakumari District (Beddome, 1877., Lawson, 1894., Bourdillon, 1908., Rama Rao, 1914.,Lawrence, 1959, 1960., Nayar, 1959., Sharma *et al.*, 1973., Rao *et al.*1975, Henry and Saminathan, 1981).Very few published research work on survey of medicinal plants in Kanyakumari district was available. Selvi (2004) made a survey of medicinal plants in Marunduvalmalai hills and she reported that about 113 medicinally important plants are used by local people to cure more than 59 ailments from Marunduvalmalai hills, Kanyakumari District. In recent years some workers (Kingston and Raj, 2000; Jeeva *et al.*, 2006; Sukumaran *et al.*, 2008; Chenthurapandy *et al.*, 2010; John De Britto *et al.*, 2010 and Kumari Subitha *et al.*, 2011) have reported various medicinal plants used in traditional medicines by local and tribal's in Kanyakumari District. However, no published research work was available from the present study area.

Considering all the above aspects, the present research work was undertaken to survey of medicinal plants in Ponmalai hills and to document data on traditional knowledge of the rural peoples in Chunkankadai village, Kalkulam Taluk of Kanyakumari District.

Material and Methods

Description of Study Area

The Kanyakumari district is located in the southernmost part of the Indian Peninsula. It is the smallest district in Tamil Nadu, covering an area of 1684 Km^2 (Gopalakrishnan, 1995). Kanyakumari district lies between in 77° 05′–77° 36′ E longitude and 8° 5′– 8° 35′ N latitude. The present study was conducted in Ponmalai hill areas of Kanyakumari District. The District is bounded by the Tirunelveli district on the north and the south and south west are surrounded by the Indian Ocean and the Arabian Sea and the west and northwest by Kerala state (Figure 13.1). The District may be divided in to three geographical regions, first the mountainous region with a number of hill tops, next the coastal region and the undulating valleys and finally the plains between the mountainous terrain and the sea coast. The minimum and maximum temperature is 14.2° C and 32 ° C December and May, respectively. Monthly

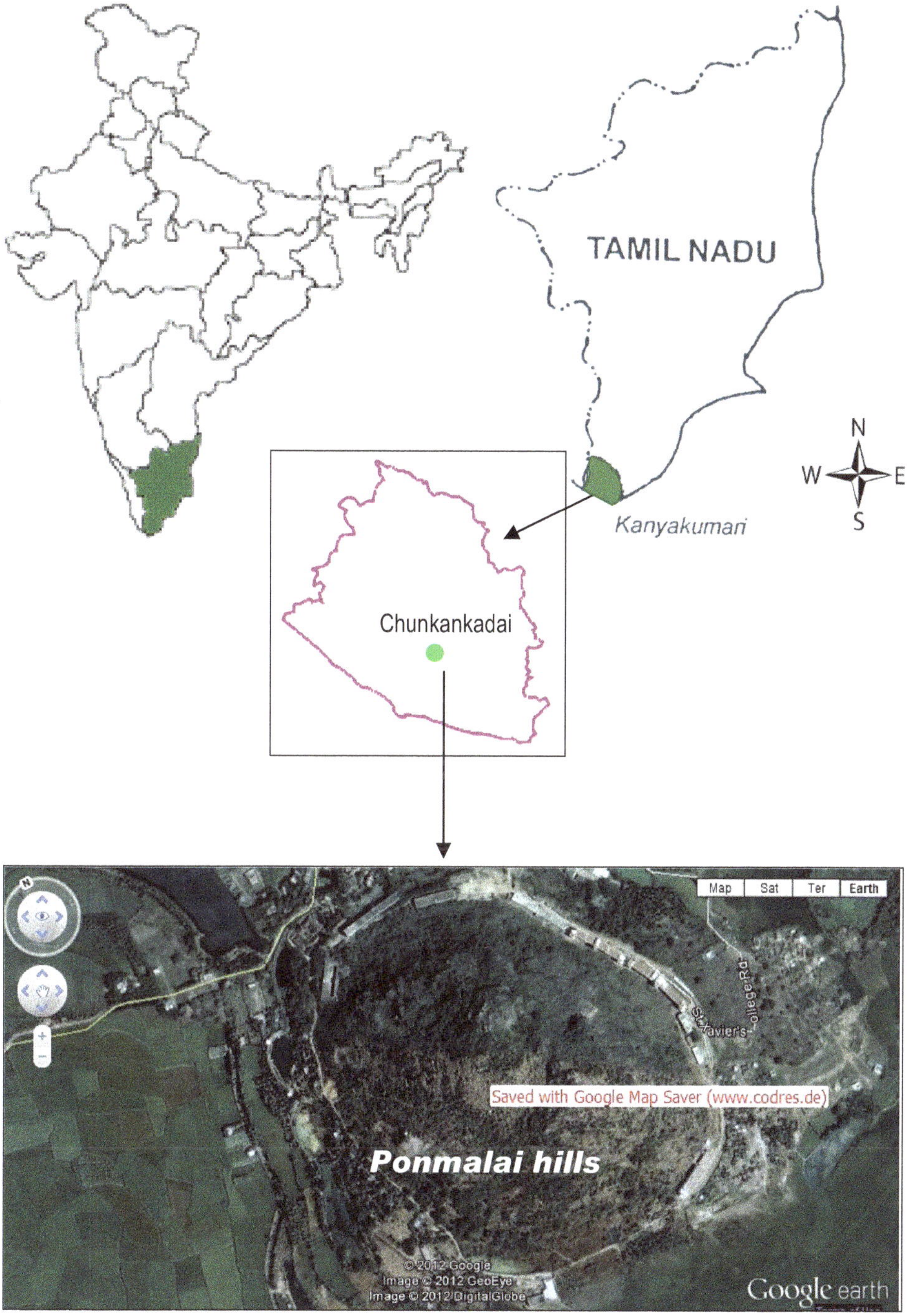

Figure 13.1: Map of the Ponmalai Hills.

rainfall ranges between 20.59 mm and 206.58mm. The district has a unique advantage of rainfall both during the southwest monsoon between June and September and northeast monsoon between October and middle of December.

The study area selected is Ponmalai hills, which lies 8° 14′ N latitude and 77° 22′ E longitude (Figure 13.1). It is 7 Km towards West in the NH – 47 and situated on the Northern side of Nagercoil – Trivandrum main Road. In many places the hill ranges project forwards into the land surface enclosing fertile areas. The sides of the hills are uneven and have lesser barren rocks. During rainy season, water stagnates between the depressions of the rocks. In the foot of the hill, where water is assured, *Cocos nucifera* have been cultivated. In dry tracts *Emblica officinalis, Anacardium occidentale, Mangifera indica* and *Eugenia jambolana* have been cultivated. As seen in other hills, Ponmalai hill is also exploited for quarrying. The Velimalai range, which is very close to Ponmalai hill, has mixed dry and moist deciduous forests. Due to over exploitation, the present study area, Ponmalai hills has very sparse vegetation. Another different feature observed is its proximity to human settlement. So it is likely to be more influenced by anthropogenic factors. Moreover this area was chosen by OISCA for afforestation programme financed by Japan, on July 1999 and 10,000 saplings were planted. Hence the documentation of available plants in this area has become an urgent necessity. This document will become basic information to study the afforestation effect on the vegetation.

Plant Collection

Periodical visits have been undertaken for collecting fresh plant specimens. The collected plants were tagged and brought to the laboratory for examination. Plants were collected according to the field and herbarium techniques (Matthew, 1981 and Pullaiah, 2007). The information on plants was collected by interviewing local traditional practitioners and village peoples. The collected medicinal plants were botanically identified using the Flora of Presidency of Madras (Gamble, 1935) and the Flora of Tamil Nadu Carnatic (Matthew, 1983). Plants were identified in the field with some of the local peoples, plant collectors and traditional healers. Herbarium specimens were deposited at the STHC Herbarium, Department of Botany, South Travancore Hindu College, Nagercoil for further reference.

The identified plants were verified and confirmed by The Botanical Survey of India, Southern Circle, and Coimbatore. Abbreviations were used wherever necessary, name of the plant collectors and numbers were followed at the end of the descriptions.

Result and Discussion

The qualitative medicinal floristic survey made from the Ponmalai hills of Chunkankadai village in Kanyakumari District revealed the presence of 88 species belonging to 80 genera in 43 families. Plant binomial, family, life form and voucher number are provided in (Table 13.1). Among life forms herbs dominate the hills with 35 sp. followed by shrubs (33sp.) trees (12sp.), climbers (4sp.) and Twiners (4sp.) (Figure 13.2). Out of 43 plant families Acanthaceae, Amaranthaceae, Asclepidaceae, Asteraceae, Convolvulaceae, Malvaceae, Rubiaceae, and Verbenaceae (3 species each) and Vitaceae (2 species), whereas 27 families represented by a single species which

Table 13.1: List of Medicinal Plants Collected from Ponmalai Hills in Chunkankadai Village of Kanyakumari District.

Sl.No	*Voucher No.*	*Botanical Name and Family*	*Habit*	*Mup*	*Ailments Cured*
		Acanthaceae			
1.	3163	*Barleria cristata* L.	H	L,R	Cough, inflammations, swelling of joints.
2.	3127	*Indoneesiella echioides* (L.) Sreem.	H	R	Fever.
3.	3147	*Justicia tranquebariensis* (L.) f.	H	L	Nasal disorders, small pox.
		Amaranthaceae			
4.	3144	*Achyranthes aspera* L.	S	Se, L, Wp, B	Anaemia, ear diseases, head ache, renal dropsies, piles, worm infestation.
5.	3148	*Digeria muricata* (L.) Mart.	H	F, Se	Urinary disorders.
6.	3117	*Gomphrena globosa* L.	H	R	Cough.
		Anacardiaceae			
7.	3178	*Mangifera indica* L.	T	F, B, Se	Dysmenorrhoea, urticaria, asthma.
		Annonaceae			
8.	3185	*Anona squamosa* L.	S	R, Se, Fr	Epilepsy, dysuria, diarrhea, dysentery.
		Apocynaceae			
9.	3126	*Carissa spinarum* L.	S	R	Eye and ear disorders, vomiting, body ache, throat pain.
10.	3187	*Rauvolfia tetraphylla* L.	S	R	Skin aliments.
		Asclepiadaceae			
11.	3154	*Pergularia daemia* (Forssk). Chiov.	C	R, L	Asthma, diarrhoea, carbuncles, uterine, menstrual troubles.
12.	3191	*Tylophora asthmatica* (L.f.) Wt & Arn.	S	Wp	Stomachic, diaphoretic, asthma, bronchitis, cough.
13.	3150	*Calotropis gigantea* (L)R. Br.	S	L	Diseases of nervous system, leprosy, swelling in joint, snake and rat bite poisonings, abdominal disorders, worm infestation.

Contd...

Table 13.1–*Contd...*

Sl.No	Voucher No.	Botanical Name and Family	Habit	Mup	Ailments Cured
		Asteraceae			
14.	3190	*Parthenium hysterophorus* L.	H	Wp	Dysentary, fever.
15.	3195	*Tridax procumbens* L.	H	L	Dysentery, diarrhoea, wounds.
16.	3111	*Vernonia cineria* (L.) Less	H	Wp	Malarial, infection, diarrhoea, fever, round worms, cough, leucoderma, soriasis, other skin diseases.
		Bignoniaceae			
17.	3184	*Milingtonia hortensis* L. f.	T	B	Fever, mouth disease, foot disease.
		Caesalpiniaceae			
18.	3188	*Caesalpinia pulcherrima* (L.) Sw.	S	B, L	Bronchitis, fever, dental disorders.
19.	3151	*Cassia absus* L.	H	Se, L	Skin troubles, wounds, cough, ulcers, pain.
20.	3133	*C. alata* L.	S	L, Se	Skin troubles, haematuria.
21.	3121	*C. auriculata* L.	S	F, R, L, Se	Skin troubles, polyuria, eye troubles, diabetes, leprosy, diarrhoea.
22.	3189	*C.fistula* L.	T	R, L, F.	Skin troubles, leucorrhoea, leprosy, fever, cardiac disease, Jaundice.
23.	3181	*Tamarindus indica* L.	T	Fr	Anaemia, pain, urinary disorder.
		Capparaceae			
24.	3155	*Cadaba trifoliata* (Roxb.) Wt. & Arn.	S	R, L	Swelling in joints, disorders of blood, spleenic disorders.
25.	3142	*Cleome viscosa* L.	H	L, Se	Chronic painful joints, ear and nasal disorders, abdominal disorders, indigestion.
		Combretaceae			
26.	3177	*Quisqualis indica* L.	S	F, Se	Soporific, diarrhoea, rickets, fever, skin troubles.

Contd...

Table 13.1–*Contd...*

Sl.No	*Voucher No.*	*Botanical Name and Family*	*Habit*	*Mup*	*Ailments Cured*
		Convolvulaceae			
27.	3180	*Evolvulus alsinoides* (L.) L.	H	Wp	Promoting hair growth.
28.	3193	*Ipomoea pes-tigridis* L.	H	L	Sores, pimples, carbuncles.
29.	3194	*Merremia tridentata* (L.) Hall. f.	H	R	Tooth ache, arthritis, urinary disorders, piles.
		Elaeocarpaceae			
30.	3128	*Muntingia calabura* L.	T	F	Headache, cold.
		Erythroxylaceae			
31.	3156	*Erythroxyllum monogynum* Roxb.	S	W, B	Stomach ache, diaphoresis, dyspepsia, fever.
		Euphorbiaceae			
32.	3172	*Breynia retusa* (Dennst.) Alston	S	L	Disease of nervous system, oedema, disorders of blood.
33.	3112	*Croton sparsiflorus* L.	H	L, Se	Wash sting, blood disorders, dental diseases, skin disease.
34.	3131	*Euphorbia antiquorum* L.	S	S, R, La	Nervine troubles, dropsy, urticaria warts, oedema cutaneous affections.
35.	3137	*E. thymifolia* L.	H	L, Se	Urinary disorders, dyspnoea, ringworms.
36.	3158	*E. tirucalli* L.	S	L	Warts, intestinal worms, tooth ache, skin diseases.
37.	3136	*Jatropha gossypifolia* L.	S	R, Se, La	Leprosy, stomachic, fevers, ulcers.
38.	3170	*Phyllanthus amarus* Schum. & Thonn.	H	Wp	Stomach troubles, diarrhoea, dysentery, dysepepsia, dropsy, jaundice, sores.
39.	3140	*P. maderaspatensis* L.	H	Se	Headache.
40.	3139	*Sebastiania chamaelea* (L.) Muell-Arg.	H	L	Diarrhoea, stop vomiting.
41.	3160	*Securinega leucopyrus* (Willd.) Muell. Arg.	T	L, S	Sores, digestive disorders

Contd...

Table 13.1–*Contd...*

Sl.No	*Voucher No.*	*Botanical Name and Family*	*Habit*	*Mup*	*Ailments Cured*
		Lamiaceae			
42.	3132	*Hyptis suaveolens* (L.) Poit.	H	L, R	Uterous affections, parasitical cutaneous disease, stomachic, worm infestation.
43.	3146	*Leucas aspera* (Willd.) Link	H	L, F	Psoriasis, skin eruptions, painful swellings, cough, cold, fever.
44.	3135	*Ocimum tenuiflorum* L.	H	Wp	Bronchitis, ringworms, cutaneous, dyspnoea, stomach ache, malarial fever, urinary disorders, mental disorders.
45.	3159	*Orthosiphon thymiflorum* (Roth) Sleensen	S	L	Diabetes.
		Lauraceae			
46.	3114	*Cassytha filiformis* L.	Tw	Wp	Anorexia, Worm infestation, urethritis, dysentery, eye, skin affections.
		Liliaceae			
47.	3164	*Sansevieria roxburghiana* S J. & J. Schult.	H	Rh	Cough, earache, oedema, tuberculosis.
		Loranthaceae			
48.	3171	*Dendrophthoe falcata* (L.f.) Etting.	H	B	Sterility in female, menstrual troubles, asthma, wounds, worm infestation.
		Lythraceae			
49.	3174	*Lawsonia inermis* L.	T	L	Leprosy, cardiac diseases, diarrhoea with blood, blood disorders, skin troubles.
		Malvaceae			
50.	3125	*Abutilon indicum* (Lam.) Sweet	S	Wp, R	Fever, piles, lumbago, urinary troubles, leucoderma.
51.	3134	*Pavonia zeylanica* (L) Cav.	H	Wp	Fever, eye disease.
52.	3145	*Sida acuta* Burm. f.	H	R, L	Stomachic, urinary disorders, worms, infection swellings, elephantiasis.

Contd...

Table 13.1–*Contd...*

Sl.No	*Voucher No.*	*Botanical Name and Family*	*Habit*	*Mup*	*Ailments Cured*
		Meliaceae			
53.	3124	*Azadirachta indica* A. Juss.	T	Wp	Fever, liver, disorders, dyspnoea, ulcers, eczema, cough, abdominal disorders, stomach ache, skin troubles.
		Menispermaceae			
54.	3162	*Tinospora cordifolia* (Willd). Hk. f. & Th.	Shurb	L, S, R	Leprosy, dyspepsia, jaundice, fever, urinary diseases.
		Mimosaceae			
55.	3153	*Mimosa pudica* L.	H	R, L	Urinary disorders, sores, piles.
		Moraceae			
56.	3182	*Ficus benghalensis L.*	T	La, B	Diarrhoea, dysentery, diabetes, vomiting, polyuria, dental disorder, ulcers.
		Nyctaginacea			
57.	3165	*Boerhavia erecta* L.	H	R	Urinary disorders, anaemia, stomach ache, asthma, nasal disorder, arthiritis.
		Oleaceae			
58.	3176	*Jasminum angustifolium* Willd.	S	L, R	Ring worm affections and honey bee sting poisoning.
		Papilionaceae			
59.	3179	*Abrus precatorius* L.	S	R, Se	Stiffness of shoulder joints, leprosy, paralysis pain, eye disease.
60.	3123	*Clitoria ternatea* L.	T	L, R, Se	Leucorrhoea, fever piles, epilepsy, skin diseases.
61.	3166	*Crotalaria laburnifolia* L.	S	Se	Blood disorders, skin disease, obesity.
62.	3129	*C. retusa* L.	S	Wp	Scabies.
63.	3130	*C. verrucosa* L.	H	L	Scabies.
64.	3197	*Derris scandens* (Roxb.) Benth.	S	R	Insect bite poisoning.

Contd...

Table 13.1–*Contd...*

Sl.No	*Voucher No.*	*Botanical Name and Family*	*Habit*	*Mup*	*Ailments Cured*
65.	3138	*Indigofera asphalathoides* Vahl ex DC.	S	Wp	Cancerous affections, leprosy, skin disease, abscesses.
66.	3161	*Stylosanthes fruticosa* (Retz.) Alst.	H	L	Diarrhoea and colds.
67.	3113	*Tephorsia purpurea* (L.) Pers.	S	L, F	Asthma, cough, stop vomiting, scabies, Itch, eczema, other skin eruptions.
		Passifloraceae			
68.	3186	*Passiflora foetida* L.	C	Fr, L	Asthma, hysteria.
		Pedaliaceae			
69.	3192	*Pedalium murex* L.	H	Wp	Urino-genital-disorders, dysuria, impotence.
		Plumbaginaceae			
70.	3118	*Plumbego zeylanica* L.	S	R	Dyspepsia, piles, anasarca, anoroxia diarrhoea, skin diseases, abscesses.
		Polygalaceae			
71.	3167	*Polygala arvensis* Willd.	H	Wp	Snake-bite poisoning.
		Rhamnaceae			
72.	3175	*Ziziphus oenoplia* (L.) Mill.	S	R, Fr	Ascaris infection, stomach ache, urinary disorders.
		Rubiaceae			
73.	3173	*Catunaregam spinosa* (Thunb.) Thiruv.	T	B, Se, R	Bone-ache, fever, pain, skin troubles, diarrhoea dysentery, digestive disorder.
74.	3183	*Morinda tinctoria* Roxb.	T	R, L, F, B	Eczema, fever, ulcers, glandular swellings, digestive disorders.
75.	3141	*Spermacosa hispida* L.	H	L, Se	Haemorrhoids, diarrhea, dysentery.
		Rutaceae			
76.	3120	*Toddalia asiatica* L.	S	R, B, L	Stomachic, malarial disease, antiperiodic, diarrhea, asthma.

Contd...

Table 13.1–*Contd...*

Sl.No	*Voucher No.*	*Botanical Name and Family*	*Habit*	*Mup*	*Ailments Cured*
		Sapindaceae			
77.	3143	*Cardiospermum halicacabum* L.	H	R, L	Lumbago, nerves diseases, glandular swelling, eczema, ear diseases.
78.	3116	*Dodonaea viscosa* Jacq.	T	L	Burns, swellings, wounds.
		Scorphulaceae			
79.	3198	*Russelia equisetiformis* Schl.	S	Wp	Malarial fever, cancer, Inflammatory disease.
		Solanaceae			
80.	3157	*Solanum xanthocarpum* Schrad. & Wendl.	H	R, L, Se	Cough, asthma, tuberculosis, epilepsy, sore-throat.
		Sterculiaceae			
81.	3168	*Melochia corchorifolia* L.	H	L	Dysentary, leucorrhoea, venereal diseases.
		Urticaceae			
82.	3196	*Pouzolazia zeylanica* (L.) Benn.	H	L	Ulcers, and fever, sores.
		Verbenaceae			
83.	3119	*Gmelina asiatica* L.	S	R, W	Digestive disorder, cystitis, fevers, leucorrhoea.
84.	3152	*Lantana camara* L.	S	L	Tumour infection, malarial fever, abdominal disorders.
85.	3169	*Stachytarpheta jamacensis* (L.) Vahl	S	L	Venereal disease, ulcers, erysipelas, stomach aliments, cataract, diarrhoea, dysentery, cardiac troubles.
		Vitaceae			
86.	3115	*Cayratia carnosa* (Wall.ex Wt. & Arn.) Gagnep.	S	R, L	Spleenic disorders. heart disease, abdominal disorders.
87.	3122	*Cissus quadrangularis* L.	S	S, R	Piles, diarrhoea, dysentery, leucorrhoea, scurvy disease, fractures, abdominal disorders.
		Zygophylllaceae			
88.	3149	*Tribulus terrestris* L.	H	R, Fr, L	Stomachic, calculous affections, leucorrhoea.

H: Herb; S: Shrub; T: Tree; Cl: Climber; Tw: Twiner; L: Leaf; R: Root; WP: Whole plant; Se: Seeds; Fr: Fruit; B: Bark; W: Wood; S: Stem; F: Flower; La: Latex.

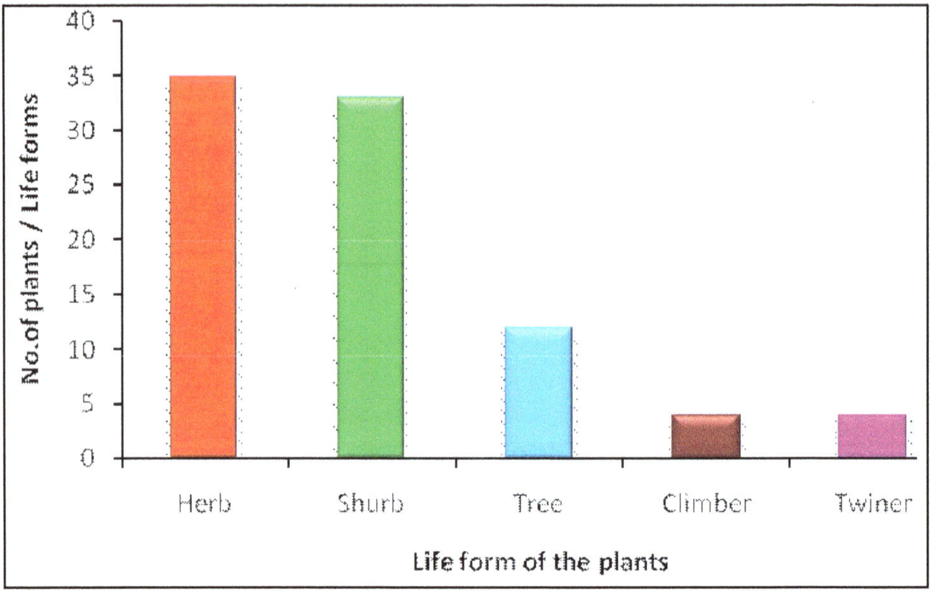

Figure 13.2: Habit of Medicinal Plants Collected from the Ponmalai Hills.

include Anacardiaceae, Annonaceae, Bignoniaceae, Combretaceae, Elaeocarpaceae, Erythroxylaceae, Lauraceae, Liliaceae, Loranthaceae, Lythraceae, Meliaceae, Menispermaceae, Mimosaceae, Moraceae, Nyctaginaceae, Oleaceae, Passifloraceeae, Pedaliaceae, Plumbaginaceae, Polygalaceae, Rhamnaceae, Rutaceae, Scorphulaceae, Solanaceae, Sterculiaceae, Urticaceae and Zygophyllaceae. (Table 13.2).

The most speciose families include Euphorbiaceae (10 sp. *viz. Breynia retusa, Croton sparsiflorus, Euphorbia antiquorum, E. thymifolia, E. tirucalli, Jatropha gossypifolia, Phyllanthus amarus, P. maderaspatensis, Sebastiania chamaelea* and *Securinega leucopyrus*) followed by Papilionaceae (9 sp.), Caesalpiniaceae (6 sp.) and Lamiaceae (4 sp.) (Figure 13.3).

A variety of plant morphological structures were utilized as medicine by the local people. Out of 88 medicinal plant species leaves of 46 plant species (31.5 per cent), roots of 33 plant species (22.6 per cent), seeds of 19 plant species (13 per cent), whole plant of 16 plant species (10.9 per cent), bark of 9 plant species (6.16 per cent), flower of 7 plant species (4.79 per cent), fruits of 6 plant species (4.1 per cent), stem of 4 plant species (2.73 per cent), latex of 3 plant species (2.05 per cent), wood of 2 plant species (1.36 per cent) and rhizome of 1 plant (0.68 per cent) were used for the treatment of various ailments by local people of present study area (Figure 13.4).

The present investigation revealed that 88 medicinal plants are used by local peoples of the study area for different diseases (Table 13.3). Skin disease and diarrhoea have the highest number of remedy where 14 medicinal plants are used. Skin disease and diarrhoea is followed by stomach ache with 13 plants and fever 12 species as remedy. Similarly dysentery is used with 11 plants. Ten species are used by the local people to treat cough.

Table 13.2: Families in Relation to Genera and Species Collected from Ponmalai Hills.

Sl.No.	Family	No. of Genera	No. of Species	Sl.No.	Family	No. of Genera	No. of Species
1.	Acanthaceae	3	3	2.	Amaranthaceae	3	3
3.	Anacardiaceae	1	1	4.	Annonaceae	1	1
5.	Apocynaceae	2	2	6.	Asclepidaceae	3	3
7.	Asteraceae	3	3	8.	Bignoniaceae	1	1
9.	Caesalpiniaceae	3	6	10.	Capparaceae	2	2
11.	Combretaceae	1	1	12.	Convolvulaceae	3	3
13.	Elaeorcarpaceae	1	1	14.	Erythroxylaceae	1	1
15.	Euphorbiaceae	7	10	16.	Lamiaceae	4	4
17.	Lauraceae	1	1	18.	Liliaceae	1	1
19.	Loranthaceae	1	1	20.	Lythraceae	1	1
21.	Malvaceae	3	3	22.	Meliaceae	1	1
23.	Menispermaceae	1	1	24.	Mimosaceae	1	1
25.	Moraceae	1	1	26.	Nyctaginaceae	1	1
27.	Oleaceae	1	1	28.	Papilionaceae	7	9
29.	Passifloraceae	1	1	30.	Pedaliaceae	1	1
31.	Plumbaginaceae	1	1	32.	Polygalaceae	1	1
33.	Rhamnaceae	1	1	34.	Rubiaceae	3	3
35.	Rutaceae	1	1	36.	Sapindaceae	2	2
37.	Socrphulariaceae	1	1	38.	Solanaceae	1	1
39.	Sterculiaceae	1	1	40.	Urticaceae	1	1
41.	Verbenaceae	3	3	42.	Vitaceae	2	2
43.	Zygophyllaceae	1	1				

Table 13.3: Number of Herbal Remedies Used to Treat Various Ailments.

Sl.No.	Name of the Disease	Total Number of Herbal Remedies	Sl.No.	Name of the Disease	Total Number of Herbal Remedies
1.	Stiffness of shoulder joints	2	2.	Leprosy	7
3.	Piles	7	4.	Urinary troubles	8
5.	Leucoderma	2	6.	Worm infestation	6
7.	Dysentary	11	8.	Diarrhoea	14
9.	Dysuria	2	10.	Stomach ache	13
11.	Ulcers	5	12.	Antiseptic	1
13.	Liver disorder	1	14.	Nasal disorders	2
15.	Asthma	4	16.	Anaemia	1
17.	Oedema	1	18.	Blood disorders	4

Contd...

Table 13.3–*Contd...*

Sl.No.	Name of the Disease	Total Number of Herbal Remedies	Sl.No.	Name of the Disease	Total Number of Herbal Remedies
19.	Swelling	4	20.	Dental disorders	2
21.	Nervous disorder	3	22.	Abdominal disorder	4
23.	Glandular swelling	2	24.	Lumbago	1
25.	Eye disorder	3	26.	Ear disorder	2
27.	Body ache	1	28.	Vomiting	3
29.	Skin troubles	14	30.	Wounds	2
31.	Cough	10	32.	Haemuturia	1
33.	Diabetes	3	34.	Polyuria	2
35.	Fever	12	36.	Cardiac disease	2
37.	Boneache	1	38.	Pain	2
39.	Heart disease	1	40.	Fractures	1
41.	Indigestion	1	42.	Leucorrhoea	6
43.	Scabies	3	44.	Scurvy disease	1
45.	Obesity	1	46.	Impetigo	2
47.	Sterility in female	1	48.	Insect- bite poisoning	1
49.	Burns	1	50.	Dyspepsia	4
51.	Warts	2	52.	Urticaria	1
53.	Tooth ache	2	54.	Ring worms	1
55.	Hair – growth	1	56.	Cystitis	1
57.	Cutaneous disease	2	58.	Cancer	2
59.	Sores	5	60.	Pimples	1
61.	Honey bee sting poisoning	1	62.	Fistulae	1
63.	Malarial infection	5	64.	Tumour infection	1
65.	Cold	3	66.	Psoriasis	2
67.	Haemorrhagre	1	68.	Mouth disorders	1
69.	Foot disorders	1	70.	Eczema	2
71.	Headache	2	72.	Malarial fever	1
73.	Bronchitis	4	74.	Mental disorders	1
75.	Urine – gental disorders	1	76.	Jaundice	2
77.	Snake – bite poisoning	1	78.	Rickets	1
79.	Soporific	1	80.	Tuberculosis	2
81.	Digestive disorder	1	82.	Elephantiasis	1
83.	Throat – pain	1	84.	Erysipelas	1
85.	Itch	1	86.	Calculous	1
87.	Ascaris infection	1			

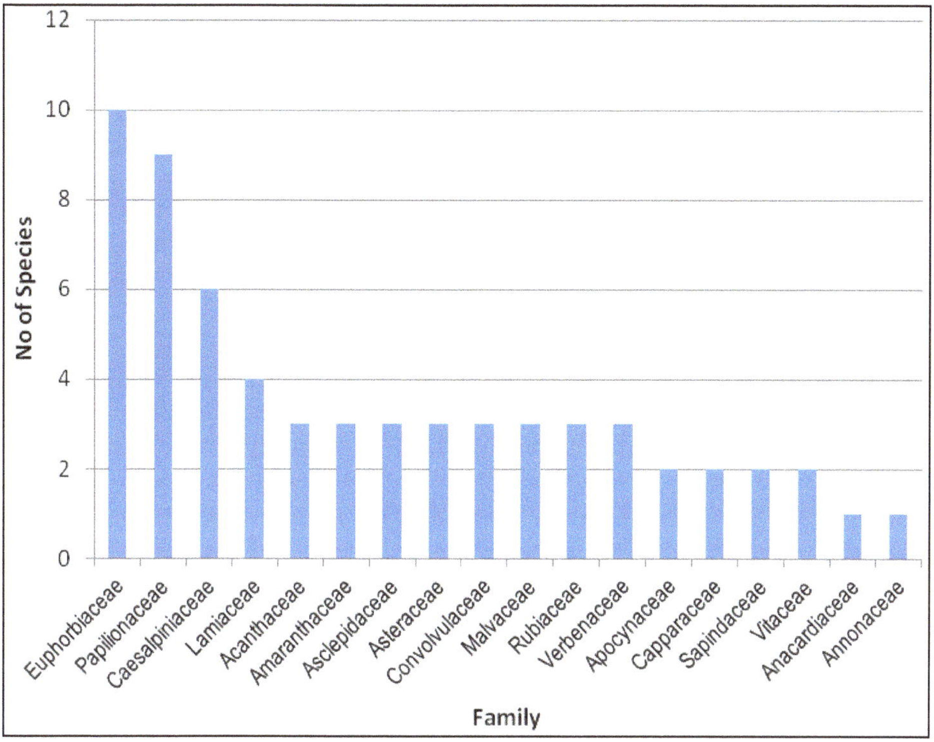

Figure 13.3: Analysis of Dominant Families with Respect to Number of Species.

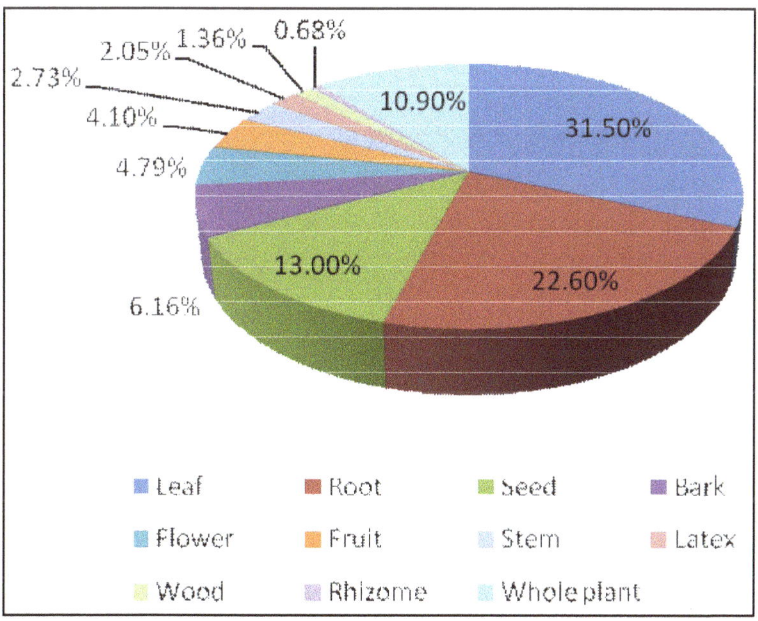

Figure 13.4: Number of Medicinal Plants and their Useful Parts Used for Rural Medicine Preparation by Local Peoples.

Leprosy and piles are treated by using 7 plants. 8 plants are used for treating urinary troubles. Leucorrhoea and worm infestation are treated by 6 plants. Malarial infection, sores, ulcers, are treated by using 5 plants by local people as medicine. For swellings, asthma, bronchitis, dyspepsia, abdominal disorders and blood disorders 4 plants are used as medicine. Cold, diabetes, scabies, eye disorder, vomiting, nervous disorder and diabetes are treated by using 3 plants each. Two plants each are used by local people as medicine for headache, cardiac disease, leucoderma, impetigo, warts, eczema, wounds, psoriasis, ear disorder, cancer, tooth ache, dental disorders, nasal disorders, polyuria, dysuria, jaundice, tuberculosis, glandular swelling, cutaneous disease and pain.

Many diseases such as bone ache, lumbago, itch, ascaris infection, anemia, erysipelas, antiseptic, ringworms, snake-bite poisoning, fistulae, hemorrhage, malarial fever, haematuria, digestive disorder, heart disease, fractures, mouth disorder, foot disorders, indigestion, promoting hair-growth, scurvy disease, obesity, throat-pain, body ache, sterility in female, cystitis, insect-bite poisoning, mental disorders, rickets, soporific edema, calculus affection, urticaria, burns, urino-gental disorders, tumour infection, liver disorder, pimples and elephantiasis, have one plant each as remedy (Table 13.3).

Table13.4: Antipoisoning Plants Collected from Ponmali Hills.

Sl.No.	*Botanical Name*	*Morphological Useful Part*	*Uses*
1.	*Calotropis gigantea*	Leaf	Snake-bite poisoning, Rat-bite poisoning
2.	*Derris scandens*	Root	Insect-bite poisoning
3.	*Jasminum angustifolium*	Leaf	Honey bee-sting poisoning
4.	*Polygala arvensis*	Whole plant	Snake-bike poisoning

Antipoisonous Plants

Out of 88 medicinal plants collected from Ponmalai hills four medicinal plants are used to treat poisonous bites. Table 13.4 Plant twig paste of *Polygala arvensis* is used to treat snake-bite poisoning. *Calotropis gigantea* leaf is also used to treat rat-bite poisoning and snake-bite poisoning. Root juice of *Derris scandens* is used as medicine for insect-bite poisoning. Leaf paste of *Jasminum angustifolium* is used to treat honeybee sting poisoning.

Nowadays, medicinal technology is well developed. This development is good for human beings to have good health and better quality of life. However, the cost of many medicines is too high for most people residing in the rural areas. Many people, therefore, still use traditional medicinal plants for curing diseases. Knowledge of medicinal plants provides people with low cost health care, and this knowledge is passed through generations (Palaniappan *et al.*, 2012).

With the passage of time, they have developed a great deal of knowledge on the use of plants and their products for curing various diseases. They have strong faith in their native folklore medicine for treatment of various diseases and they rely on medicinal plants for their own herbal treatment (Sajem and Gosai, 2006). Traditional medical knowledge of medicinal plants and their use by indigenous cultures are not only useful for conservation of cultural traditions and biodiversity but also for community healthcare and drug development in the present and future (Pei, 2006).

Conclusion

Medicinal plants play an important role in the medicinal and health care needs of rural people. This preliminary study focused on documenting most of the curative plants used by the local peoples in and around Ponmalai hills of Chunkankadai village. However, destruction of habitat through deforestation and over exploitation for commercial purposes and changes in cultural attitude threatens to constrain many of these species into extinction. From the conservational angle, the data presented here provides a basis for future studies on this ecosystem in general. In view of various resources use, habitat uniqueness and human pressure of the forest resources the need for conservation of Ponmalai hills. Conservation of valuable medicinal plants would develop the wealth of the forest and thereby the nation.

Acknowledgements

The authors are grateful to Dr. S. Karuppasamy, Asst. Prof. Dept. of Botany, Madura College, and S. Jeeva, Assistant Prof. Dept. of Botany, Scott Christian College, Nagercoil for their helps and support. The authors are also thankful to the Management and Principal of South Travancore Hindu College, Nagercoil for their support and their encouragement.

References

Anon, 1990. Tamil Nadu Forests.Tamil Nadu Forest Department, Madras.

Banerjee, R.N. and L.K. Banerjee. 1975. *Polygala raoi*. R.N. Ban & L.K. Ban. (Polygalaceae). A new species for Kanyakumari, Tamil Nadu, India. Proc. Indian. Acad. Sci. 82: 218-220.

Beddome, R.H. 1877. The forest flora of the Tirnevelly District, Indian Forester. 3: 19-25.

Bourdillion, T.F. 1908. The forest trees of Travancore. The Travancore Govt. Press. Trivandrum.

Chenthurapandy, P., V. R. Mohan and C. Kalidass. 2010. An ethnobotanical survey of medicinal plants used by the Kanikkar Tribe of Kanyakumari District of Western Ghats, Tamil Nadu for the treatment of skin diseases. J. Herbal. Med. Toxicol. 4 (1): 179-190.

Gamble, J.S. 1935. The Flora of Presidency of Madras. Adlard and son, Ltd.London. Vol I-III.

Gilbert,L.E, 1980. Food web Organisation and conservation of Neotropical diversity. Soule, M.E. and B.A. Wilcox. Conservation Biology, Sunderland, U.K. Sinaver. Pp. 11-33.

Gopalakrishnan,M. 1995. Gazetteers of India, Tamil Nadu state, Kanyakumari District, Govt. of Tamil Nadu. Chennai.

Henry,A.N. and M.S, Swaminathan.1981. Observations on the vegetation of Kanyakumari District. Tamil Nadu. Bull.Bot. Surv. India. 23: 135-139.

Jain, S.K. 1987. The problem of endangered species its study and solution.Presidential address, Biological Sciences, 57th session of National Academy of Sciences, Allahabad.

Jeeva,S. et. al. 2006. Weeds of Kanyakumari District and their value in rural life. Indian J. Trad. Knowledge 5 (4): 501-509.

John De Britto, A., R. Mary Sujin., R. Mahesh and K. Dharmar. 2010. Ethnomedicinal wisdom of the Manavalakuruchi people in Kanyakumari District, Tamil Nadu. Inter. J. Biol. Tech. 1 (2): 25-30.

Kingston,S and A.D.S. Raj. 2000. Plants used in ethnomedicine by the habitat people of Kanyakumari District. Proce. Mano. Res. Forum. B20-23.

Kothari, A. 1994. Widening perspectives on biodiversity. (Ed.) Krattiger, A.F. pp. 137-145.

Kumari Subitha., T. Ayyanar, M. Udayakumar and T. Sekar. 2011. Ethnomedicinal plants used by Kani tribals in Pechiparai forests of Southern Western Ghats, Tamil Nadu, India. Inter. Res. J. Pl. Sci. 2 (12) 349-354.

Lawrence, C.A. 1959. Observation on the flora of Marunduvalmalai. Cape Comerin. J. Bombay Nat. Hist. Soc. 56: 95-100.

Lawrence, C.A. 1960.The vegetation of Kanyakumari District (Cape Comorin). J. Bombay Nat. Hist. Soc. 57: 184-195.

Lawson, M.A. 1894. Notes of a tour in Travancore etc. Rec. Bot. Surv.India. 1 (4) 58-60.

LSPSS, 1993. Micro studies carried out by members of All India Network of NGO's (LSPSS, Coimbatore) involved in the revitalization of local health traditions.

Matthew, K.M. 1981.Materials for a flora of the Tamil Nadu Carnatic. The Rapinat Herbarium, St. Josephs College, Tiruchirapalli, India. Pp. 73-85.

Matthew, K.M. 1983. The flora of the Tamil Nadu Carnatic. Vol I-III. The Rapinat Herbarium St. Josephs College, Tiruchirapalli, India.

Ministry of Health. 1991. Statistics of ministry of Health, Government of India Report, 1991.

Nayar,M.P. 1959. The vegetation of Kanyakumari, kanyakuamri Disrtrict. Bull.Bot.Surv. India. 1: 122-126.

Palaniappan, P., M. Pandian., S. Natarajan and C. Pitchairamu. 2012. Ethnomedicinal wisdom of Alagar Hills in Eastern Ghats, Tamil Nadu, India. Inter. J. Appl. Biores. 6: 28-34.

Pei,J.S. 2001. Ethnobotanical approaches of traditional medicinal studies: some experiences from Asia. Pharmaceutical Biology. 39: 74-79.

Pullaiah, T. 2007. Field and Herbarium methods. In: Taxonomy of Angiosperms, Regency Publications, New Delhi. Pp.130-145.

Rama Rao, M. 1914. The flowering plants of Travancore. Trivandrum.

Rao, T.A., L.K. Banerjee and A.K. Mukeherjee. 1975. A contribution to the coastal flora and vegetation of Tamil Nadu (India). Indian Forester. 101: 460-475.

Sajem, A.L. and K. Gosai. 2006. Traditional use of medicinal plants by the Jaintia tribes in North Cachar Hills District of Assam. Northeast India. J.Ethnobiol. Ethnomed. 2: 33.

Samydurai,P., S. Jagatheshkumar., V. Aravinthan and V. ThangaPandian. 2012. Survey of wild aromatic ethnomedicinal plants of Velliangiri hills in the Southern Western Ghats of Tamil Nadu, India. Inter, J. Med. Arom. Plants. 2 (2): 229-234.

Selvi, V.2004. Survey, documentation and enumeration of the medicinal plants of the Marunduvalmalai hills, Kanyakumari District. M.Phil. Dissertation, M.S. Univ. Thirunelveli. P.114

Sharma,B.D., B.V. Shetty., S. Karthikeyan and M. Chandrabose. 1973. Studies on the vascular Flora of Mahendragiri Hills and the surrounding regions, Kanyakumari and Triunelveli districts. Tamil Nadu. Bull.Bot.Sur. India. 15: 45-70.

Srivastava, R. 2000. Studying the information needs of medicinal palants stakeholders in Europe. Traffic Dispatches. 15: 5.

Sukumaran, S., S. Jeeva., A.D.S. Raj and D. Kannan. 2008. Floristic diversity, conservation status and economic value of miniature Sacred Grooves in Kanyakumari District, Tamil Nadu, Southern Penisular India. Turk. J. Bot. 32: 185-199.

Whittmore, T.C. 1990. Tropical rain forests, Oxford, Clarendon Press.

Chapter 14

Wild Edible Plants of Tribals of Khammam District, Andhra Pradesh

*R. Ratna Manjula, J. Koteswara Rao and T.V.V. Seetharami Reddi**

Department of Botany, Andhra University, Visakhapatnam – 530 003

Introduction

India has a variety of tribal population, reflecting its great ethnic diversity. Majority of tribal population of Andhra Pradesh lives in forest ecosystems and has its own socio-cultural pattern, traditions and typical food practices. Settled agriculture does not provide them sufficient food due to indifferent land situation and as most of them are farmers practicing *podu* cultivation they depend on natural food resources. Many wild plant species are used for edible purpose since ancient times by tribals inhabiting forested areas. In India, it is estimated that, around 800 species of plants are consumed as food, chiefly by tribal inhabitants (Singh and Arora, 1978). But studies on such wild edible plants in our country are limited (Jain, 1964; Gour, 1977; Arora, 1991; Patole and Jain, 2002; Rekha Sinha and Valeria Lakra, 2007; Mohanty, 2010; Selva Kumari *et al.*, 2012). Andhra Pradesh is inhabited by 33 tribal communities representing 6.59 per cent of the total population (Census 2001). In Khammam district [Plate 14.1(1)] there are six tribal communities, *viz.*, Koya, Lambada [Plate 14.1(2)],

---

* Corresponding Author: E-mail: reddytvvs@rediffmail.com

Gond/Naikpod [Plate 14.1(4)], Yerukula, Nayak and Konda Reddi [Plate 14.1(3)] constituting 26.47 per cent of the total population in the district. Most of them depend on forest resources for their livelihood due to lack of agricultural land and take edible forms of flowers, roots, fruits, tubers, rhizomes, leaves, etc. for food. There are some sporadic reports only on wild foods from few places of Andhra Pradesh (Pal and Banerjee, 1971; Krisahna Prasad *et al.*, 1999; Sudhakar and Vedavathy, 1999; Reddy *et al.*, 2007; Suneetha *et al.*, 2013). The present paper highlights some of the important wild food plants of Khammam district which need to be documented for future food security. Khammam district lies between 16 45′ and 18 35′ N latitude and 79 47 to 81 47′ E longitude occupying an area of 16, 029 sq. km and the forests occupies 52.6 per cent.

Methodology

The study was conducted in Khammam district of Andhra Pradesh from 2008-2012. A total of 46 mandals covering 116 villages inhabited by six types of tribal groups were selected randomly for data collection. It was collected through a combination of tools and techniques of questionnaire, PRA, focused group interview and discussion. Plant specimens were collected and identified with regional floras (Gamble, 1928; Pullaiah and Chennaiah, 1997; Pullaiah and Moulali, 1997; Pullaiah, 1997). All the specimens were deposited in the herbarium of the Department of Botany, Andhra University, Visakhapatnam. The plants are enumerated alphabetically with their botanical name, family, vernacular name, English name, parts used, uses and mode of uses.

Enumeration

Food

***Bambusa arundinacea* (Retz.) Roxb.** Bambusaceae VN: Bongu veduru E: Thorny bamboo

Cooked grains are eaten.

***Oryza sativa* L.** Poaceae VN: Buvva E: Paddy

Seeds are used to prepare food.

Vegetables

***Acacia sinuata* (Lour.) Merr**. Mimosaceae VN: Sikaya E: Soap pod

Leaves are used as vegetable [Plate 14.1(5)].

***Allium cepa* L.** Liliaceae VN: Yerragadda E: Onion

Bulb is used as vegetable.

***Allium sativum* L.** Liliaceae VN: Tellagadda E: Garlic

Bulb is used as vegetable.

***Alternanthera sessilis* (L.) R. Br. ex DC.** Amaranthaceae VN: Ponnaganti E: Sessile joyweed

Leaves are used as vegetable.

***Amaranthus viridis* L.** Amaranthaceae VN: Erra thotakura E: Green amaranth

Leaves are used as vegetable.

***Bauhinia purpurea* L.** Caesalpiniaceae VN: Devakanchanam, E: Pink bauhinia

Tender leaves are used for curries [Plate 14.2(8)].

***Bauhinia racemosa* Lam**. Caesalpiniaceae VN: Arechettu E: Sweta kanchan

Flowers are used for making pickles.

***Colocasia esculenta* (L.) Schott** Araceae VN: Chamadumpa E: Taro *Arum*

Leaves and tubers are used as vegetable.

***Cucumis sativus* L.** Cucurbitaceae VN: Dosakaya E: Common cucumber

Fruits are used as vegetable.

***Cucurbita maxima* Duch. ex Lam.** Cucurbitaceae VN: Gummadi E: Red pumpkin

Fruits are used as vegetable [Plate 14.2(9)].

***Cyamopsis tetragonaloba* (L.) Taub.** Fabaceae VN: Goruchikkudu E: Cluster bean

Fruits are used as vegetable.

***Daucus carota* L.** Apiaceae VN: Carrot E: Carrot

Tubers are used as vegetable.

***Dioscorea bulbifera* L.** Dioscoreaceae VN: Adavi dumpa E: Bulb bearing yam

Tubers are used as vegetable.

***Ipomoea batatas* (L.) Lam.** Convolaceae VN: Silkada dumpa E: Sweet potato

Tubers are used as vegetable.

***Lablab purpureus* (L.) Sweet** Fabaceae VN: Chikkudu E: Indian butter bean

Fruits are used as vegetable.

***Lycopersicon esculentum* Mill.** Solanaceae VN: Thakkali E: Tomato

Fruits are used as vegetable.

***Murraya koenigii* (L.) Spreng.** Rutaceae VN: Karivepaku E: Curry leaf

Leaves are used as vegetables.

Fruits

***Anacardium occidentale* L.** Anacardiaceae VN: Jeedimamidi E: Cashew nut

Fruits are edible.

***Ananas comosus* L.** Bromeliaceae VN: Anasa E: Pineapple

Fruits are edible.

***Annona muricata* L.** Annonaceae VN: Laxamana phalam E: Soursop tree

Fruits are edible.

***Annona reticulata* L.** Annonaceae VN: Ramapalam E: Bullocks heart of India

Fruits are edible [Plate 14.1(6)].

***Annona squamosa* L.** Annonaceae VN: Seethaphalam E: Custard apple

Fruits are edible.

***Anthocephalus chinensis* (Lam.) Rich. ex Halp.** Rubiaceae VN: Kadimi E: Parvaty's tree

Fruits are edible.

***Artocarpus heterophyllus* Lam.** Moraceae VN: Panasa E: Jack fruit

Fruits are edible [Plate 14.2(7)].

***Capparis zeylanica* L.** Capparaceae VN: Tella uppi E: Copper plant

Fruits are edible.

***Carica papaya* L.** Caricaceae VN: Boppayi E: Papaya tree

Fruits are edible.

***Citrus limon* (L.) Burm.** Rutaceae VN: Adavi nimma E: Lemon

Fruits are edible.

***Citrus medica* L.** Rutaceae VN: Maadiphalam E: Wild lemon

Fruits are edible.

***Cocos nucifera* L.** Arecaceae VN: Kobbari chettu E: Coconut

Fruit is edible.

***Cordia dichotoma* Forst. *f*.** Boraginaceae VN: Eriki E: Sebesten plum

Fruits are edible.

***Diospyros melanoxylon* Roxb**. Ebenaceae VN: Tunikaku E: Ebony

Fruits are edible.

***Grewia hirsuta* Vahl** Tiliaceae VN: Jibilika E: Veronix calolia

Fruits are edible.

***Holoptelea integrifolia* (Roxb.) Planch.** Ulmaceae VN: Nemali nara chettu E: South Indian elm

Fruits are edible.

***Lantana camara* L.** Verbenaceae VN: Lavangipulu E: Wild sage

Fruits are edible.

***Limonia acidissia* L.** Rutaceae VN: Velakkaya E: Elephant apple

Fruits are edible.

***Madhuca indica* Gmel.** Sapotaceae VN: Ippa chettu E: Indian butter tree

Fruits are edible [Plate 14.2(10)].

***Mangifera indica* L.** Anacardiaceae VN: Mamidi chettu E: Mango

Fruits are edible.

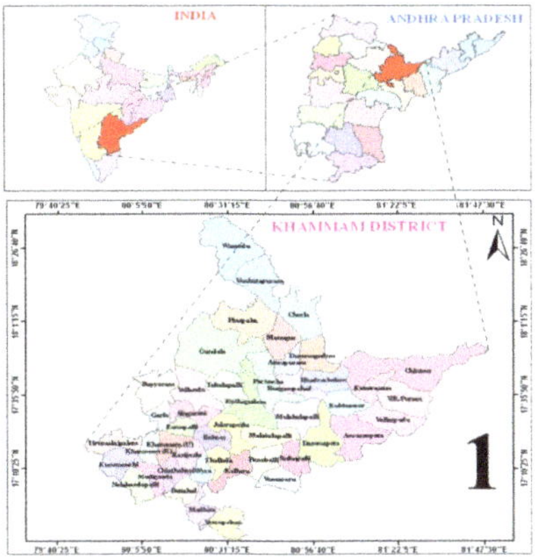

Study Area

Lambada Family

Kondareddi Family

Naikpod Family

Acacia sinuata

Annona reticulata

Plate 14.1(1–6).

Artocarpus heterophyllus

Bauhinia purpurea

Cucurbita maxima

Madhuca indica

Mimusops elengi

Morus alba

Plate 14.2(7–12).

***Manilkara zapota* (L.) P. Royen**. Sapotaceae VN: Sapota E: Sapota

Fruits are edible.

***Mimusops elengi* L. Sapotaceae** VN: Pogada E: Spanish cherry

Fruits are edible [Plate 14.2(11)].

***Morinda pubescens* Sm.** Rubiaceae VN: Togaru E: Togari wood of Madras

Fruits are edible.

***Morus alba* L.** Moraceae VN: Pattaku E: Mulberry

Fruits are edible [Plate 14.2(12)].

***Musa ornata* Roxb.** Musaceae VN: Adavi arati E: Wild banana

Fruits are edible.

***Musa paradisiaca* L.** Musaceae VN: Arati E: Plantain

Fruits are edible.

***Pavetta indica* L.** Rubiaceae VN: Papidi E: White pavetta

Fruits are edible.

***Phoenix sylvestris* (L.) Roxb.** Arecaceae VN: Eatha E: Date sugar palm

Fruits are edible.

***Phyllanthus acidus* (L.) Skeels** Euphotbiaceae VN: Rasa usiri E: Country gooseberry

Fruits are edible.

***Phyllanthus emblica* L.** Euphorbiaceae VN: Usiri E: Indian gooseberry

Fruits are edible.

***Physalis minima* L.** Solanaceae VN: Buddakasi E: Sun-berry

Fruits are edible.

***Psidium guajava* L.** Myrtaceae VN: Jama chettu E: Guava

Fruits are edible.

***Raphanus sativus* L.** Brassicaceae VN: Mullangi E: Radish

Tubers used as vegetable.

***Saccharum officinarum* L.** Poaceae VN: Cheruku E: Sugarcane

Stem is edible.

***Solanum melongena* L. var. *insanum* (L.) Prain** Solanaceae VN: Mulluvankaya E: Bitter apple

Fruits are used as vegetables.

Solanum surattense* Burm. *f. Solanaceae VN: Vakudu E: Dwarf lid brinjal

Fruits are used as vegetables.

***Solanum torvum* Sw.** Solannaceae VN: Kondusti, E: Turkey berry

Fruits are edible.

***Sorghum vulgare* (L.) Pers.** Poaceae VN: Jonnalu E: Sorghum

Seeds are used as food.

***Syzygium cumini* L.** Myrtaceae VN: Adavii jinne E: Indian cherry

Fruits are edible.

***Terminalia catappa* L.** Combretaceae VN: Badam E: Indian almond tree

Fruits and seeds are edible.

***Trigonella foenum- graecum* L.** Fabaceae VN: Tagiresa E: Fenugreek

Leaves are used as vegetables.

***Zea mays* L.** Poaceae VN: Mokkajonna E: Corn

Roasted and boiled grains are eaten.

Results and Discussion

The present study yielded 64 plants species used for food (4), vegetables (24) and fruits (36). The food plants all belong to the family Poaceae. The 24 species of vegetables belongs to 21 genera and 14 families. Herbs are the dominant ones (13), followed by climbers (4), shrubs and trees (2 each) and one underground corm. In 9 species leaf is used as vegetable followed by fruit (8), tuber (4), bulb (2) and flower (1). Fabaceae, Cucurbitaceae, Apiaceae, Solanaceae are the dominant families each represented by 3 species followed by Caesalpiniaceae, Amaranthaceae and Liliaceae each represented by 2 species and the rest of the families each represented by one species. Leaves are collected during different seasons, cooked and eaten with their staple food. Most of the tribal people of the district are using leafy vegetables as part of their food. They are either collected from forest areas or found as weeds in moist areas of cultivated, open fields and on the edges of canals. Women are collecting and preparing the curries. Fruit yielding species are 36 belonging to 31 genera and 22 families. Trees are the dominant ones (26) followed by herbs (5), shrubs (4) and a lone climber. Family- wise Annonaceae, Rubicaeae and Sapotaceae are the dominant ones represented by 3 species each followed by Anacardiaceae, Rutaceae, Myrtaceae, Solanaceae, Euphorbiaceae, Moraceae, Musaceae and Arecaceae, each with 2 species and the rest of the families represented by one species each. Fruits are eaten as raw as well as in ripen form. Fruits of *Anacardium occidentale, Ananas comosus, Annona squamosa, A. muricata, A. reticulata, Artocarpus hererophyllus, Carica papaya, Citrus limon, C. medica, Cocos nucifera, Limonia acidissima, Mangifera indica, Manilkara zapota, Musa paradisiaca, Phoenix sylvestris, Phyllanthus acidus, P. emblica, Psidium guajava, Saccharum officinarum* and *Terminalia catappa* are eaten depending upon their ripening in different seasons. They sell most of them in weekly shandies for their living. Raw fruits of *Anacardium occidentale, Artocarpus hererophyllus, Carica papaya, Citrus limon, C. medica, Cocos nucifera, Limonia acidissima, Mangifera indica, Musa paradisiaca, Phyllanthus acidus, P. emblica* and *Solanum torvum* are used in the preparation of curries and pickles.

Use of *Allium cepa*, and *Colocasia esculenta* as vegetables by Valaiyan tribe of Madurai district, Tamil Nadu was reported by Subramanian *et al.* (2011). *Alternanthera sesssilis, Dioscorea bulbifera, Ipomoea batatas* and *Trigonella foenum-graecum, Diospyros melanoxylon, Lantana camara, Musa paradisiaca, Phoenix sylvestris, Physalis minima* and *Solanum torvum* are used as vegetables and fruits in Gaya district of Bihar (Pratap Aditya *et al.*, 2009). *Alternanthera sesssilis, Amaranthus viridis, Bauhinia racemosa* and *Murraya koenigii* are used as vegetables in Tuticorin district of Tamil Nadu (Selva Kumari *et al.*, 2012). The first two are also reportedly used for the same purpose by the tribals of Jharkhand, Orissa and West Bengal (Sinha and Lakra, 2007). Thus most of the wild plants used as vegetables and fruits reported in the present study are also used for the same purpose by the tribals in different parts of India.

Though the detailed nutritional value of these plants is not known, they are consumed by the tribals most commonly. The wild relatives of edible plants propose yet another approach to conserve these genetic resources on the assumption that the cultivated races are best preserved *in situ* under traditional agricultural practices. The search for novel high quality but inexpensive sources of food has always remained a major concern of all agencies involved in providing adequate food and improving nutritional status of population. In this context, these indigenous foods of tribals will play a vital role towards food and nutritional security of the nation, as their economic value is beyond dispute.

Summary

A total of 4 species of food plants, 24 species of vegetables belonging to 21 genera and 14 families and 36 species of fruits belonging to 31 genera and 22 families used by the tribals of Khammam district were identified. Most of the edible parts are fruits followed by leaves, seeds, tubers, flowers, t ender shoots, and root. Most of the edible plants are used mainly by the tribals both during normal and at the time of famine. Safe conservation and sustainable use of this diversity is the need of the hour for meeting the present and future food needs of the tribals during the periods of food scarcity.

Acknowledgements

Authors are grateful to the tribal people of Khammam district for sharing their valuable information on wild food plants and cooperation during field trips.

References

Arora, R.K. 1991. Native food plants of the tribals of Noetheastern india. Pp. 137-152. *In:* S.K.Jain (Ed.). Contributions to Ethnobotany in India. Scientific Publishers, Jodhpur.

Gamble, J.S. 1915-35. Flora of Presidency of Madras. (Adlard and Son Ltd., London).

Gour, R.D. 1977. Wild edible fruits of Garhwal Hills. Himalaya **1**: 66-70.

Jain, S.K. 1964. Wild plant foods of tribals of Bastgar. Khadi Gramodyog **10**: 557-561.

Krishna Prasad, V., Rajagopal, T., Yogesh Kumar and Badarinath, K.V.S. 1999. Food plants of Konda Reddis of Rampa Agency, East Godavari district, Andhra Pradesh- A case study. Ethnobotany **11**: 92-96.

Mohanty, R.B. 2010. New wild edible plants from some pockets of Dhenkanal district, Odisha. Ethnobotany **22**: 111-113.

Pal, D.C. and Banerjee, D.K. 1971. Some less known plant foods among the tribals of Andrhra Pradesh and Orissa state. Bull. Bot. Surv. Ind. **13**: 221-223.

Patole, S.N., and Jain, A.K. 2002. Some wild edible plants of Pachmarhi Biosphere Reserve (M.P.). Ethnobotany **14**: 48-51.

Pratap Aditya, Somya Yadav, Anil Kumar and Yadav, D.K. 2009. Indigenous knowledge of wild plant resources used as food in Gaya district, Bihar. J.Non-Timber Forest Products **16**: 85-88.

Pullaiah, T. and Chennaiah, E. 1997. Flora of Andhra Pradesh, Vol I (Scientific Publishers, Jodhpur).

Pullaiah, T. and Moulali, D.A. 1997. Flora of Andhra Pradesh, Vol II (Scientific Publishers, Jodhpur).

Pullaiah, T. 1997. Flora of Andhra Pradesh, Vol III (Scientific Publishers, Jodhpur).

Reddy, K.N., Chiranjibi Patnaik, Reddy, C.S. and Raju, V.S. 2007.Traditionhal knowledge on wild food plants in Andhra Pradesh. Indian J. Trad. Knowl. **6**: 223-229.

Selva Kumari, R., Raja Kumar, T.J.S., Murugesan, S. and Chellaperumal, N. 2012. Wild edible plants of Kudiraimozhi Theri in Tuticorin District, Southern India. J. Non-Timber Forest Products **19**: 245-246.

Singh, H.B. and Arora, R.K. 1978. Wild edible plants of India. ICAR, New Delhi.

Sinha, R and Lakra, V. 2007. Edible weeds of tribals of Jharkhand, Orissa and West Bengal. Indian J. Trad. Knowl. **6**: 217-221.

Subramanian, A., Mohn, V.R. and Maruthupandian, A. 2011. Little known wild edible plants used by Valaiyans from Madurai district, Tamil Nadu. J.Non-Timber Forest Products **18:** 333-336.

Sudhakar, A. and Vedavathy, S. 1999. Wild edible plants used by the tribals of Chittoor district (Andhra Pradesh), India. J.Econ. Taxon. Bot. **23:** 321-329.

Suneetha, J., Koteswara Rao, J., Prabhakara Rao, P. and Seetharami Reddi, T.V.V. 2013. Wild edible plants used by the tribals of East Godavari District, Andhra Pradesh pp. 271-301. *In*: T.Pullaiah and K.Jaganmohan Reddy (eds.) Biodiversity in India. Vol **6** Regency Publications, New Delhi.

Chapter 15

Ex situ Conservation of Wild Medicinal Plants at Botanical Garden, Yogi Vemana University, Kadapa

A. Madhusudhana Reddy

Department of Botany, Yogi Vemana University, Kadapa – 516 003, Andhra Pradesh

Introduction

Medicinal plant conservation strategies need to be understood and planned for based on an understanding of indigenous knowledge and practices. Many drugs contain herbal ingredients, and it has been said that 70–80 per cent of the world's population relies on some form of non-conventional medicine and around 25–40 per cent of all prescription drugs contain active ingredients derived from plants in the United States alone. Many countries rely on these medicinal plants for the health and well being of its population, but the market demand has led to an increased pressure on the natural resources that lend to the production of some of these plants. The most serious proximate threats when extracting medicinal plants generally are habitat loss, habitat degradation, and over harvesting. Developing markets for natural products, particularly those that are harvested from the wild, can trigger a demand that cannot be met by available or legal supplies and demands a conservation initiative so the local populations are not exploited, causing more damage to their resources. Many times populations are taken advantage of for their resources and knowledge, which can often be for financial gain. Conservation of medicinal plants in its biocultural perspective not only implies conservation of biodiversity but also places an equal emphasis on conservation of cultural diversity.

India is one of the 12-mega biodiversity centers in the country with three hotspots of biodiversity, *viz.*, Western Ghats, the Himalayas and the Indo-Burma region. From ages plants are used as source of medicine and are closely associated with cultural traditions. Most of common medicinal plants are found in the forest. The use of medicinal herbs is still a tradition, continued by ethnic communities who are living in undulating planes and at foothills of dense forests. From ancient times in India, they are using a large number of wild and cultivated plants for the treatment of various ailments thus; a considerable amount of information on medicinal plants is available with these communities. The plants growing around form an integral part of their culture. These people are largely dependent on their traditional healing system for their healthcare and the information is passed on from generation to generation through the word of mouth. There is little or nothing in written. In earlier day many of the plants used by tribal communities are at the verge of extinction due to over exploitation and for many plants there is no information about the identity and the proper usage and there will be miss-identification for many species. In order to maintain a plant repository of species in danger, which are largely used by tribal communities from ages and also to make proper identification of all those plants with information on uses, the garden has established a dedicated medicinal plants block with aim to conserve and disseminate the information about the usage and the status of the plants. Some important plants conserved in Botanical Garden is given in Table 15.1. *Ex situ* conservation is one of the important aspect in conserving the species under threat.

Methodology

This Botanical Garden was established in 2009 with a total area of 19 acres. The garden is divided into 25 sectors with one sector earmarked for Medicinal plants. Medicinal plants were collected from all over Andhra Pradesh and their identification was confirmed with the help of local floras and literature. Information on use of medicinal plants was collected from local elder persons of the family. The vernacular names given by the informers were confirmed with local flora and also their botanical name, family and local name were identified with the help of available literature.

Results

Total of 120 plants species belonging to 70 genera and 40 families have been conserved in medicinal plants sector of Botanical Garden. The data on ethno-medicinal uses with method of drugs administration in different ailments were presented. These plants are being used by various ethnic groups for the treatment of various diseases ranging from simple injuries, wounds, cuts, fever, diarrhea, ulcers, swelling, bone fractures, potency, antidote, skin care, night blindness, toothache, asthma, cough and cold. Based on the data collected by consulting various ethnic groups and local people an analysis has been done on the use of different parts of plants as source of medicine. The analysis showed that leaves, bark and root are the

Table 15.1: List of some Important Medicinal Plants.

Sl.No.	*Name of the Species*	*Family*	*Telugu Name*
1.	*Aegle marmelos* (L.) Corr.	Rutaceae	Maredu, Bilvamu
2.	*Aloe vera* (L.) Burm.f.	Liliaceae	Manchikalabanda
3.	*Aristolochia indica* L.	Aristolochiaceae	Nalla eswari
4.	*Arundo donax* L.	Poaceae	Kakiveduru
5.	*Balanites aegyptiaca* (L.)Del.	Balanitaceae	Gara chettu
6.	*Boswellia ovalifoliolata* Bal. & Henr.	Burseraceae	Konda Sambrani
7.	*Callistemon citrinus* (Curtis) Skeels	Myrtaceae	Bottle Brush
8.	*Centella asiatica* (L.)Urban.	Apiaceae	Swarasvathiaku
9.	*Cochlospermum religiosum* (L.)Alston	Cochlospermaceae	Koda pathi
10.	*Commiphora caudata* (Wt. & Arn.) Engl.	Burseraceae	Konda mamidi
11.	*Costus speciosus* (Koen.) Smith	Zingiberaceae	Vanavasa
12.	*Crateva magna* (Lour.) DC.	Capparaceae	Ulimidi
13.	*Cymbopogon citratus* (DC.) Stapf.	Poaceae	Nimma gaddi
14.	*Curculigo orchioides* Gaertn.	Hypoxidaceae	Nelathati
15.	*Decalepis hamiltonii* Wt. & Arn.	Asclepiadaceae	Nannari gaddalu
16.	*Dioscorea pentaphylla* L.	Dioscoreaceae	Yerra teega
17.	*Dioscorea oppositifolia* L.	Dioscoreaceae	Eseru gaddalu
18.	*Diospyros melanoxylon* Roxb.	Ebenaceae	Tumki
19.	*Euphorbia milii* Der.	Euphorbiaceae	Kuchu chettu
20.	*Euphorbia nivulia* Buch.-Ham.	Euphorbiaceae	Errakalli
21.	*Euphorbia tirucalli* L.	Euphorbiaceae	Manchikalli
22.	*Ficus benghalensis* L.	Moraceae	Marri
13.	*Ficus hispida* L.f.	Moraceae	Kakimedi
24.	*Ficus racemosa* L.	Moraceae	Medi
25.	*Ficus religiosa* L.	Moraceae	Ravi
26.	*Gardenia gummifera* L.f.	Rubiaceae	Bikki
27.	*Gardenia resinifera* Roth	Rubiaceae	Pedda bikki
28.	*Givotia moluccana* (L.) Sreem.	Euphorbiaceae	Tella poliki
29.	*Gloriosa superba* L.	Cochlaceae	Nabhi
30.	*Gymnema sylvestre* (Retz.) Schult.	Asclepiadaceae	Podapatri
31.	*Gyrocarpus americanus* Jacq.	Hernandiaceae	Poliki
32.	*Hemidesmus indicus* (L.) R.Br.	Asclepiadaceae	Sugandipala
33.	*Holoptelea integrifolia* (Roxb.) Planch.	Ulmaceae	Tapase
34.	*Hymenodictyon orixense* (Roxb.) Mabb.	Rubiaceae	Dudippa
35.	*Justicia adhatoda* L.	Acanthaceae	Addasaramu
36.	*Leptadenia reticulata* Schult.	Asclepiadaceae	Palateega

Contd...

Table 15.1–*Contd...*

Sl.No.	Name of the Species	Family	Telugu Name
37.	*Limonia acidissima* L.	Rutaceae	Velaga
38.	*Maerua apetala* (Roth) Jacobs	Capparaceae	Pilli adugu
39.	*Maytenus emarginata* (Willd.) Ding.	Celastraceae	Danthi
40.	*Mitragyna parviflora* (Roxb.) Korth.	Rubiaceae	Battaganapa
41.	*Morinda pubescens* J.E.Smith	Rubiaceae	Maddi
42.	*Naringi crenulata* (Roxb.) Nicols.	Rutaceae	Kukkavelaga
43.	*Ocimum americanum* L.	Lamiaceae	Kukkatulasi
44.	*Ocimum tenuiflorum* L.	Lamiaceae	Tulasi
45.	*Oroxylum indicum* (L.)Vent.	Bignoniaceae	Mokka Vepa
46.	*Pandanus fascicularis* Lam.	Pandanaceae	Mogali
47.	*Pavetta tomentosa* Roxb.	Rubiaceae	Tellapapidi
48.	*Piper trioicum* Roxb.	Piperaceae	Adavi miriyalu
49.	*Pterocarpus santalinus* L.f.	Fabaceae	Yerrachandanam
50.	*Pterocarpus marsupium* Roxb.	Fabaceae	Yogisa
51.	*Pterospermum xylocarpum* (Gaertn.) S.-W.	Sterculiaceae	Chinna thada
52.	*Salacia oblonga* Wt. & Arn.	Hippocrateaceae	Garika musti
53.	*Sansevieria roxburghiana* Schult.	Asparagaceae	Chaga
54.	*Schefflera stellata* (Gaertn.) Harms	Araliaceae	Marrimamidi
55.	*Soymida febrifuga* (Roxb.)A.Juss.	Meliaceae	Somidi
56.	*Strychnos nux-vomica* L.	Loganiaceae	Mushti
57.	*Sterculia foetida* L.	Sterculiaceae	Adavi badam
58.	*Sterculia urens* Roxb.	Sterculiaceae	Errapoliki
59.	*Terminalia catappa* L.	Combretaceae	Badham chettu
60.	*Terminalia chebula* Retz.	Combretaceae	Nallakaraka
61.	*Terminalia arjuna* Roxb. ex DC.	Combretaceae	Tellamaddi
62.	*Terminalia bellirica* (Gaertn.) Roxb.	Combretaceae	Thandra
63.	*Terminalia alata* Roth	Combretaceae	Nallamaddi
64.	*Terminalia pallida* Brandis	Combretaceae	Tella karaka
65.	*Tylophora indica* (Burm.f.) Merr.	Asclepiadaceae	Kukkapala
66.	*Vitex altissima* L.f.	Verbenaceae	Nemaliadugu
67.	*Vitex negundo* L.	Verbenaceae	Vavili
68.	*Withania somnifera* (L.) Dunal	Solanaceae	Aswagandha
69.	*Wrightia tinctoria* R.Br.	Apocynaceae	Palvareni
70.	*Ximenia americana* L.	Olacaceae	Konda nakkeru

Plate 15.1

A: ***Barleria*** sp.; B: ***Caesalpinia bonduc***; C: ***Caralluma dscendens***;
D: ***Datura metel*** seeds; E: ***Drimia nagarjunae***; F: ***Gardenia gummifera***.

Plate 15.2

A: *Pterospermum* sp.; B: *Solanum seaforthianum*; C: *Soymida febrifuga*; D: *Terminalia* sp.

Plate 15.3

A: *Breynia retusa*; B: *Phoenix* sp.; C: *Crateva* sp.; D: *Terminalia arjuna*; E: *Ficus racemosa*; F: *Pandanus* (Mogali).

major source of medicine. The medicinal plants with root as medicinal part need to be given more attention for conservation as collection of such plants will lead to the endangeredness of those plants. Those information need to be disseminated in the public for sustainable utilization.

Acknowledgements

I am Thankful to Ministries of Industries and Commerce, Govt. of Andhra Pradesh for sponsoring the Agri Science Park Project.

References

Gamble, J.S. and Fischer, C.E.C. 1915-1936. Flora of the Presidency of Madras. Adlard and Son, London.

Madhusudan Rao, A. 1989. A floristic studies on the Flora of Kadapa District, A.P. India, Ph.D., Thesis, Sri Venkateswara University, Tirupathi.

Chapter 16

A Review on Biotechnological Approaches for the Rescue of Endangered Genus *Ceropegia* L. in India

K. Sri Rama Murthy[1]*, *M. Chandrasekhara Reddy*[1], *R. Kondamudi*[1], *D. Raghuramulu*[2], *S. Karuppusamy*[3] and *T. Pullaiah*[4]

[1] *School of Conservation Biology and Plant Biotechnology,*
[2]*Department of Botany and Biotechnology,*
Montessori Mahila Kalasala, Vijayawada – 520 010, Andhra Pradesh
[2]*Department of Botany,*
S.M.L. Govt. Degree College, Yemmiganur – 518 360, Andhra Pradesh
[3]*Department of Botany, The Madura College,*
Madurai – 625 011, Tamil Nadu
[4]*Department of Botany, Sri Krishnadevaraya University,*
Anantapur – 515 003, Andhra Pradesh

ABSTRACT

In vitro propagation and restoration into their natural habitat has increasingly become a significant biotechnological tool for rescue of rare, vulnerable and endangered taxa. The present review gives a summary on the protection of the

* Corresponding Author: E-mail: drksrmurthy@yahoo.com

genus *Ceropegia* L. employing various techniques of micropropagation. In *Ceropegia* fruit and seed setting is very low due to limited reproductive capacity. Therefore micropropagation is the only viable alternative technique for raising these plant species and to reintroduce into their natural wild environment. Factors that affect *in vitro* propagation of different *Ceropegia* species are mentioned. The influence of plant growth regulators (auxins and cytokinins), either alone or in combinations, on the morphogenesis of different *Ceropegia* spp. are summarized. In most of the species cytokinin (BAP) at different concentrations play a significant role in induction and proliferation of multiple shoots. Whereas the auxin (IBA) at different concentrations play a significant role in root induction. Different concentrations of 2,4 -D either alone or in combination with cytokinins produce maximum amount of callus which further leads to either somatic embryogenesis or organogenesis. Higher concentrations of cytokinins (BAP and Kn) influence the *in vitro* tuber induction.

Keywords: *Ceropegia spp., Conservation, Micropropagation, Plant growth regulators, In vitro Tuber induction, In vitro flowering.*

Introduction

The genus *Ceropegia* belonging to the family Apocynaceae, subfamily Asclepiadoideae and tribe Ceropegieae comprises of more than 200 species distributed in tropical and subtropical Asia, Africa, Australia and in Canary and Pacific Islands (Anonymous, 1992; Bruyns, 2003). In India majority of the *Ceropegia* species are tuberous erect herbs, non succulent twiners, leafless succulent twiners and rarely sub shrubs (Surveswaran *et al.*, 2009). Ansari (1984) revised the Indian *Ceropegia* and reported 44 species, of which 28 are said to be endemic to India. Yadav and Mayur (2008) and Kullayiswamy *et al.* (2012) added 5 more novelties to the list. The existence of most of the Indian *Ceropegia* species has become restricted to remote pockets in the Himalayans and the Western Ghats, two biodiversity hot spots (Yadav, 1996; Malpure *et al.*, 2006; Nautiyal *et al.*, 2009). Majority of these species are facing different threats (Patil *et al.*, 2005) and are placed under the categories of rare, endangered, vulnerable, extinct and threatened plants (Gadgil, 2004). Many of the species are now listed in the Red Data Book of Indian plants (Nayar and Sastry, 2000), IUCN Red list category (Gaikwad and Yadav, 2003) and in the recently published list of Rare, endangered and threatened (RET) plants by Botanical Survey of India (BSI, 2009).

In recent years due to the recognition of importance of conservation of rare plants, the genus has attracted attention of several biologists because many species are rare, some are known only from their specific locations and at the verge of extinction (Yadav *et al.*, 2004). The major causes for reduction of natural population of this genus are restriction of some species to a specific habitat, habitat modification, industrialization, pollution, forest fires and climatic changes. The starchy tubers are edible and some of the wild animals and tribals feed on tubers (Mabberley, 2001). The tubers get decay due to bacterial and fungal infections. On the other hand improper pollination, scarcity of the pollinators, low fruit and seed setting, delicate seeds, dormancy of the seeds, low seed viability and poor germination rate of the seeds are the other major problems (Karuppusamy and Pullaiah, 2009: Murthy *et al.*, 2012b).

Conservation of genus *Ceropegia* is a challenge to scientific community because propagation and maintenance of these species in botanic gardens is very difficult.

The genus has excellent economic and medicinal importance. The tuberous roots of many *Ceropegia* species are edible, boiling or roasting can eliminate the bitterness of the tubers (Mabberley, 1987). The tubers of this genus contain sugars, starch, fats, albuminoids, gums and crude fibers which are useful as nutritive tonic (Kirtikar and Basu, 1935).The species of *Ceropegia* can be used as a source of soma a plant based drug of the Ayurvedic medicine with a wide variety of uses (Asolkar *et al.*, 1992; Jagtap and Singh 1999). The fleshy stems are used as a raw material for the traditional and folk medicines for the treatment of diseases like diarrhea and dysentery (Nadkarni, 1976), stomach and gastric disorders (Jain and Defillips, 1991). The pharmacological importance of the genus *Ceropegia* was mainly due to the presence of a pyridine alkaloid cerpegin (Sukumar *et al.*, 1995). The tuberous roots are the official source for cerpegin (Nadkani, 1976). The alkaloid cerpegin is relatively rare in nature and it exhibits promising hepatoprotective, antipyretic, analgesic, local anesthetic, anti ulcer, anti inflammation, mast cell stabilizing and hypertensive activities without side effects (Adibatti *et al.*, 1991).

Cerpegin, a furopyridine alkaloid isolated from the chloroform extract of *Ceropegia juncea* was found to be toxic at doses above 400 mg/kg and the mice showed excitation, irritability, convulsions and respiratory paralysis (Sukumar *et al.*, 1995).While, studying the cicatrizant, haemostatic and chemotherapic activities of *Ceropegia fusca* in curing the wounds, elevated levels of hydroxyproline in regenerated tissue suggests enhanced collagen synthesis, an important constituent of extracellular matrix (Darias *et al.*, 1996). *Ceropegia rupicola* is useful for the skin diseases when it was applied externally (Rawiya and Mathias, 2008). Apart from some metabolites of pyridoxal to the best of our knowledge cerpegin is the only naturally occurring bicyclic furo (3, 4-C) pyridine ring system. The structural study of this cerpegin has been undertaken to verify the molecular structure proposed from the spectral data (Sivakumar *et al.*, 1990). Due to the lack of natural source of this alkaloid several workers (Kelly and Walsh, 1992; Adibatti *et al.*, 1991; Matsuo and Arase, 1995) attempted studies for chemical synthesis and synthesized cerpegin within five steps and compared with the isolated standard cerpegin from the *C. juncea* respectively. Whereas, one pot four steps synthesis of cerpegin was carried out by Lazaur *et al.* (2008).

So, it is important to prevent the extinction of the genus *Ceropegia* for its taxonomical and ethnobotanical importance. However *in situ* conservation efforts has had limited impact on healing the decline in the natural population. Poor seed germination and delayed rooting of seedlings made it unsuitable for propagation through seeds. Vegetative propagation through root tubers is also not an appropriate method because in one season the tuber can produce a single plant. It is therefore necessary to study *ex situ* conservational methods like micropropagation. Propagation by tissue culture represents an important potential to effectively propagate them (Nautiyal *et al.*, 2009). In recent years there is increasing interest for exploit tissue culture for propagation of these endangered taxa.

In vitro Seed Germination Studies

In nature normally the seed germination was very low in genus *Ceropegia*. It is due to improper pollination, low fruit and seed setting and high dormancy of the seeds. *In vitro* seed germination is an appropriate method for development of successful micropropagation protocol. Aseptic seedlings developed *in vitro* can used as a good source of explants. *In vitro* seed germination can also be helpful to study the physiological dormancy of the seeds and to study the developmental biology of the seedlings etc. Raghuramulu and Pullaiah (1999) studied the induction of *in vitro* seed germination in *C. bulbosa*. They found that Hydrogen peroxide (H_2O_2) was effective for raising the aseptic seedlings. The maximum percentage of seed germination, seedling emergence and growth (in terms of length of seedlings and number of nodes) was achieved on Gamborg's (B_5) medium supplemented with 2.22 micro mole per liter (µM/l) Benzyl amino purine (BAP). Occurrence of twin seedlings in *C. bulbosa* was reported by Raghuramulu *et al.* (1999). In another species *C. sahyadrica* germination of the seeds (collected from surface sterilized follicles) on Murashige and Skoog's (MS) basal medium containing 3 per cent sucrose was reported by Nikam and Savanth (2007). In *C. fantastica* 90 per cent of seed germination was achieved on ½ strength MS basal medium where as 75 per cent on full strength MS basal medium (Chandore *et al.*, 2010).

In *C. odorata* releasing of physiological dormancy of the seeds by using Gibberellic acid ($GA_{3)}$, H_2O_2 and Potassium nitrate ($KNO_{3)}$ was evaluated by Srinivasrao *et al.* (2010). Normally slight release of dormancy (7 per cent) after 80 days of storage and complete loss of dormancy (100 per cent) following one year of storage was observed in untreated seeds. Among the three chemicals tested for breaking the dormancy of the seeds 0.1 per cent H_2O_2 promoted germination up to 62 per cent followed by 51 per cent with 150 ppm of GA_3. KNO_3 was unable to break the dormancy in fresh seeds, but was effective in seeds air dried for 35 days. In *C. juncea* aseptic seedlings were induced on ½ strength MS basal medium (Krishnareddy *et al.*, 2011). Recently Rajput *et al.* (2012) reported the occurrence of tricotyledony seedlings in *C. mahabalei* while conducting experiments for *in vitro* propagation of this critically endangered plant species.

Axillary Bud Proliferation and Multiplication

In vitro propagation of axillary and/or adventitious buds is used as an advanced biotechnological system for the production of elite plants. It is now recognized as valuable technique for propagation, *in vitro* conservation and reestablishment of threatened plant species back into the wild (Bapat *et al.*, 2008). Patil (1998) first attempted the axillary bud culture and shoot multiplication in 3 different species of *Ceropegia*. *C. jainni* a rare plant of Indian subcontinent, *C. bulbosa* var. *bulbosa* and *C. bulbosa* var. *lushii* are two conmon wide distributed species. He found that ½ strength MS medium with 8.88 µM/l BAP induced axillary buds from nodal cultures. Subsequent subculture on to the same medium (1/2 MS) supplemented with 2.22 µM/l BAP produced multiple shoots in all the 3 species. *In vitro* shoot multiplication experiments in *C. bulbosa* var. *bulbosa* (using B5 medium supplemented with different concentrations of BAP and adenine sulfate (Ads) in combination with α- Naphthalene

Table 16.1: Tissue Culture Studies in the Genus *Ceropegia* L.

Sl.No..	Species Name	Source of Explant	Result	Reference
1.	*Ceropegia attenuata* Hook.	Nodes	Micropropagation and *in vitro* flowering	Chavan *et al.,* 2011a
2.	*Ceropegia barnesii* Bruce & Chatterjee	Different explants	Micropropagation	Ananthan, 2003
3.	*Ceropegia bulbosa* Roxb.	Seeds	Polyembryony	Raghuramulu *et al.,* 1999
		Seeds	Seed germination, Multiple shoot induction, plantlet regeneration and Callus cultures	Raghuramulu and Pullaiah, 1999
		Nodes	Micropropagation	Rathore *et al.,* 2010
		Nodes	*In vitro* propagation	Goyal and Bhadauria, 2006
		Nodes	Synthetic seeds and genetic fidelity evaluation of *in vitro* regenerated plants	Dhir and Shekhawat, 2013
		Seeds	Seed germination, Callus induction, multiple shoot initation and *ex vitro* rooting	Phulwaria *et al.,* 2013
4.	*Ceropegia bulbosa* var. *bulbosa* Roxb.	Node	Multiple shoot formation, *in vitro* flowering and microtuber induction	Britto *et al.,* 2003
		Seeds	Seed germination, Multiple shoot induction and *In vitro* flowering	Nair *et al.,* 2007
		Nodes	Multiple shoot induction, *in vitro* and *ex vitro* rooting and microtuberization	Patil, 1998
		Leaf and stem segments	Callus culture, somatic embryogenesis and plantlet regeneration	Patil, 1998
5.	*Ceropegia bulbosa* var. *lushii* Hook.f.	Nodes	Multiple shoot induction, *in vitro* and *ex vitro* rooting and microtuberization	Patil, 1998
		Leaf and stem segments	Callus culture, somatic embryogenesis and plantlet regeneration	Patil, 1998
6.	*Ceropegia candelabrum* L.	Node	Axillary bud multiplication and *in vitro* rooting	Beena *et al.,* 2003
		Leaf and internodes	Callus induction, somatic embryogenesis and plantlet regeneration	Beena and Martin, 2003

Contd...

Table 16.1–*Contd...*

Sl.No..	*Species Name*	*Source of Explant*	*Result*	*Reference*
7.	*Ceropegia elegans* Wall.	Nodes	Shoot multiplication and rooting	Krishnareddy and Pullaiah, 2012
8.	*Ceropegia fantastica* Sedgw	Seeds	Seed germination. Shoot induction, multiplication and *in vitro* flowering	Chandore *et al.,* 2010
9.	*Ceropegia hirsuta* Wight & Arn.	Seeds	Induction and multiplication of microtubers	Pandit *et al.,* 2008
		Seeds	Seed germination, Multiple shoot induction and *In vitro* flowering	Nair *et al.,* 2007
		Nodes	Micropropgation	Nikam *et al.,* 2008
10.	*Ceropegia intermedia* Wight	Nodes	Micropropagation	Karuppusamy *et al.,* 2009
11.	*Ceropegia jainii* Ansari & Kulk	Node	Multiple shoot induction, *in vitro* and *ex vitro* rooting, microtuberization and *in vitro* flowering	Patil, 1998
		Leaf and stem segments	Callus culture, somatic embryogenesis and plantlet regeneration	Patil, 1998
12.	*Ceropegia juncea* Roxb.	Nodes	Multiple shoots, callus culture and cerpegin (secondary metabolite) estimation.	Nikam and Savanth, 2009
		Seeds	seed germination and Multiple shoot induction	Krisnhareddy *et al.,* 2011
13.	*Ceropegia lawii* Hook.	Seeds	Seed germination, Multiple shoot induction and *In vitro* flowering	Nair *et al.,* 2007
		Seeds	Induction and multiplication of microtubers	Pandit *et al.,* 2008
14.	*Ceropegia maccannii* Ans.	Seeds	Seed germination, Multiple shoot induction and *In vitro* flowering	Nair *et al.,* 2007
		Seeds	Induction and multiplication of microtubers	Pandit *et al.,* 2008
15.	*Ceropegia mahabalei* Hem. and Ans.	Seeds	Tricotyledonous seedlings	Rajput *et al.,* 2012
16.	*Ceropegia noorjahani* Ansari	Axillary buds	Multiple shoot formation and *in vitro* rooting	Kedage *et al.,* 2006
17.	*Ceropegia oculata* Hook.	Nodes	Shoot multiplication and *in vitro* rooting	Nikam and Savanth, 2007a

Contd...

Table 16.1–*Contd...*

Sl.No..	*Species Name*	*Source of Explant*	*Result*	*Reference*
18.	*Ceropegia panchganiensis* Blatter & McCann	Nodes explant	Multiple shoots, induction of callus, *in vitro* flowering, micro tubers, rooting , genetic stability of micro propagated plants	Chavan *et al.*, 2013
		Seeds	Seed germination, Multiple shoot induction and *In vitro* flowering	Nair *et al.*, 2007
		Seeds	Induction and multiplication of microtubers	Pandit *et al.*, 2008
19.	*Ceropegia pusilla* Wight & Arn.	Different explants	Multiple shoot induction	Ananthan, 2003
		Node, TCLS and internodes	Morphogenetic callus and multiple shoots	Kondamudi *et al.*, 2010
		Node, inter nodes and callus	Micropropagation and *in vitro* flowering	Kondamudi and Murthy, 2011
		Nodes, shoot tips and callus	Synthetic seeds	Murthy *et al.*, 2013
		Nodes	Callus cultures and Multiple shoot induction	Prabakaran *et al.*, 2013
		Nodes	Tuberization	Murthy *et al.*, 2012a
20.	*Ceropegia sahyadrica* Ans. et Kulk.	Seeds	Induction and multiplication of microtubers	Pandit *et al.*, 2008
		Seeds	Callus culture and multiple shoot induction	Nikam and Savanth, 2007b
		Seeds	Seed germination, Multiple shoot induction and *In vitro* flowering	Nair *et al.*, 2007
21.	*Ceropegia santapaui* Wight & Arn.	Seeds	Seed germination, Direct and indirect shoot organogenesis, analysis of genetic status of *in vitro* raised plantlets and their biochemical characterization	Chavan *et al.*, 2013
22.	*Ceropegia spiralis* Wight	Nodes, TCLs and internodes	Micropropagation andmultiple shoot induction	Murthy *et al.*, 2010a
		Callus	Somatic embryogenesis	Murthy *et al.*, 2010b

Contd...

Table 16.1–*Contd...*

Sl.No..	*Species Name*	*Source of Explant*	*Result*	*Reference*
		Node and callus	*In vitro* flowering	Murthy and Kondamudi, 2010
		Nodes	Tuberization	Murthy *et al.*, 2012a
		Nodes	Micropropagation	Chavan *et al.*, 2011b
		Nodes, shoot tips and callus	Synthetic seeds	Murthy *et al.*, 2013
23.	*Ceropegia thwaitesii* Hook.	Nodes	Shoot multiplication and rooting	Muthukrishnan *et al.*, 2012
		Nodes	Multiple shoot induction and rooting	Muthukrishnan *et al.*, 2013

Figure 16.1: A. *Ceropegia juncea*; B. *Ceropegia pusilla*, C. *Ceropegia pullaiahii*; D. *Ceropegia spiralis.*

Figure 16.2: *In vitro* Propagation of *Ceropegia juncea*

A: Multiple shoots from nodal explants cultured on MS medium supplemented with BAP 8.87 μM; B: Nodal explants cultured on MS medium supplemented with TDZ 4.54 μM; C: Multiple shoots form nodal explants cultured on BAP 8.87 μM + TDZ 4.54 μM; D: *In vitro* rooting on half strength MS medium supplemented with IBA 4.90 μM + NAA 1.27 μM.

Figure 16.3: Micropropagation of *Ceropegia pusilla*.

A. Induction of callus on the MS medium supplemented with BAP 4.44-22.2 µM along with NAA 2.26 µM; B. Different stages of somatic embryos on the media containing BAP and different auxins from TCLs; C. Multiple shoots regenerated on the medium augmented with BAP 2.22 µM; D. Regeneration of shoots from TCLs on the MS medium supplemented with BAP 13.32 µM+IBA 0.49 µM; E. *In vitro* flowering on the medium containing BAP 13.32 µM+IAA 11.54 µM; F. TDZ (40.86 µM) induced shoot regeneration from the nodes on MS medium; G. Basal microtubers induced in MS medium containing BAP 13.32 µM+2,4-D 0.11 µM; H. Shoots with well developed roots in the medium supplemented with KN 0.45 µM+IBA 0.49 µM; J. Acclimatized plantlets after 15 days.

Figure 16.4: Induction of Callus and Somatic Embryogenesis from the TCLs Excised from the Nodes and Internodes of *Ceropegia spiralis*.

A: Friable embryogenic callus bearing loosely attached granular bipolar somatic embryos; B: Different stages of somatic embryos (Globular to heart); C: Unipolar somatic embryos grown in to root like structures lacking shoot pole; D: Different stages of somatic embryos (Globular, torpedo with secondary somatic embryos, heart stage and an embling, E: An embling; complete plant from the somatic embryo; F: Longitudinal section of the somatic embryo showing small projections of leaf primordia and development of vasculature. Bar = 180 µm; G: Base of the somatic embryo showing vascularisation into the cotyledons at torpedo to heart staged somatic embryo. Bar = 180 µm; H: Longitudinal section through somatic embryo of the cotyledonary stage showing xylery element differentiation. Bar =180 µm. I: Emblings in the mist chamber getting acclimatized.

Figure 16.5: *In vitro* Propagation of *C. elegans*.

A: Natural Habit; B: Multiple shoot induction form BAP 13.37 µM; C Multiple shoot induction production form Kn 23.20 µM; D: Multiple shoot induction on MS medium supplemented with Kn23.20 µM+IAA 5.71 µM; E: *In vitro* rooting on ½ MS +IBA 4.90 µM after 30 days; F: Acclimatized plant let ready to transfer in to soil.

Figure 16.6: Micropropagation of *Ceropegia intermedia* through Axillary Bud Multiplication.

A: Axillary bud initiation on MS + BA 6.66 μM after 15 days culture; B: Axillary bud multiplication on MS + BA 6.66 μM after 30 days culture; C: Axillary bud multiplication on MS + KN 6.07 μM after 30 days culture; D: Axillary bud multiplication on MS + Zeatin 2.28 μM after 30 days culture; E: Axillary bud growth on MS + TDZ 2.27 μM after 30 days culture; F: *In vitro* rooting on ½ MS + NAA 5.37 μM after 40 days culture; G: Normal growth of micropropagated plant after one month of growth in green house condition.

acetic acid (NAA) 0.26 µM) were conducted by Britto *et al.* (2003). Maximum number of multiple shoots (up to 12 shoots/node) was achieved in the presence of 13.32 µM/l BAP and 0.26 µM/l NAA. The high frequency of shoot formation (85 per cent) in *C. bulbosa* was achieved on MS medium fortified with 4.44 µM/l BAP, 0.58 µM/l GA_3 and 0.27 µM/l NAA (Goyal and Bhadauria, 2006). In *C. candelabrum* MS medium supplemented with 8.88 µM/l BAP and 2.46 µM/l Indole-3-butyric acid (IBA) was best suitable for axillary bud proliferation (Beena *et al.*, 2003). Further the subsequent sub culture on to the same medium enhanced the rate of shoot multiplication. MS medium supplemented with 7.5 µM/l BAP induced more number of multiple shoots in *C. oculata* a rare endangered plant (Nikam and Savanth, 2007a). Induction of multiple shoots from axillary buds of *C. hirsuta* was achieved on MS medium with 7.5 µM/l BAP (Nikam *et al.*, 2008)

In *C. intermedia* MS medium fortified with 6.66 µM/l BAP was best suitable for induction of more number of shoots and further enhancement of shoot number was observed in subsequent sub cultures up to third generation (Karuppusamy *et al.*, 2009). MS medium with 2.22 µM/l BAP induced more number of multiple shoots in *C. spiralis* an endemic and endangered plant of south India (Murthy *et al.*, 2010a). *C. fantastica* is a critically endangered plant which is under possible extinction. Chandore *et al.* (2010) reported that MS medium supplemented with 4.66 µM/l BAP induced more number of multiple shoots in *C. fantastica*. In contrast to the above Nikam and Savanth (2009) reported that in *C. juncea* induction of multiple shoots were achieved on MS medium supplemented with 7.5 µM/l Kinetin (Kn). However, Krishnareddy *et al.* (2011) found that the maximum numbers of multiple shoots in *C. juncea* were produced on MS medium supplemented with 8.87 µM/l BAP and 4.54 µM/l Thidiazuron (TDZ). High frequency of multiple shoots were induced on MS medium supplemented with 13.31 µM/l BAP in *C. attenuata* a rare and endangered plant of Maharashtra (Chavan *et al.*, 2011a).

Krishnareddy and Pullaiah (2012) attempted *in vitro* conservation studies of *C. elegans* using nodal explant cultures. They found that MS medium supplemented with 23. 20 µM/l Kn and 5.71 µM/l IAA induced maximum shoot multiplication. Similar type of report *i.e.*, the effect of Kn and IAA on *in vitro* shoot multiplication was proposed by Muthukrishnan *et al.* (2012). They reported that in *C. thwaitesii* high frequency shoot bud proliferation and multiplication was achieved on MS medium supplemented with 13.94 µM/l Kn and 28.54 µM/l IAA. Later they studied the effect of agar concentration on multiple shoot induction of *C. thwaitesii* (Muthukrishnan *et al.*, 2013) and found that reduced concentration of agar than normal and its complete elimination favoured multiplication of shoots. Chavan *et al.* (2013) attempted experiments on direct shoot organogenesis of *C. santapaui* by using cotyledonory node explants and found that high frequency shoot multiplication occurred when cultured on MS medium supplemented with 2.0 mg/l 2ip.

Callus Cultures, Somatic Embryogenesis and Organogenesis

Micropropagation through callus mediate regeneration is an efficient method for mass propagation and conservation of endangered plants. Normally in nature the callus is formed in response to injury. The callus is defined as an unorganized mass of parenchymatous tissue induced by a dedifferentiation mechanism from an inoculated explant *in vitro*. In genus *Ceropegia* the first attempt of callus induction and regeneration was reported by Patil (1998). He observed that 9.04 µM/l 2,4-dichlorophynoxy acetic acid (2,4- D) was optimum for callus induction from both stem and leaf explants in all the three species tested (*C. jainii, C. bulbosa* var *bulbosa* and *C. bulbosa* var *lushii*). The callus derived from leaf explants was not maintained beyond the second subculture. However the callus derived from the stem segments when sub cultured onto the fresh ½ MS medium with 2.22 µM/l BAP induced somatic embryos in all the 3 *Ceropegia* genotypes. An efficient protocol for shoot organogenesis and plantlet regeneration from callus derived from seedling explants in *C. bulbosa* has been developed by Phulwaria *et al.* (2013). Among different explants tested epicotyls were found to be the most suitable explants for establishment of callus cultures and further shoot regeneration. They reported that maximum callus proliferation occurred on MS medium supplemented with 4.52 µM/l 2,4-D and 2.22 µM/l BAP and further shoot induction from callus was achieved on MS medium supplemented with 4.44 µM/l BAP and 0.53 µM/l NAA. In *C. candelabrum* when internodes and leaf explants were inoculated on MS medium supplemented with 4.52 µM/l 2,4-D induced friable callus which latter undergone somatic embryogenesis (Beena and Martin, 2003). Further when this callus was sub cultured into fresh ½ or ¼ strength MS medium fortified with 0.23 or 0.45 µM/l 2,4 -D induced the maximum number of somatic embryos.

Nikam and Savanth (2007b) found that in *C. sahyadrica* of 195 combinations of plant growth regulators tested, extensive callus proliferation took place on MS medium supplemented with 1.0 µM/l 2,4–D and 5.0 µM/l BAP. In *C. spiralis* induction of embryogenic callus was achieved on MS medium fortified with 13.32 µM/l BAP and 4.52 µM/l 2, 4 -D. However when this callus was sub cultured on to fresh MS medium fortified with 4.56 µM/l Kn and casein 250 milligrams per liter (mg/l) maturation and germination of somatic embryos were noticed (Murthy *et al.*, 2010b). In *C. spiralis* extensive embryogenic callus was induced from thin cell layer explants on MS medium with 13.32 µM/l BAP and 1.10 µM/l 2,4-D (Murthy and Kondamudi, 2011). Kondamudi *et al.* (2010) induced organogenic callus and shoot multiplication in *C. pusilla*. They reported that 13.32 µM/l BAP in combination with 0.45 µM/l 2,4-D induced maximum amount of organogenic callus. However morphogenesis took place when this callus was sub cultured on to MS supplemented with 22.7 µM/l TDZ. Instead of somatic embryogenesis and organogenesis on the other hand induction of callogenesis can also be helpful in studying the production and estimation of phytochemicals (secondary metabolites). Induction of callus and estimation of alkaloid cerpegin accumulation in callus cultures of *C. juncea* was investigated by Nikam and Savanth (2009). Maximum amount of callus induction and proliferation was observed on MS medium in combination with 1 µM/l 2,4-D and 5 µM/l BAP. However maximum cerpegin content (470 ug/g dry weight) was recorded in dried callus derived on MS medium with 10 µM/l Indole-3-acetic acid (IAA) and 5 µM/l BAP. Chavan *et al.* (2013) cultured cotyledon explants of *C. santapaui* for callus induction and subsequent shoot organogenesis. Maximum response for callus induction and proliferation was achieved on MS medium supplemented with

6.21 µM/l picloram. The high frequency of shoot regeneration from callus was noticed when callus transferred onto MS medium supplemented with 11.1 µM/l BAP and 1.8 µM/l IBA.

In vitro Flowering and Fruiting

Flowering involves the conversion of the apical meristematic initials into floral meristem. In nature signals that change the fate of apical meristem into floral meristem include the age of the plant, climate, temperature and photoperiod etc. However in *in vitro* conditions flowering depends on explant, nutrients, signaling molecules (hormones) and additives along with temperature and photoperiod. Therefore it is very difficult to present complete picture on the *in vitro* flowering process. However induction of *in vitro* flowering is an important study in this genus. It can be helpful in studying the reproductive physiology of these endemic and endangered species. *In vitro* flowering can also helpful in developing micro breeding strategies in supportive of conservation.

In *C. jainii* induction of *in vitro* flowering was noticed on MS medium supplemented with BAP 2.22 µM/l and 0.25 µM/l spermine (Patil, 1998). Induction of *in vitro* flowering in *C. bulbosa* var. *bulbosa* on B5 medium containing 2.22 µM/l BA and 1.0 mg/l GA_3 and fruit, seed setting after 10 days was reported by Britto *et al.* (2003). However they also found that the seeds from *in vitro* developed follicles showed 65 per cent of germination in the soil and the seedlings showed normal growth in the green house. High frequency *in vitro* flowering in 4 rare and endemic species (*C. lawii, C. maccannii, C. oculata* and *C. sahyadrica*) and in 2 wide distributed species (*C. bulbosa* var. *bulbosa* and *C. hirsuta*) from *in vitro* regenerated shoots or *in vitro* raised seedlings was reported by Nair *et al.* (2007). However in all the six species maximum frequency of flowering was achieved on MS medium supplemented with either 26.6 µM/l BAP or 175 µM/l sucrose.

In vitro flowering from thin cell layer cultures of *C. spiralis* was induced on MS medium supplemented with 13.32 µM/l BAP and 5.77 µM/l IAA (Murthy and Kondamudi, 2011). MS medium fortified with 2.27 µM/l TDZ induced *in vitro* flowering in *C. fantastica* (Chandore *et al.*, 2010). In *C. spiralis* ½ strength MS medium with 10.74 µM/l NAA induced flowering, whereas in *C. pusilla* maximum frequency of flowering was induced on ½ strength MS with 0.012 µM/l IBA (Murthy and Kondamudi, 2010). Chavan *et al.* (2011) noticed induction of flowering and fruiting in *C. attenuata* on MS containing 4.14 µM/l picloram

In vitro Tuberization

In vitro microtuberization would be an ideal strategy for the conservation of these species, if micro tubers could be yielded throughout the year (Fay, 1992). Micro tubers are easy to acclimatize and reintroduce into the wild because they get established fast in soil. So, *in vitro* tuberization studies were carried out by a number of biologists in different species of *Ceropegia*. Induction of micro tubers in three species of *Ceropegia* (*C. jainii, C. bulbosa* var. *bulbosa* and *C. bulbosa* var. *lushii*) was achieved on MS medium supplemented with either 22.2 µM/l BAP or 23.2 µM/l Kn (Patil, 1998). However in

C. bulbosa var. *bulbosa* 85 per cent of shoots formed micro tubers on B5 medium containing 4 per cent sucrose, 0.32 µM/l Kn and 9.84 µM/l IBA (Britto *et al.*, 2003).

Pandit *et al.* (2008) reported high frequency microtuber production from the seedling apical buds in five different *Ceropegia* species (*C. hirsuta, C. lawii, C. oculata* and *C. sahyadrica*). They found that MS medium containing 17.76 to 26.64 µM/l BAP and 3 to 8 per cent sucrose with continuous illumination was optimum for microtuber induction. Further sub culturing of these microtubers induced the cluster of secondary microtubers. In *C. pusilla* MS medium with 13.32 µM/l BAP and 0.11 µM/l 2,4-D induced basal tuberization (Kondamudi and Murthy, 2011). Cultures treated with 13.32 µM/l BAP along with the combination of 2.68 µM/l NAA showed 84 per cent of microtuber formation in *C. spiralis* (Murthy *et al.*, 2010).

Synthetic Seeds

Production of artificial seeds is the best method for large scale clonal propagation, conservation, short term storage, cryopreservation of germplasm and germplasm exchange and distribution. Artificial seed technology is also an alternative technique for the cost-effective mass propagation of elite plant genotypes (Reddy *et al.*, 2012). In genus *Ceropegia*, for the first time an attempt for the production of seed analogues was made by Dhir and Shekhawat (2013). In *C. bulbosa* they encapsulated nodal explants and reported 100 per cent plantlet conversion. They also found that encapsulated nodal segments could be stored at low temperature (4 °C) for up to 60 days with a survival frequency of 50 per cent. Latter Murthy *et al.* (2013) encapsulated different explants for production of synthetic seeds in *C. spiralis* and *C. pusilla*. They found that among different explants encapsulated, nodal explants encapsulated with MS medium supplemented with 13.32 µM/l BAP + 3 per cent sucrose + 3 per cent sodium alginate sprouted well and induced maximum number of multiple shoots.

Rooting and Acclimatization

Rooting is the last and most important step for complete plantlet regeneration during *in vitro* propagation of any plant species. Acclimatization is the end and most critical step in micropropagation of plants. When the plant is growing in *in vitro* conditions it shows heterotrophic type of growth. Immature chloroplasts are present in the leaves and the plantlets are too sensitive. During the time of acclimatization the plantlet wants to develop their complete photosynthetic system to prepare their own food materials. Patil (1998) reported both *in vitro* and *ex vitro* rooting in three *Ceropegia* genotypes. He found that ¼ strength MS with 2.46 µM/l IBA and 60 g/l sucrose was optimum for root induction. In *C. bulbosa* var. *bulbosa* the shoots of 3 to 4 cm length were rooted in MS medium supplemented with 9.84 µM/l IBA (Britto *et al.*, 2003). Successful *ex vitro* rooting of *in vitro* derived shoots in *C. bulbosa* was reported by Pulwaria *et al.* (2013). Where as in *C. candelabrum* MS medium along with 0.49 µM/l IBA was optimum for root induction (Beena *et al.*, 2003). However Chandore *et al.* (2010) reported that the *C. fantastica* was best rooted on MS medium free of $CaCl_2$ containing 4.92 µM/l IBA. ½ strength MS medium with 5.37 µM/l NAA produced more number of roots in *C. intermedia* (Karuppusamy *et al.*, 2009).

In *C. pusilla* MS medium in combination with 13.32 µM/l BAP and 0.49 to 1.23 µM/l IBA induced maximum number of roots (Kondamudi *et al.*, 2010). Nikam and Savanth (2007) found that 60 µg/l spermine along with 5 per cent sucrose favored root induction in *C. sahyadrica*. In *C. spiralis* induction of maximum number of roots was achieved on ½ MS supplimented with 10.74 µM/l NAA (Murthy and Kondamudi, 2011). ½ MS medium fortified with 0.5 to 2.0 µM/l IAA or NAA favored root induction in *C. juncea* (Nikam and Savanth, 2009). In *C. bulbosa* the shoots were best rooted on MS medium containing 11.42 µM/l IAA and 3.55 µM/l BAP (Goyal and Bhadauria, 2006). However Nikam *et al.* (2008) found that in *C. hirsuta* ½ strength MS in combination of 2 µM/l IAA along with 5 per cent sucrose showed highest rooting percentage. In *C. attenuata* best rooting was observed on ½ MS with 2.46 µM/l IBA (Chavan *et al.*, 2011), where as in *C. juncea* ½ MS in combination with 4.90 µM/l IBA and 1.27 µM/l NAA induced more number of roots (Krishnareddy *et al.*, 2011). The acclimatization and survival percentage of plantlets varied from 70 to 100 per cent for every individual species.

Genetic Stability of *in vitro* Derived Plants

Maintenance of genetic stability is more important in micropropagation of medicinal plants. Somaclonal variations have been frequently reported in many micropropagated plants. The regenerated plants did not show any detectable variation in morphology when compared with mother plants. However sometimes some changes can be induced among *in vitro* cultures in their genome level. So, it is important to study the genetic fidelity of *in vitro* regenerated plantlets. In recent years DNA marker technology has been developed for analysis of their genetic status. In genus *Ceropegia* the genetic stability was first studied in *C. bulbosa* by using random decamer primers (RAPD) (Dhir and Shekhawat, 2013). They reported that morphologically identical plants generated *in vitro* were genetically identical to the mother plants. Chavan *et al.* (2013) studied the molecular profiling of direct shoot regenerated plants from nodes (DSO) and indirect shoot regenerated plants from callus (ISO) in *C. santapaui* by using random amplified polymorphic DNA (RAPD) and inter simple sequence repeats (ISSR) markers. They reported that RAPD and ISSR marker based profiling reveled uniform banding pattern in DSO-derived plantlets which was similar to mother plant. However in ISSR fingerprints of ISO derived plants there is slight variation. In *C. spiralis* of the two PCR based assays tested ISSR marker were more effective in detecting polymorphism among micro propagated plants (Chavan *et al.*, 2013).

Conclusion

In recent days biotechnology is playing an important role in plant protection through various methods like micropropagation. Micropropagation plays an important role in propagation of many endangered plants if conventional propagation methods are very slow. This review can be helpful to develop efficient cost effective reproducible protocols to other *Ceropegia* species. It can also be helpful to forest and other biodiversity conservation departments to conserve these red listed species more easily and effectively. This can also be helpful in restoration and conservation of other endemic, endangered, rare and threatened plants too.

Acknowledgements

The receipt of financial assistance from the Council of Scientific and Industrial Research (CSIR) and University Grants Commission (UGC), New Delhi, are gratefully acknowledged.

References

Adibatti, N.A. Thirugnanasambantham, P. and Kuilothungan C, 1991. A pyridine alkaloid from *Ceropegia juncea*, Phytochemistry **30:** 2449-2450.

Ananthan, R. 2003. *In vitro* approaches for the propagation and conservation of the endemic *Ceropegia barnesii* Brance *et* Chattarjee and *Ceropegia pusilla* Wight et Arn. Asclepiadaceae. Ph.D. Thesis, Bharathiyar University, Coimbatore, India.

Anonymous, 1992. *Ceropegia* Linn. (Asclepiadaceae), In: The Wealth of India. Vol 3. (CSIR, New Delhi). pp. 448–449.

Ansari, M. 1984. Asclepiadaceae: Genus *Ceropegia*–Fascicles of Flora of India. Botanical Survey of India, Calcutta. 16: 1-34.

Asolkar, L.V. Kakkar, K. K. and Chakre, O. J. 1992. 2nd supplement to the Glossary of Indian Medicinal plant with Active Principle. Part 1 (A-K), Publication and Information Directorate, CSIR, New Delhi. pp. 193.

Bapat, V.A., Yadav, S.R. and Dixit, G.B. 2008. Rescue of endangered plants through biotechnological applications. Nat. Acad. Sci. Lett., **31:** 201-210.

Beena, M.R. and Martin, K.P. 2003. *In vitro* propagation of the rare medicinal plant *Ceropegia candelabrum* L. through somatic embryogenesis, In Vitro Cell Dev. Biol.-Plant, **39:** 510-513.

Beena, M.R., Martin, K. P., Kirti, P. B. and Hariharan, M. 2003. Rapid *in vitro* propagation of medicinally important *Ceropegia candelabrum*, Plant Cell Tiss. Org. Cult., **72:** 285-289.

Britto, S.J., Natarajan, E. and Arockiasamy, D.I. 2003. *In vitro* flowering and shoot multiplication from nodal explants of *Ceropegia bulbosa* Roxb. var. *bulbosa*, Taiwania, **48:** 106-111.

Bruyns, P.V. 2003. Three new succulent species of Apocynaceae (Asclepiadoideae) from Southern Africa, Kew Bull. **58:** 427- 435.

BSI, 2009. Botanical Survey of India. http://envfor.nic.in/bsi/research.html accessed on August 25th 2009.

Chavan, J.J., Gaikwad, N.B. and Yadav, S.R. 2013. High multiplication frequency and genetic stability analysis of *Ceropegia panchganiensis*, a threatened ornamental plant of Western Ghats: Conservation implications Scientia Horticulturae. 161.

Chavan, J.J., Gaikwad, N.B., Kshirsagar, P.R., Umdale, S.D., Bhat, K.V., Dixit, G.B. and Yadav, S.R. 2013. Application of molecular markers to appraise the genetic fidelity of *Ceropegia spiralis*, a threatened medicinal plant of South India. Curr. Sci. **105:** 1348-1350.

Chandore, A. N., Nimbalkar, M.S., Gurav, R.V., Bapat, V.A. and Yadav, S.R. 2010. A protocol for multiplication and restoration of *Ceropegia fantastica* Sedgw.: A critically endangered plant species, Curr. Sci. **99:** 1593-1596.

Chavan, J.J., Mansingraj, S.N., Avinash, A.A., Sharad, S.K., Nikhil, B.G., Ghansham, B.D., Rajaram,V.G., Bapat, V.A. and Yadav, S.R. 2011a. Micropropagation and *in vitro* flowering of endemic and endangered plant *Ceropegia attenuata* Hook, J Plant Biochem. Biotechnol. **20:** 276–282.

Chavan, J.J., Nimbalkar, M.S., Gaikwad, N.B., Dixit, G.B. and Yadav, S.R. 2011b. *In vitro* propagation of *Ceropegia spiralis* Wight–an endemic and rare potential ornamental plant of peninsular India, Proc Nat Acad Sci India Sect B, **81**: 120-126.

Chavan, J. J., Gaikwad, N. B., Umdale, S. D., Kshirsagar, P. R., Bhat, K. V. and Yadav, S. R. 2013. Efficiency of direct and indirect shoot organogenesis, molecular profiling, secondary metabolite production and antioxidant activity of micropropagated *Ceropegia santapaui*, Plant Growth Regul, *link.springer.com/content/pdf/10.1007 per cent 2Fs10725-013-9830-7.pdf*

Darias, V., Martin, D. and Abdala, S. 1996. Preliminary study of the cicatrizant, haemostatic and chemotherapic activities of *Ceropegia fusca* Bolle. Phytother. Res, **10:** S_1-S_2.

Dhir, R. and Shekhawat, G.S. 2013. Production, storability and morphogenic response of alginate encapsulated axillary meristems and genetic fidelity evaluation of *in vitro* regenerated *Ceropegia bulbosa*: A pharmaceutically important threatened plant species, Industrial Crops and Products, **47:** 139-144.

Fay, 1992. Conservation of rare and endangered plants using *in vitro* methods. In vitro Cell Dev Biol-Plant, **28**: 1- 4.

Gadgil, M. 2004.ENVIS Technical Report No. 16, Environmental Information, Bangalore, http://www.ces.iisc.ernet.in/hpg/envis. pp. 96-98.

Gaikwad, S.P. and Yadav, S. R. 2003. Endemic flowering plant species of Maharashtra and their possible utilization. In: Biodiversity in India. Vol II, edited by T. Pullaiah (Regency Publications, New Delhi). pp.28-52.

Goyal, D. and Bhadauria, S. 2006. *In vitro* propagation of *Ceropegia bulbosa* using nodal segments, Indian J. Biotechnol., **5:** 565-567.

Jagtap, A. and Singh, N. P. 1999. Fasc. Flora of India 24, Botanical Survey of India, Kolkata, pp. 211-241.

Jain, S. K. and Defillips, R. A. 1991. Asclepiadaceae. In Medicinal plants of India, Vol 1. Reference Publications Inc., Michigan, USA. Pp.144-152.

Karuppusamy, S. and Pullaiah, T. 2009. Pollination system and *ex-situ* fruit set in *Ceropegia juncea* Wight (Apocynaceae)–an endemic species of India, Acad. J. Plant Sci., **2:** 242-245.

Karuppusamy, S., Kiranmai, C., Aruna, V. and Pullaiah, T. 2009. *In vitro* conservation of *Ceropegia intermedia*-an endemic plant of South India, Afr. J Biotechnol., **8:** 4052-4057.

Kedage, V. V., Mhatre, M. and Dixit, G.B. 2006. *In vitro* propagation of *Ceropegia noorjahani* Ans.: A critically endangered, endemic medicinal plant of Maharashtra. International symposium on frontiers in Genetics and Biotechnology–Retrospect and Prospects. Pp. 162.

Kelly, T. R. and Walsh, J. J. **1992.** The synthesis of cerpegin, *J. Org. Chem,* **57:** 6657-6658.

Kirtikar, K.R. and Basu, B.D. 1935. In Indian Medicinal Plants Vol.3 (Bishen Singh Mahendrapal, New Delhi, India). Pp. 1638.

Kondamudi, R. and Murthy, K.S.R. 2011. Micropropagation and *in vitro* flowering of *Ceropegia pusilla.* J. Trop. Med. Plants, **12:** 41- 47.

Kondamudi, R., Vijayalakshmi, V. and Murthy, K. S. R. 2010. Induction of morphogenetic callus and multiple shoot regeneration in *Ceropegia pusilla* Wight & Arn, Biotechnology, **9:** 141-148.

Krishnareddy, P. V., Karuppusamy, S. and Pullaiah, T. 2011. *In vitro* propagation of *Ceropegia juncea* Roxb., Afr. J Plant Sci., **3:** 345-357.

Krishnareddy, P.V. and Pullaiah, T. 2012. *In vitro* conservation of *Ceropegia elegans*, an endemic plant of South India, Afr. J Biotechnol., **11:** 12443-12449.

Kullayiswamy, K. R., Sandhyarani, S. and Karuppusamy, S. 2012. *Ceropegia pullaiahii* sp. nov. (Apocynaceae, Asclepiadoideae) from India. Nordic J. Bot. **30:**11-14

Lazaur. J., Hoarau. C., Mougin, F., Trecourt, F., Godard, A., Queguiner, G. and Marsais, F. 2008. One pot four step synthesis of cerpegin, Tetrahedron Letters, **46:** 3811-3813.

Mabberley, D. J. 1987. The Plant Book. Cambridge University Press, Cambridge. 114-115.

Mabberley, D. J. 2001.The Plant Book. Cambridge University Press, Cambridge. Pp. 143.

Malpure, N.V., Kamble, M.Y. and Yadav, S.R. 2006. A new species of *Ceropegia* L. (Asclepiadaceae) from the Western Ghats of India with a note on series *Attennuat* Huber, Curr. Sci., **91:** 1142.

Matsuo, K. and Arase, T. 1995. Synthesis of a pyridone alkaloid, cerpegin, Chemical and Pharmaceutical Bulletin, **43:**2091-2094.

Murthy, K. S. R. and Kondamudi, R. 2010. Effect of cytokinins and auxins on *in vitro* flowering of endangered *Ceropegia spiralis* Wight and *C. pusilla* Wight & Arn, Phytomorphology, **60:** 32-37.

Murthy, K.S.R. and Kondamudi, R. 2011. Rapid shoot regeneration from thin cell layer explants of an endangered Asclepiad *Ceropegia spiralis.* Plant Tissue Cult. and Biotechl. **21:** 63-73.

Murthy, K.S.R., Kondamudi, R. and Vijayalakshmi, V. 2010a. Micropropagation of an endangered medicinal plant *Ceropegia spiralis* Wight & Arn. J. Agr. Tech., **6:** 179-191.

Murthy, K.S.R., Kondamudi, R. and Pullaiah, T. 2010b. High frequency somatic embryogenesis in *Ceropegia spiralis* Wight.–An endemic and endangered medicinal plant, Indian J. Biotechnol. **9:** 414-418.

Murthy, K. S. R., Kondamudi, R. and Karuppusamy, S. 2012a. Microtuberization of *Ceropegia spiralis* Wight and *Ceropegia pusilla* Wt. & Arn., Afr. J. Plant Sci. **6:** 321-327.

Murthy, K. S. R., Kondamudi, R., Reddy, M. C., Karuppusamy, S. and Pullaiah, T. 2012b. Check list and conservation strategies of the genus *Ceropegia* in India. Int. J. Biodvers. Conserv, 4: 304-315.

Murthy, K. S. R., Reddy, M. C. and Kondamudi, R. 2013. Synthetic seeds–a novel approach for the conservation of endangered *Ceropegia spiralis* Wt. and *C. pusilla* Wt. & Arn., Bangladesh J. Sci. Ind. Res. **48:** 39-42.

Muthukrishnan, S., Benjamin, J. H. F. and Rao, M. V. 2013. Influence of agar concentration and liquid medium on *in vitro* propagation of *Ceropegia thwaitesii* Hook- an endemic species, J. Agr. Tech., **9:**1013-1022.

Muthukrishnan, S., Benjamin, J. H. F., Muthukumar, M., Sheriff, N. A., Kumar, T. S. and Rao, M. V. 2012. *In vitro* propagation of *Ceropegia thwaitesii* Hook- an endemic species of Western Ghats of Tamil Nadu, India. Afr. J Biotechnol. **11:**12277-12285.

Nadkarni, K. M. 1976. Indian Materia Medica Vol.1 Pp.303-304 Popular Prakashan, Bombay, India.

Nair, A. K., Dilip, N. D. and Subhash, P. S. 2007. High-frequency *in vitro* flowering in six species of *Ceropegia,* J. Plant Biol., **50:** 374-377.

Nautiyal, D.C., Sharma, S. K. and Pandit, M. K. 2009. Notes on the taxonomic history, rediscovery and conservation status of two endangered species of *Ceropegia* (Asclepiadaceae) from Sikkim Himalaya. J. Bot. Res Inst. Texas, **3:** 815-822.

Nayar, M. P. and Sastry, A. R. K. (eds.) 2000. Red Data Book of Indian Plants, BSI, Kolkatta, India.

Nikam, T.D. and Savanth, R.S. 2007a. Conservation of *Ceropegia oculata* Hook. (Asclepiadaceae) by micropropagation using nodal explants. Plant Cell Biotech. Mol. Biol., **8:**167-172.

Nikam, T. D. and Savanth, R. S. 2007b. Callus culture and micropropagation of *Ceropegia sahyadrica* Ans. et Kulk: An edible starchy tuberous rare asclepiad. Indian J. Plant Physiol. **12:**108-114.

Nikam, T. D., Savanth, R. S. and Parage, R. S. 2008. Micropropagation of *Ceropegia hirsuta* Wt. & Arn., a starchy tuberous asclepiad, Indian J Biotechnol, **7:** 129-132.

Nikam, T. D. and Savanth, R. S. 2009. Multiple shoot regeneration and alkaloid ceropegin accumulation in callus culture of *Ceropegia juncea* Roxb, Physiol Mol. Biol Plant, **15:** 71-77.

Pandit, S. S., Aneeshkumar, N. and Naik, D. D. 2008. Towards conservation of threatened *Ceropegia* species endemic to a biodiversity hot spot: *in vitro* micro tuber production and proliferation, a novel strategy. J. Plant Sci. **24:** 79-88.

Patil, C. R., Rahangdale, S. R., Rahangdale, S. S. and Patil, S.C. 2005. Occurrence of tricarpy in *Ceropegia attenuata* Hook. (Asclepiadaceae). Curr. Sci. **88:** 1215 -1215.

Patil, V. M. 1998. Micropropagation of *Ceropegia* spp. *In Vitro* Cell. Dev. Biol.-Plant, **34:** 240-243.

Phulwaria, M., Shekhawat, N. S., Rathore, J. S. and Singh, R. P. 2013. An efficient *in vitro* regeneration and ex *vitro* rooting of *Ceropegia bulbosa* Roxb.–a threatened and pharmaceutical important plant of Indian Thar Desert. Industrial Crops and Products **42:** 25-29.

Prabakaran, R., Sasikala, T. and Kalimuthu, K. 2013. Regeneration of shoots from callus of *Ceropegia pusilla* Wight & Arn. British Biotechnology Journal, **3:** 416-423.

Raghuramulu, D. and Pullaiah, T. 1999. *In vitro* propagation of *Ceropegia bulbosa* Roxb. from the seedling explants.In: Kavi kishore, P.B. (ed.) Plant Tiss. Cult. Biotechnol. Emerging Trends: pp. 202-210.

Raghuramulu, D., Pullaiah, T. and Murthy, K. S. R. 1999. Occurrence of twin seedlings in *Ceropegia bulbosa* Roxb, Indian For., **125:** 835-837.

Rajput, B. P., Ghate, V. S., Upadhye, A. S. and Datar, M. N. 2012.Tricotyledony in critically endangered plant, *Ceropegia mahabalei* Hemadri et Ansari (Apocynaceae), Curr. Sci,**103:** 24-25

Rathore, M.S., Gehlot, H. S. and Shekhawat, N. S. 2010. Biotechnological approaches for propagation and prospecting of important medicinal plants from Indian Thar Desert. Int. J. Plant Prod., **4:** 67-72.

Rawiya, A. and Mathias, M. 2008. The *in vitro* inhibition of Human neutrophil elastate activity by some Yemeni medicinal plants. Sci. Pharm., **76:** 471-483.

Reddy, M. C., Murthy, K. S. R. and Pullaiah, T. 2012. Synthetic seeds: A review in agriculture and forestry. Afr. J Biotechnol., **11:** 14254-14275.

Sivakumar, K., Eswaramurthy, S., Subrahmanian, K. and Natarajan, S. 1990. Structure of cerpegin, a new alkaloid, Acta Cryst., **46:** 839-841.

Srinivasarao, C., Soni, A., Dulloo, M. E. and Naithani, S. C. 2010. Overcoming physiological dormancy in *Ceropegia odorata* seeds, an endangered rare species, with GA_3, H_2O_2 and KNO_3. Seed Sci. and Technol. **38:** 341-347.

Sukumar, E., Hamsaveni, G. R., Bhima, R. R., Viswanathan, S., Thirugnanasambantham, P. and Vijayasekaran, V. 1995. Pharmacological actions of cerpegin, a novel pyridine alkaloid from *Ceropegia juncea*. Fitoterapia. **66:** 403-406.

Surveswaran, S., Kamble, M, Y., Yadav, S. R. and Sun, M. 2009. Molecular phylogeny of *Ceropegia* (Asclepiadoideae, Apocynaceae) from Indian Western Ghats. Plant Syst. Evol., **281:** 51- 63.

Yadav, S.R. and Mayur, Y. K. 2008. Threatened *Ceropegias* of the Western Ghats and strategies for their conservation pp. 123-134. In: Rawant G.S. (Ed). 2008. Special habitats and threatened plants of India. ENVIS Billetin: Wildlife and protected Areas Vol. 11. Wildlife Institute of India, Deharadun, India.

Yadav, S. R. 1996. Flytrap flowers of Western Ghats, Hornbill, 1: 1-7.

Yadav, S. R., Sardesai, M. M. and Gaikwad, S. P. 2004. *Ceropegia anantii* (Asclepiadaceae), a new species from Western Ghats, India, J Bombay Nat. Hist. Soc. **101:** 141-143.

Chapter 17

Snakes of Sri Krishnadevaraya University Campus and its Surroundings, Andhra Pradesh

Y. Amarnath Reddy[1], *B. Sadasivaiah*[2]*, *P. Indira*[1] *and T. Pullaiah*[3]

[1]*Department of Zoology, Sri Krishnadevaraya University, Anantapur, Andhra Pradesh*
[2]*Department of Botany, Govt. Degree and PG College, Wanaparthy, Mahabubnagar, Andhra Pradesh*
[3]*Department of Botany, Sri Krishnadevaraya University, Anantapur, Andhra Pradesh*

Introduction

The Planet Earth was born 4.5 billion years ago and life was originated 3.5 billion years on it. There is no certainty in predicting the species number existing on earth at present. Hawksworth and Arroyo (1995) estimated 13.6 million species and May (1992) said 5 to 30 million species present on earth. India occupies 2.4 per cent of the world's land area and it is one of the world's mega biodiversity centers, contributing 8 per cent of the world's species diversity.

Snakes are among the animals that have most influenced the human psyche since ancient times. Images, myths and beliefs regarding these reptiles are present in

* Corresponding Author: E-mail: chumsada@gmail.com

many societies due to their culturally salient biological and behavioral characteristics, such as their speed and agility, the bifid tongue, unblinking lidless eyes, their ability to periodically renew their skin, their mode of mating, and the ability to produce and inject a deadly substance (Cardoso *et al.,* 2003). In ancient times, many people worshiped snakes to try to appease them in order to avoid the evil they could cause. Attributes such as wisdom, cunning, power, fertility, sexuality and renewal of life have been attributed to them by different people in India, Egypt, ancient Europe, ancient Persia, Mexico and most of Africa (Marcum, 2007). Among the snakes, four of them are considered the most dangerous to humans, besides causing more aversion and fear in the population. In general, local people have strong negative behavior towards snakes, killing them whenever possible. Until the anti venom was present and available, the locals used only charms, prayers and homemade remedies to treat or protect themselves and others from snake bites (Fita *et al.,* 2010).

At some point of our lives we have come across the words 'Mapila', 'Ahatulla' or 'Malsara'; *the feared, mysterious snakes that come in a group of seven and suck the blood of sleeping people; the eye-plucking monsters......* the stories go on (Somaweera, 2004). But these snakes are more innocent and helpful to human beings than we think. Snakes are the 'youngest' reptile group on earth, the last big invention of nature. Although the first 'snakelike' creatures are considered to have appeared on earth about 135–180 million years ago in the Jurassic period of the Mesozoic era (Mattison, 1995), when other tetra pod reptiles have already gained ground, the earliest known snake fossils belong only to the Cretaceous period about 95 million years ago. Today nearly 3000 species of snakes, belonging to 18 snake families, inhabit most parts of the Earth excluding some countries like Ireland, New Zealand, Bermuda, Antarctic region etc.

Ever since Adam's early encounter with the 'Serpent', snakes have been the receivers of a bad press. Snakes have an image problem due to which most people fear and loathe them. To many people, they are worth more dead than alive. Due to this negative impression of these snakes, which arose from the early days, several snakes are ruthlessly over-killed through fear, misidentification, poor knowledge and as precautionary measure against snakebite, despite their being ecologically important as pest controllers (Somaweera, 2004). According to de Silva (1990) an average of five snakes are killed daily in Mahaweli settlement areas alone. Based on many misbelieves on the snakes among the people the Government agencies, non-governmental organizations and conservationists of these snakes in any country must be backed by strong educational programmes and public awareness campaigns on the importance of these animals in the ecology of nature. Other than merciless-killing, habitat loss and human interference are the two most common threats faced by all snakes in the country. Lack of consistent studies on the population dynamics and distribution etc. for most species precludes confident statements or even inference concerning the reduction in populations. Deforestation or the reduction of wild land habitats to less than the critical amount necessary for the survival of the species has become a serious environmental issue. Especially podu (shifting) cultivation, an ancient practice that still continues in many parts of the country, the main geographical distribution of some snakes, does a major damage. Over 1000 hectares of forests and grasslands are known to be set fire annually, damaging most snakes and their eggs etc in a small country like Sri Lanka (Somaweera, 2004). The increasing of human population and real estate are also some of the most serious threats to the diversity of snakes. Secondary effects of deforestation, such as soil erosion, climatic changes and exposure to predators also affect the snake fauna. Fragmentation of habitats and victimization to road kills, due to the above situations, are also significant factors.

Snakes are an important part of the natural food chain and play a vital role in the ecosystem balance. An example, snakes are helpful in reducing the population of rodents, and in turn they are prey to the most of the serpents and other animals. The snake poison is used to cure many dangerous diseases like cancer etc. There is an urgent need to search the diversity of snakes in any part of the developing country like India. Based on the above importance of the snakes, the present work has been taken.

Nguyen *et al.* (2009) described reptiles of Vietnam. Rasmussen (2011) listed 25 species of sea snakes of Vietnam and arranged keys based on external characters to identify them. Peters (1964) provided meaning of the terms used in herpetology. Schleich and Kastle (2002) listed 130 species of reptiles of Nepal and described them with details of distribution, ecology, behavior and reproduction.

The important publications on Indian snakes are Deoras (1965), Khaire (1996), Sharma (1998). Shebberare and Barker (1998) described snakes of Sikkim and Bengal. Mathew (1999) described venomous snakes of north-east India.

Regarding to the reptile fauna of Andhra Pradesh, very few reports were found and these are scattered, not covering the whole state (Murthy, 2010), Rayalaseema and Telangana regions of Andhra Pradesh were least explored. Earlier reports on reptiles of Andhra Pradesh include Beddome (1870) and Annandale (1906). McCann (1945) reported reptiles of Vizagapatnam and neighbouring Ghats. Sharma (1971) reported reptiles of Nagarjuna sagar dam area. Biswas (1984) reported on turtles of coast of Andhra Pradesh. Thulsi Rao *et al.* (2005) reported 48 species, 34 genera, 12 families from Nallamalaih hills, of which 5 species were reported for the first time from Andhra Pradesh.

Study Area

Sri Krishnadevaraya University campus lies at about NL 14°38' and EL 77°42'; an altitude of +350m MSL and 10 km away from Anantapur city, Andhra Pradesh. It is extended to over 482 acres. The campus is surrounded by three villages namely Akuthotapalli, Itukulapalli and Chinthakunta. The University was established in 1981 as University. The campus areas is divided into two halves by National Highway (No. 205), one side is with all residential quarters, women hostels, campus school, Vice-chancellor's bungalow, forest nursery, Education department etc bordered with cultivated fields and the other side is with all Departments, Stadium, administrative buildings and men hostels (Rao, 2012).

The soil in the campus is red and the climate is tropical with erratic rainfall. The maximum temperature is 42°C recorded in the month of April. Before establishment of the campus, much of the area is open and some parts with exposed rocks and supporting grass dominating vegetation. It was, at the time, the campus devoid of any trees or shrubs except few Tamarind trees along the highway. After widening the highway, all the tamarind trees were cut down. A good number of trees were planted by the University authorities and many of them are naturalized. The ground vegetation is dominated by the grasses like *Heteropogon contortus*, *Aristida* spp., *Chrysopogon* spp. and *Eragrostis* spp. Shrubby species are confined to corners and fencing area.

All the three villages are surrounded by cultivated fields where crops like Groundnut, Rice, Jowar, and vegetable crops are cultivated.

Materials and Methods

Field explorations were conducted intensively and extensively for a period of 4 years, during 2009-2013 covering all the seasons. All the snakes encountered in the study area were recorded. Snakes killed on the road or by the people were collected and preserved according to methodology described by Director, ZSI (1990). The specimens were labeled with relevant information. All the specimens are kept in the museum, Department of Zoology, Sri Krishnadevaraya University at Anantapur. Identification of specimens was done following standard faunas. The survey was conducted during day and night time. Snakes were collected by hand picking and with the help of snake hook. Shields on the head, number of upper and lower labial, number of ventral scales, subcaudals, and dorsal scales around mid body of the snakes were recorded with care. They were identified following Smith (1931, 1935,1943), Murthy (1990, 2010), Daniel (2002), Whitaker and Captain (2004) and Das (2008). Dichotomous keys were provided for easy identification of order, family, genus and species of the snakes.

Results

A total of 820 individuals of snakes belonging to 21 species and 5 families were identified in the University campus and its vicinity. Of these, 14 snakes are Non-Venomous (NV), 3 are Semi-Venomous (SV) and 4 are Venomous (V). Among the 21 snakes Vine snake and Common Bronze Back snakes are found commonly on trees and sometimes Indian rat snake and cobra also found on trees. The Cat snakes are generally nocturnal, but we found this in day time in resting stage. Some of the snakes were presented in Plates 17.1 and 17.2.

Key to the Families of Ophidia

1A). Blind snakes, eyes covered with scales .... **Typhlofidae**

1B) Snakes not as above .... **(2)**

2A) Head triangular, distinct from neck; covered with small scales; fangs to 1-2cm .... **Viperidae**

2B) Head not as above; covered with shields or small scales; fangs absent or if present not more than 1cm .... **(3)**

3A) Head covered with small scales; supra labials not in contact with eyes, body stout, eyes are small .... **Boidae**

Eryx conicus

Lycodon striatus

Naja naja

Oligodon arnensis

Oligodon taeniolatus

Ptyas mucosa

Plate 17.1

Ahaetulla nasutus

Argyrogena fasciolatus

Boiga trigonata

Dendrelaphis tristis

Echis carinatus

Elaphe helena

Plate 17.2

3B) Head covered with large shields; supra labials in contact with eyes, body moderate-slender, eyes moderate-large .... **(4)**

4A) Tail length moderate, not slender; fangs are present at front side .... **Elaphidae**

4B) Tail long, slender; fangs absent or present; if so present on rear side .... **Colubridae**

Family: Typhlofidae

Rhamphotyphlops braminus (Daudin, 1803)

Nonvenomous. Body worm like, smooth and with shiny scales. Head and tail are blunt. Eyes indistinct and covered by scales.

Scalation – Rostral large. Upper labials – 4, of them last two in contact with ocular. scales are 20 round the body.

Size and colour: Maximum total length of the body is 14cm. Dorsal side is brown or blackish brown, ventral side is lighter than dorsal side.

Habit and Habitat: Lives underground and in termite nests. This species comes out during night times and feeds on ants, termites and their larvae.

Common name: Common Worm Snake; *Local name*: Yeddu mukkulo purugu.

Distribution: *Outside India*: Oriental region, Africa, Mexico and New Guinea

India: Throughout

Andhra Pradesh: Throughout

SKU: Common

Family: Boidae

Key to the species of Boidae

1A) Tail blunt and resembles to head; mental grove present .... *Eryx johni*

1B) Tail pointed; mental grove absent .... *Eryx conicus*

***Eryx conicus* (Schneider, 1801)**

Non venomous. Body stout, short and heavy. Head moderately elongated, rounded evenly from side to side. Snout long, overhanging the chin. Eyes small and with vertical pupil. Nostrils slit like. Tail very short, conical in shape. A curved-claw like rudimentary hind limbs present just above the vent.

Scalation: The head is covered with small scales but nasal and inter nasals are enlarged. Supra labials 11-13, infra labials 12-17, mental groove and chin shields are absent. Eyes surrounded by 10-15 small scales. Dorsal body scales are keeled, more rigid on the posterior side of the body and on tail. Middle of the body with 40-55 rows of scales. Ventral scales 161-196, anal entire, subcaudals 16-24.

Size and Colour: Females are longer than males. The maximum total length of the body is 96cm. Dorsum of body is brownish and with a continuous dorsal series of dark brown spots. Ventral surface is yellowish to white. A dark stripe present behind the eye.

Habit and Habitat: Nocturnal and feeds on small mammals, birds, reptiles and amphibians and large insects. Kills its prey by constriction. It is a strong burrower in loose sand or soil. The snake is viviparous and gives birth to 6-11 young at a time.

Common name: Common Sand Boa; *Local name*: Daasari pamu

Distribution: *Outside India*: Afghanistan, Pakistan and Sri Lanka

India: Throughout; *Andhra Pradesh*: Nalgonda district

SKU: Occasional.

***Eryx johni* (Russell, 1801)**

Non venomous. Body stout, cylindrical and heavy, size from head to tail uniform. Head not distinct from neck. Eyes small, round with vertical pupil. Nostril is slit like. Tail short, blunt, superficially similar to head in shape.

Scalation: Rostral shield is wide, broader than its vertical width. Usually two pairs of broad scales present behind the rostral. Eyes are completely surrounded by 9-12 small scales. Supra labials 9-12, infra labials 13-18. Mental groove is present. Chin shields are absent. Dorsal scales are smooth in anterior region but keeled in posterior region. Scale row at the middle of the body 51-67. Ventral scales in males 193-212, females 198-215, anal entire, subcaudals 28-40 in males and 29-34 in females.

Size and Colour: Maximum total length of the body is 98 cm and 110 cm of males and females respectively. Adults uniform light brown, ruddy brown. Ventral surface lighter than back. Young brick-red or sandy red. Back with dark blotches or ruddy brown cross bars. Ventral surface buff.

Habit and Habitat: Nocturnal and feeds on small mammals, birds and insects. It is sluggish and strong burrower and prefers to live in loose clay soil with sparse grass.

Common name: Red sand Boa; *Local name*: Poodu pamu

Distribution: *Outside India*: Afghanistan, Iran, Pakistan and Sri Lanka

India: Throughout; *Andhra Pradesh*: Guntur district

SKU: Common.

Family: Colubridae

Key to the genera and species of Colubridae

1A) Snout projecting and pointed, pupil horizontal .... **Ahaetulla *(A. nasutus)***

1B) Snout not projecting; pupil round .... **(2)**

2A) Fangs present at rear .... **(3)**

2B) Fangs absent .... **(4)**

3A) Body compressed laterally; head triangular and distinct from neck, 'Y' shaped mark present on head .... **Boiga *(B. trigonata)***

3B) Body round-cylindrical; head not triangular and slightly distinct from neck, no marks on head .... **Enhydris (*E. enhydris*)**

4A) Vertebral scales slightly enlarged, narrower than surrounding scales, ventral shields entire with a suture like lateral keel, a notch on each side .... **Dendrolepis (*D. tristis*)**

4B) Vertebral scales and ventral shields not as above .... **(5)**

5A) 3-6 anterior teeth longer than others and inwardly bent, separated by toothless gaps from the rest of teeth .... **Lycodon**

5B) Teeth are not as above .... **(6)**

6A) Head very short, not distinct from neck .... **Oligodon**

6B) Head long, distinct from neck .... **(7)**

7A) Nostril is between the two nasals .... **Elaphe (*E. helena*)**

7B) Nostril is between two nasals and first supra labials .... **Ptyas (*P. mucosa*)**

***Ahaetulla nasutus* (Lacepede, 1789)**

Mild venomous. Body slender, laterally compressed. Head distinct from neck. Snout very long, ends as a pointed tip. Snout bearing a median groove above, formed usually entirely by the rostral. Eyes large and with horizontal pupil.

Scalation: Preocular-1, Postoculars-2, Temporals-7, Upper labials-8 in them 5th touching the eye. Inter nasals and prefrontals are in the contact with the labials. Middle of the body with 15 rows of scales. Vertebral scales are larger than surrounding scales. Body scales smooth. Ventral shields 166-207, Anal divided and Subcaudals 135-180 paired.

Size and Colour: Maximum total length of the body is 2m. Chin and throat white, with light sky blue and yellow mottling. Remaining total body bright green but belly lighter than above. A thin white or yellow line separating upper body scales from belly scales. Mouth pink in colour.

Habit and Habitat: Diurnal and feeds on lizards, small mammals, birds, small snakes and amphibians. When provoked, open the mouth widely and inflate the body, at this time black and white inter scale colouration is exhibited. It has the habit of striking on eye of opponent. The arboreal snake has wonderful turnoff of speed over foliage. As the body is very slender and light weight, it can twine small stems and takes support of small twigs. Viviparous, give birth to 3-22 individuals at a time from March-December.

Common name: Common Vine Snake; *Local name*: Pasarika pamu

Distribution: *Outside India*: Burma, Cambodia, Complete Indo-Chinese region, Sri Lanka, Thailand and Vietnam; *India*: Throughout (except North West and Gangetic basin)

Andhra Pradesh: Kurnool and Nalgonda districts

SKU: Rare

Boiga trigonata **(Schneider, 1802)**

Mild venomous. Body long, thin and compressed laterally. Head distinct from neck and triangular in shape. Eyes large and with vertical pupil. Tail very long. Maxillary teeth 8-10, last one or two are enlarged fangs.

Scalation: Preocular-1, Temporals-5, Upper labials-8 in them 3rd, 4th and 5th touching the eye, Lower labials 10-11. Middle of the body with 21 rows of scales. Body scales smooth, vertebral scales slightly enlarged. Ventral shields 206-256, Anal entire and Subcaudals 75-96 paired.

Size and Colour: Maximum total length of the body is 123cm. Body ground colour is yellowish brown, sandy or fawn, uniform or mottled with dark shades. Dorsally a series of dark Y-shaped mark which meet at the centre and resemble arrow heads. On top of head Y-shpaed mark present. A narrow dark streak from behind eye to gape.

Habit and Habitat: Nocturnal and feeds on agamid lizards, small mammals, birds and their eggs and large insects. The prey is killed by constriction. Inhabits in bushes, thatched roofs, under tree bark or stone. An excellent climber. When disturbed coil tightly vibrate tail and strike repeatedly.

Common name: Cat snake

Distribution: Outside India: Iran, Pakistan, Saudi Arabia, Sri Lanka and Uzbekistan

India: Throughout; *Andhra Pradesh:* Kurnool district

SKU: Rare

Dendrelaphis tristis **(Daudin, 1803)**

Non venomous. Body elongated, slender and smooth. Head distinct from neck. Snout bluntly rounded. Eyes large with round pupil. Tail very long.

Scalation: Loreal-1, Supra labials-9 in them 5th and 6th touching the eye. Vertebral scales slightly narrower than surrounding scales. Middle of body with 15 rows of scales. Ventral shields 163-205, Anal divided and subcaudals 105-150, paired.

Size and Colour: Maximum total length of the body is 172cm. Bronze-brown or purplish brown back with a dark brown black stripe on either side of body. Indistinct black streak behind eye. Under side usually white or gray.

Habit and Habitat: Diurnal and feeds on frogs, lizards, insects and eggs of birds. This is arboreal snake and jumps from one to tree to another. Inhabits open areas like disturbed forests, forest clearings and around human habitations in rural areas especially on thatched roofs of houses.

Common name: Common Bronze back Tree Snake; *Local name:* Ielliriki pamu

Distribution: Outside India: Pakistan and Sri Lanka

India: Gujarat, Maharashtra, Orissa, Southern India and West Bengal

Andhra Pradesh: Kurnool district

SKU: Common

***Elaphe helena* (Daudin, 1803)**

Non venomous. Body elongated and slender. Head distinct from the neck and snout long. Eyes large with round pupil. Tail long.

Scalation: Pre frontal longer, upper labials 8-11, in them 5, 6 and 7 touching the eye. Preocular-1, postoculars-2 (rarely 1). Temporals 2+2, 2+3 or rarely 2+4. Body scales are keeled posterior of body and tail. Scales around mid body are 23-27 or 29 rows of scales. Ventrals 210– 44, angulate laterally, anal undivided, subcaudals 73-100 paired.

Size and Colour: Maximum total length of the body 172cm. Dorsal surface light to dark brown and ventral side white. An oblique streak extended from eye to lip and two black stripes on neck and light bands on fore body; hind body has two prominent stripes and extends to tail.

Habit and Habitat: Nocturnal and feeds on small mammals, birds, lizards, snakes and amphibians. It takes shelter in crevices, tree holes, termite mounds, dense bushes and frequently enters into houses.

Common name: Common Trinket Snake; *Local Name*: Tati Jerri

Distribution: Outside India: Pakistan and Sri Lanka

India: Throughout; *Andhra Pradesh:* Guntur and Kurnool districts

SKU: Occasional

***Enhydris enhydris* (Schneider, 1799)**

Mild venomous. Body moderate size, stout and cylindrical. Head seems to be depressed and slightly distinct from neck. Snout rounded. Eyes placed high on the face and with vertical pupil. Tail abruptly tapered.

Scalation: Inter nasal single, touching loreal. Upper labials-8, in them 4th touching the eye. Postoculars-2. Body scales smooth and rows of scales at the middle-21. Ventrals 134-177, anal divided, subcaudals 46-79 paired.

Size and Colour: Maximum total length of the body is 98cm. Dark-olive or olive brown above and light yellow below. A dark lateral line present on both sides of the body. A continuous or interrupted dark line often present in the middle of the belly.

Habit and Habitat: Diurnal and feeds on fish mainly and occasionally on tadpoles, frogs and lizards. When handled, usually not attempt to bite. Inhabits freshwater bodies such as lakes, ponds, rice fields and slow moving streams.

Common name: Common Smooth scaled water snake; *Local name:* Neeruguttu pamu

Distribution: Outside India: Bangladesh, Cambodia, China, Laos, Malay, Myanmar, Peninsula, Thailand and Vietnam.

India: Assam, Bihar, Madhya Pradesh, Uttar Pradesh and West Bengal

Andhra Pradesh: Kurnool and Nalgonda districts

SKU: Rare

Key to the Species of Lycodon

1A) Ventral shields angulate laterally .... *L. aulicus*

1B) Ventral shields not angulate laterally .... *L. striatus*

***Lycodon aulicus* (Linnaeus, 1754)**

Non venomous. Body slender, cylindrical. Head flattened, broader than neck, projects beyond lower jaw. Eyes black.

Scalation: Loreal-1, Preocular-1, Temporals-2+3, Supra labials 9 in them 3rd, 4th and 5th touching eye. Body scales smooth and middle of the body with 17 rows of scales. Ventral shields 172-224, strongly angulated laterally, Anal divide or single, Subcaudals 57-80 paired.

Size and Colour: Maximum total length of the body is 93cm. Body glossy gray brown or black above and usually with 10-20 narrow white or yellow bands that fork on sides of body. Neck usually has whitish or yellowish collar. Bands most distinct on fore body and either faint or entirely absent on hind body. Upper lip white or spotted with brown. Ventral surface white.

Habit and Habitat: Strictly nocturnal and feeds on geckos and skinks. Found in and around caves, wells and hollow trees, often enters into houses. Climbs vertical walls with ease.

Common name: Common Wolf Snake

Distribution: Outside India: Bangladesh, Hong Kong, Indonesia, Malaysia, Maldives, Myanmar, Nepal, Pakistan, Philippines, South China and Sri Lanka

India: Throughout; *Andhra Pradesh:* Anantapur and Vizayanagaram districts

SKU: Common

***Lycodon striatus* (Shaw, 1802)**

Non venomous. Body cylindrical, slender. Head slightly distinct from neck. Snout blunt. Eyes entirely black.

Scalation: Preocular-1, Postocular-2, Temporals-5, Supra labials-8 in them 4th touching eye, Lower labials 9-10. Dorsal scales smooth. Middle of the body with 17 rows of scales, Ventral shields 172-214 and strongly angulated laterally. Anal divided, subcaudals 57-80, paired.

Size and Colour: Maximum total length of the body is 42cm. Dorsum black to dark brown with a series of white or yellow transverse bands, which widen on sides and enclose triangular spots. Bands may be less distinct towards tail. A white bar on the nape is generally present. Upper lip and ventral side are white.

Habit and Habitat: Nocturnal, terrestrial and feeds on geckos and skinks. Inhabits dry regions, including forest edges and semi deserts, agriculture lands, hiding under stone during the day and emerging at night to forage.

Common name: Barred Wolf Snake

Distribution: Outside India: Pakistan and Sri Lanka

India: Throughout; *Andhra Pradesh:* Kurnool district

SKU: Occasional.

Key to the Species of Oligodon

1A) Head with 3 chevron shaped marks; 2nd, 3rd or 4th supra labial touching the eye .... *O. arnensis*

1B) Head with other type of marks; only 3rd and 4th supra labials touching the eye .... *O. taeniolatus*

***Oligodon arnensis* (Shaw, 1802)**

Non venomous. Body cylindrical, short, smooth and even sized. Head depressed, snout short and blunt. Eyes have a round pupil. Tail short.

Scalation: Prefrontal-1, Preocular-1, Postoculars-2, Upper labials 6-7 in them 2nd and 3rd, sometimes 4th also touching the eye. Middle of the body with 17 rows of scales. Ventral shields 164-202, anal shield divided, sub caudals 40-59 paired.

Size and Colour: Maximum total length of the body is 70cm. Dorsum is brown or occasionally purple. Back with black bars which have pale white margins. Head with three conspicuous black marks, first between eyes to below them, second and third are arrow shaped. Ventral surface white.

Habit and Habitat: Nocturnal or crepuscular. Feeds on insects, spiders, reptile eggs, geckos, skinks and small mice. Found in termite mounds, caves, crevices, tree holes and abandoned houses.

Common name: Banded Kukri Snake

Distribution: Outside India: Nepal Pakistan and Sri Lanka

India: Throughout; *Andhra Pradesh:* Kurnool district

SKU: Occasional.

***Oligodon taeniolatus* (Jerdon, 1853)**

Non venomous. Body slender with uniform diameter from neck to vent. Head slightly or not broader than neck. Eyes moderate, round pupil. Tail short and with pointed tip.

Scalation: Rostral large, preocular-1; postoculars- 2; upper labials-7 in them 3rd and 4th touching eye. Middle of the body with 15-17 rows of scales. Ventral scales 158-218, anal divided, sub caudals 20-56, paired.

Size and Colour: Maximum total length of the body is 63cm. Dorsum of body is light brown, blackish brown or buff and with narrow cross bars or transverse spots. Nape with W-shaped dark collar and usually there is a dark band across the head at level of eyes.

Habit and Habitat: Nocturnal and feeds on small lizards and eggs of reptiles. It is found in gardens, bushes, under stones and wooden logs and in tree holes and rocky areas, clay and sandy places.

Common name: Streaked Kukri Snake

Distribution: Outside India: Pakistan and Sri Lanka

India: Throughout; *Andhra Pradesh:* Guntur and Prakasam districts

SKU: Occasional.

***Ptyas mucosa* (Linnaeus, 1758)**

Non venomous. Body robust, long compressed. Head long and distinct from neck. Snout obtuse, nostrils large. Eyes large with round pupil. Tail slender, long, round and prehensile.

Scalation: Loreals 2-4; preoculars-2; postoculars 2-3; temporal-4; upper labials-8; in them 4th and 5th touching eye; lower labials 9-10. Chin shields 4. Middle of the body with 16 to 17 rows of scales. Anterior body scales smooth, posterior slightly keeled. Ventral shields 190-213, anal divided, subcaudals 100-146, paired.

Size and Colour: Maximum total length of the body is 285cm. Dorsal surface of body yellowish or olive green or brown, ventral surface pale white to cream. Labials are separated by vertical stripes. Scales on the posterior part is irregularly margined with black and forming a reticulated pattern.

Habit and Habitat: Diurnal, feeds mainly on rats and also consumes reptiles, birds and their eggs and amphibians. It is most probably found in agriculture fields and around human habitations, also found in grass lands and plains.

Common name: Indian Rat Snake; *Local name:* Jerripothu

Distribution: Outside India: Iran, Java, Malaysia, Pakistan, South China, Sri Lanka, Sumatra, Taiwan and Vietnam

India: Throughout; *Andhra Pradesh:* Guntur, Kadapa and Nalgonda districts

SKU: Common.

Family: Elapidae

Key to the Family Elapidae

1A). Sankes with hood .... *Naja naja naja*

1B). Snakes not hooded .... *Bungarus caeruleus*

***Bungarus caeruleus* (Schneider, 1801)**

Venomous. Body long, moderate size and cylindrical. Head flattened, not distinct from neck. Snout blunt and round. Eyes small, entirely black, round pupil indistinguishable. Tail short.

Scalation: Preocular-1; Postoculars-2; Loreal absent; Temporals-3; Supra labials-7; in them 3rd and 4th touching eye; Lower labials 8. Scales of body smooth. Middle of the body 15 rows of scales. Vertebral scales hexagonal. Ventrals 194-234, Anal entire, Subcaudals 42-52 entire.

Size and Colour: Maximum total length of the body 150cm. Dorsal surface of body black or dark brown and with narrow white bands generally in pairs. The fore body lacking this bands but with white vertebral spots. Underside glossy white or yellowish.

Habit and Habitat: Strictly nocturnal. Feeds on snakes, if not available, prefer rodents, lizards and frogs. The krait is quiet and inoffensive in disposition, bites only under severe provocation. When alarmed in wild, it makes no attempt to escape or defend but lies quietly and conceals the head in the coils of the body. Wide spread in the plains, in woodlands, agricultural fields and in and around human habitations.

Common name: Common Krait; *Local name:* Kattla pamu

Distribution: Outside India: Pakistan and Sri Lanka

India: Throughout; *Andhra Pradesh:* Nalgonda and Kurnool districts

Rare: Common.

***Naja naja naja* (Linnaeus, 1758)**

Venomous. Body dorsiventrally flattened in front and sub cylindrical behind. Head depressed, slightly distinct from neck. Snout rounded. Eyes moderate, round pupil. Neck wide dilates to form a hood.

Scalation: Preoculars 10-12, Postoculars-3, Temporals 5-6, Supra labials-7 in them, 3rd largest and touching the eye. Lower labials 7-9. A small triangular shield is present at oral margin and between 5th and 6th lower labial scales. Scales of the body smooth and glossy. Middle of the body with 21-25 rows of scales. Ventral shields 176-200, Subcaudals 48-75 paired.

Size and Colour: Maximum total length of the body 210cm. Body above dark olive to dark brown, ventral side is pale gray to light yellow. Hood with a spectacle mark on dorsal side and transverse black marks on ventral side.

Habit and Habitat: Active during day and night. Feeds on rodents, toads, frogs, birds and snakes. On agitation, raises the fore body and expands hood, makes hissing sound and strikes forcefully. It is usually found in crop fields, granaries, near streams, occasionally enters into houses.

Common name: Spectacled Cobra; *Local name:* Nagupamu

Distribution: Outside India: Bangladesh, Bhutan, Nepal, Pakistan and Sri Lanka

India: Throughout (except Northeast); *Andhra Pradesh:* Throughout

SKU: Common.

Family: Viperidae

Key to the Genera and Species of Viperidae

1A) Subcaudals 40-64; maximum length of the body to 180cm .... *Vipera rasselli*

1B) Subcaudals less than 40; maximum length of the body to 80cm ..... *Echis carinata*

***Echis carinatus* (Schneider, 1801)**

Venomous. Body slender, more or less flattened dorsoventrally. Head short, triangular, distinct from neck. Snout rounded, wider than high. Eyes large, iris golden yellow, pupil vertical. Tail short.

Scalation: Top of head with small, elongated, imbricate, strongly keeled scales and devoid of shields. Upper labials 10-12 none of them touching eye, 1-2 series of scales are present between upper labials and eye. Body scales strongly keeled and rough in appearance. Middle of the body with 27-37 rows of scales. Ventrals 132-185, Anal entire or divided, Subcaudals 23-39 entire.

Size and Colour: Maximum total length of the body 80cm. Dorsum of the body is light or dark brown, brick red, gray or sand colour with dark brown or blackish spots in vertebral series. Lateral sides with narrow, undulating, white line. A light arrow mark on top of head. Under side white with or without brown spots.

Habit and Habitat: Nocturnal. Feeds on amphibians, mice, squirrels, lizards and insects. When it is threatened tries to escape quickly, if not possible then produce the hissing sound by rubbing side scales of the body and attempts to strike quickly. Inhabits in sandy, rocky and semi arid or arid environment with grass, herbs, shrubs, xerophytes and other vegetation.

Common name: Saw-scaled Viper; *Local name:* Thuttapusuku

Distribution: *Outside India:* Arabia, Iran, Iraq, Kenya, Nigeria, Pakistan, Russia, and Sri Lanka

India: Throughout (except West Bengal and North East India)

Andhra Pradesh: Chittoor, Guntur, Nalgonda, and Prakasam districts

SKU: Rare.

***Vipera russelli* (Shaw, 1797)**

Venomous. Body stout, massive and long. Head flattened, triangular, distinctly separated from neck. Nostril is very large, pierced in nasal. Eyes large with vertical pupil. No sensory pit between eye and nostril. Snout obtuse. Tail short thin.

Scalation: Head above covered with small, imbricate, strongly keeled scales, shields are absent. Upper labials 10-12, none contact with eyes. Middle of the body with 27-33 rows of scales. Ventrals 153-180, Anal entire, Subcaudals 41-64 paired.

Size and Colour: Maximum total length of the body 180cm. Dorsum brown with a bold and distinct pattern consisting of three series of large, dark oval spots. Head with two large black spots at base and a light V-shaped mark with its apex on top of snout. Lower parts yellowish-white.

Habit and Habitat: Nocturnal and feeds on rodents and frogs. It is sluggish animal and moves slowly. On provocation makes loud hissing sound, not readily strikes. This ground dwelling viper inhabits grasslands, scrub forests and open forests, plains, foot hills, uplands and cultivated fields.

Common name: Russell's Viper; *Local name:* Rakthapinjari

Distribution: Outside India: Burma, Pakistan, Sri Lanka, and Thailand

India: Throughout; *Andhra Pradesh:* Guntur and Kurnool districts

SKU: Occasional.

Conclusions

Based on the study good number of snakes were recorded in the campus and vicinity. It indicates the diversity of snakes is more, at the same time there is a little bit of danger is also exhibited. Presence of four common dangerous snakes in the campus indicates an urgent need of recruitment of snake catchers to avoid snakebites. It is better to establish a snake park for the conservation of snakes and to create awareness on snakes to the people and students.

Acknowledgements

Authors are thankful to Mr. U. Kaalicharan, who gave training to catch the snakes, BARC for financial support in the form of project (Flora and Fauna of Thummalapalle Uranium Mining Area, Kadapa, Andhra Pradesh) and local people for their cooperation in rescuing and conservation of snakes.

References

Annandale, N. 1906. Notes on the fauna of a Desert tract in Southern India Part 1. Batrachians and Reptiles. Mem. Asiat. Soc. Bengal, **1:** 183-202.

Beddome, R.H. 1870. Descriptions of new Reptiles from Madras Presidenncy. Madras Month. J. Med. Sci. **9:** 169-176.

Cardoso J.L.C., F.O.S. França, H.W. Fan, C.M.S. Malaque and V. Haddad 2003. Animais peçonhentos no Brasil: biologia, clínica e terapêutica dos acidentes São Paulo: Savier.

Das, I. 2008. A Photographic guide to Snakes and Other reptiles of India. Om Books International, New Delhi, India.

Deoras, P.J. 1965. Snakes of India. National Book Trust, New Delhi.

De Silva A. 1990. Color guide to the snakes of Sri Lanka. R and A Publishing Limited, England.

Director 1990. Collection and Preservation of Animals. Zoological Survey of India, Calcutta, India.

Fita, S.D., Eraldo M Costa Neto and Alexandre Schiavetti 2010. 'Offensive' snakes: cultural beliefs and practices related to snakebites in a Brazilian rural settlement. Journal of Ethnobiology and Ethnomedicine 6: 1-13.

Hawksworth, D.L. and M.T. K. Arroyo (1995). In: Global Biodiversity Assessment (ed. V.H. Heywood), Cambridge University, Cambridge.

Henderson, R.W. and R. Powell 2009. Natural History of West Indian Amphibians and Reptiles. University Press, Florida.

Khaire, N. 1996. Indian Snakes. Indian Herpetological Society, Pune, India.

Marcum, H. 2007. Living with animals: snakes and humans. Encyclopedia of human-animal relationships: A global exploration of our connections with animals Westport: Greenwood Press Bekoff M, 1181-1184.

Mathew, R. 1999. Venomous snakes of north-east India. In: Indian wildlife resources, ecology and development. (Ed. B.D. Sharma): 23-24, Daya Publishing House, Delhi, India.

Mattison C. 1995. The encyclopedia of Snakes. Dah Hua Printing Press C. Ltd. Hong Kong.

May, R.M. 1992. How many species inhabit the earth?. Scientific American. **267 (4):** 42-48.

Mc Cann, C. 1945. Reptiles and amphibians of Vizagapatnam and neighbouring Ghats. Journal of Bombay Natural History Society. **45:** 436.

Murthy, T.S.N. 1981. Checklist and key to the Uropeltid snakes of India. The Snake. **13:** 142-150.

Murthy, T.S.N. 1990. A Field book of the lizards of India. Rec. zool. Surv. India, Occ. Paper No.115. Zoological Survey of India, Calcutta.

Murthy, T.S.N. 2010. The Reptile Fauna of India. B.R. Publishing Corporation, Delhi, India.

Nguyen, V.S., T.C. Ho and Q.T. Nguyen 2009. Herpetofauna of Vietnam. Edition Chimaira, Frankfurtam Main.

Peters, J.A. 1964. Dictionary of Herpetology. Hafner Publishing Co., New York.

Rao, Ravi Prasad, B 2012. Flora of Sri Krishnadevaraya University Campus. Sri Krishnadevaraya University Publication, Anantapur.

Rasmussen, A. R., J. Elmberg and P. Gravlund 2011. Sea snake (Serpentes: Subfamilies Hydrophiinae and Laticaudinae) in Vietnam: a comprehensive checklist and an updated identification key. Zootaxa. **2894:** 1-20.

Schleich, H.H. and W. Kastle 2002. Amphibian and Reptiles of Nepal. Distributors: Koeltz Scientific Books, Germany.

Sharma, B.D. 1998. Snakes in India: A Source Book. Asiatic Publishing House, Delhi, India.

Sharma, R.C. 1971. The reptile fauna of Nagarjuna sagar dam area (Andhra Pradesh, India). Rec. Zool. Surv. India. **63 (1-4):** 77-93.

Shebberare and Barker 1998. The snakes of Sikkim and Bengal. Asiatic Publishing House, Delhi, India.

Smith, M.A. 1931. The fauna of British India including Ceylon and Burma. Reptilia and Amphibia (Vol. I): Loricata and Testudines. Taylor and Francies, London.

Smith, M.A. 1935. The fauna of British India including Ceylon and Burma. Reptilia and Amphibia (Vol. II): Sauria. Taylor and Francies, London.

Smith, M.A. 1943. The fauna of British India including Ceylon and Burma, including the whole of the Indo-Chinese Sub-region. Reptilia and Amphibia (Vol. III): Serpents. Taylor and Francies, London.

Somaweera Ruchira 2004. Sri Lankan Colubrid Snakes. Sri Lanka Naturalist **6 (3-4):** 32-46.

Stidworthy, J. 1972. Snakes of the world. Bantam books, Toronto, USA.

Thulsi Rao, K., H.V. Ghate, M. Sudhakar, S.M. Maqsood Javed and Siva Rama Krishna 2005. Herpetofauna of Nallamalai Hills with eleven new records from the region including ten new reports for Andhra Pradesh. Zoos' Print Journal. **20 (1):** 1737-1740.

Whitaker, R. and A. Captain 2004. Snakes of India: The field guide. Draco Books, Chennai, India.

Chapter 18

Loranthaceae and Viscaceae of Karnataka

M.C. Thriveni*, G.R. Shivamurthy** and K.N. Amruthesh***

Department of Studies in Botany, University of Mysore, Manasagangotri, Mysore – 6, Karnataka

ABSTRACT

Mistletoes (members of both Loranthaceae and Viscaceae) are highly specialized and the most diverse group of parasitic flowering plants belong to the order Santalales with worldwide distribution. They occur over a wide range of habitats all over the world. One can encounter these mistletoes in each and every terrestrial ecosystem. They have become more fascinating through their peculiar habitat *i.e.*, aerial parasitism. Their parasitic growth form is critical to understand their biology, growth and distribution. They exploit or parasitize a wide range of host plants. From the taxonomic point, the family Loranthaceae has been usually divided into two subfamilies Loranthoidae and Viscoidae. Based on the embryological data, both these families have been separated and raised to the rank of family.

In Karnataka mistletoes are represented by 24 species, out of which Loranthaceae comprises seven genera and 16 species and Viscaceae comprising one genus and eight species. The mistletoes are differentially distributed throughout the State. They can attack a large number of taxonomically unrelated hosts. The mistletoes are recorded on a wide range of host plants in each locality

E-mail: *thrivenimc@gmail.com, **grsgyarahally@gmail.com and ***dr.knamruthesh@gmail.com

visited. Most of the monoculture plantations like Teak, Rose wood, Nandi, Honne, Mathi, Eucalyptus, Rubber, Silver oak and horticultural trees like Mango, Orange, Sapota, Pomegranate, Guava and Custard apple are severely infested by these mistletoes. A complete list of hosts for each mistletoe species is provided.

Mistletoes fail to attack a specific host and are considered as host generalists. Host specificity is rare in mistletoes. The selection of hosts is assisted by birds which disperse the seeds. The fleshy pulp of mistletoe berry is most favourite diet for the birds. They play an important role in the distribution of mistletoes. These birds are also involved in the pollination. As it is observed for several years, birds casually visit the flowers in search of food, often primarily to seek insects concealed in inflorescences although they will take nectar if it is available. The Loranthacean flower demonstrates the classical complement of adaptation to bird assisted self-pollination. The birds peck the flower at the base with the help of its beak to take the nectar from the corolla tube and while doing so, it triggers the corolla lobes tip to recurve back explosively resulting in self pollination.

Apart from the host specificity, many mistletoes show host preferences where they prefer to grow on certain host species. The host preference may vary from one place to another and also between the species.

Mistletoes also mimic their hosts. Number of mistletoe species and their hosts look similar in their vegetative characters. In such cases it is very difficult to distinguish the host and the parasite. It is said that the mimicry serves as a protective adaptation from the herbivores. The wide host range and predation pressure from herbivores made mistletoes to mimic their hosts.

The most amazing observation is hyperparasitism, where a mistletoe species parasitize on other mistletoe species. Viscaceae members are frequently recorded as hyperparasitic on Loranthaceous mistletoes. Another fascinating observation is double parasitism which involves the parasitisation of two species of mistletoes on the same host.

Mistletoes are widely recognised as damaging agents. They cause localized hypertrophy on the host tissue at the site of infection. Over a period of time, the host branch distal to mistletoe infection dries up and dies. In many horticultural crops and fruit trees, mistletoes reduce the fruit yield. The effect of mistletoes include reduced vigor and growth rates, poor fruit or seed set, malformation of the woody tissues, sparse foliage, top dying, predisposition to insect and other disease attack and premature death.

Apart from the damaging agents, many species of mistletoes are used as food and fodder for many domestic animals. Various mistletoes are cut from their hosts and fed to cattle during drought. They are also used in the preparation of various medicines and in the treatment of a number of ailments. The control of mistletoes involves the cutting and burning of host branch beyond the point of infection.

Introduction

The plant, animal and human pathogenic fungi, bacteria, viruses and nematodes are probably the first living beings that come to our mind when we think of pathogens. These organisms certainly do cause damage to plants of economic importance, but one may be surprised to know that parasitic flowering plants are also important pathogens (Nickrent and Musselman, 2004).

Just as insectivorous plants captured the imagination of humans for many generations, parasitic flowering plants have also been the subject of much curiosity since ancient times. These two groups of plants share common features that they both have evolved nutritional modes for a green plant *i.e.*, photosynthetic, motionless and oblivious to other organisms around them.

These phanerogamic parasites are among the most bizarre group of organisms and have enthralled the naturalist for centuries. They steal their requirements from unsuspecting hosts and some adroitly entice their hosts to assist in obtaining their nutrients. Because of their exotic way of life and mode of nutrition, the parasitic flowering plants have fascinated biologists since time immemorial. The above features make the study of parasitic flowering plants all the more interesting not only as objects of intrinsic botanical and physiological interest but also because of increased awareness of their importance as pests of Agriculture, Horticulture and Forestry.

The initial curiosity would have gradually given way to fear on realization of the problems that some of the parasites can pose. These are quite distinct from other weeds in forming an intimate attachment with their host crops, whether to roots or the shoots. Through this attachment they absorb water, minerals, varying proportions of their carbohydrate and nitrogen requirements. As a result of this parasitic feeding habit, hosts become weakened or suffer more readily from water stress.

The parasitic flowering plants are highly diverse and constitute a remarkable set of Angiosperm taxa that are common members of most plant communities. One can encounter parasitic plants in nearly every habitat type found throughout the world and range from polar regions to the dry and humid tropics.

In 1822, the Botanical world was astonished by the description, of a most incredible flowering plant. An inhabitant of the jungles of Sumatra, the plant was nothing less than a flower of gigantic dimensions emerging from the thin root of a liana. The plant was named *Rafflesia arnoldii* R.Br. and regarded as "Queen of Parasites" because it has the largest known flower (nearly a meter in diameter) (Kuijt, 1969).

Although the precise number is difficult to determine, it has been estimated that about 1 per cent of flowering plants, approximately 4,100 species in total exhibit parasitic mode of nutrition. The total number of parasite species is not distributed evenly among the 277 genera of 18 families of flowering plants (Nickrent and Musselman, 2004).

Classification

Based on the host organ with which they establish haustorium, they are classified into

- ☆ Root parasites and
- ☆ Stem parasites

Root parasites belong to the families Apodanthaceae, Balanophoraceae, Cynomoriaceae, Hydnoraceae, Krameriaceae, Lennoaceae, Mitrastemonaceae, Myzodendraceae, Olacaceae, Opiliaceae, Orobanchaceae, Rafflesiaceae, Santalaceae and Scrophulariaceae.

The stem parasites belong to the families Loranthaceae, Viscaceae, Convolvulaceae and Lauraceae. Among these, the members of Loranthaceae and Viscaceae are commonly called as mistletoes (Parker and Riches, 1993). Approximately there are 60 per cent root and 40 per cent shoot (stem) parasites.

Based on the presence/absence of chlorophyll, they are classified as

- ✰ Hemi or Semiparasites: chlorophyll is present and
- ✰ Holoparasites: chlorophyll is absent

Hemiparasites can be further divided into two types, facultative and obligate, depending on their degree of dependence upon host. Facultative hemiparasites do not require a host to complete their life cycle but are photosynthetic and when presented with host roots, invariably form haustorial connections and grow luxuriantly, whereas obligate hemiparasites must attach to host plants to complete their life cycle.

The exact origin of the word mistletoe is far from clear and several views have been made in this regard. This name was derived from the old German 'Mist' means dung and 'Tan' means twig which describes the observable origin of these plants from the droppings of birds that have been feeding on mistletoe fruits (Calder, 1983). In Sanskrit mistletoes are called as 'Vrikshabhaksha' meaning that they eat trees which indicate the damage done by them and suggest that they were known in India from ancient times (Prakash, 2004).

Mistletoes have long been a source of fascination to humans, and references to these parasitic plants can be found among the legends and superstitions of people throughout the world. Some cultures believed that mistletoes were endowed with mystical powers because they grow from the branches of other plants and because many species fruit in winter when other temperate zone plants are dormant (Aukema, 2003).

Examples of parasitism amongst Angiosperms are found in a number of life forms, including trees (*e.g.*, Santalaceae), shrubs (*e.g.*, Mistletoes), vines (*e.g.*, *Cuscuta* and *Cassytha*) and herbs (*e.g.*, Scrophulariaceae). Root parasitic vines are infrequent (*e.g.*, *Buttonia* of Scrophulariaceae; *Cansjera* of the Opiliaceae). It is difficult to categorize clearly the habit of some Orobanchaceae, Hydnoraceae and Balanophoraceae that have species with fleshy aerial portions and subterranean parts that live for many years (Press and Graves, 1995).

From the available literature on parasitism, the first fact we encounter is that no parasites are known from any vascular Cryptogams. In the Gymnosperms there is a solitary report from New Caledonia (De Laubenfels, 1959) *Parasitaxus ustus* (Vieill.) Laub. (Podocarpaceae) is always parasitic on the host *Falcatifolium taxoides* Brongn. & Gris de Laub. (also of Podocarpaceae). Except this there are no known parasitic

Gymnosperms. Perhaps more surprising is the absence of parasitism from the monocotyledons. It is quite a startling and unexplained fact that the known parasitic vascular plants are limited only to the Dicotyledons (Kuijt, 1969).

Haustorium

The parasitic plants form a close connection with the vascular system of their host plant (s) through a specialized structure known as haustorium. It is the penetrating organ into the host tissue, which is the most common phenotypic expression of this highly diverse group of organisms originating in different species at their shoot or root portions. It is a remarkable unit, unique both in structure and function. When mature, it provides for a vascular conduit of water and nutrients and form the morphological and physiological bridge between the parasite and host plants. Mature haustoria often appear as swollen, round structures firmly attached to or clasping the host surface. The portion that penetrates into the host vascular tissue is referred to as the endophyte (Riopel and Timko, 1995). As such the haustorium becomes the nodal point in identifying the parasitic habit of a species. Its mode of development and penetration into the host organ, its vascular organization, reported absence of phloem and occurrence of inclusions have continued to baffle the botanists to this day.

There is an extraordinary diversity in the number of host species that parasitic Angiosperms may infect and it is often said that a good parasite does not kill its host (Nickrent, 2002). Most parasitic plant species can potentially attack a large number of different hosts indicating that they have a broad host range (Gibson and Watkinson, 1989; Nilsson and Svensson, 1997; Pennings and Callaway, 2002; Westbury, 2004). In this regard, most parasitic plants are considered as generalists.

Mistletoes are found on a wide range of woody plants, preferentially parasitizing trees and shrubs. Their greatest diversity is found in forests and woodlands (Kuijt, 1969; Calder, 1983; Hawksworth, 1983).

Examples of wide host range are documented for both root and shoot parasites. In the former case, *Castilleja* spp. are known to parasitize more than 100 different hosts from a variety of plant families (Press, 1998), whereas *Rhinanthus minor* L. has 50 different host species from 18 different families (Gibson and Watkinson, 1989). Although shoot parasites tend to have a smaller host range than root parasites (Norton and Carpenter, 1998), broad host ranges are still apparent such as *Cuscuta* sp. (dodders) with hundreds of different host species (Kelly *et al.*, 1988; Musselman and Press, 1995). Many individual mistletoe species are capable of parasitizing a large number of host species. For example, *Dendrophthoe falcata* has 411 known host species recorded so far (Thriveni and Shivamurthy, 2010).

There is no clear relationship between the parasitic mode and habitat exists. Different parasites may be found at different stages of succession within the community. As a generality, with the exception of mistletoes, parasitic angiosperms are frequently characteristic of disturbed areas (Press and Graves, 1995).

Among the parasitic flowering plants, the families Loranthaceae and Viscaceae (mistletoes) have fascinated the Botanists from time immemorial because of their peculiar habit, host preference, distribution pattern, taxonomy, morphology, structure of the haustorium, floral biology, phenology, pollination mechanism, fruit dispersal, germination and establishment of parasite on hosts and so on. Taxonomically mistletoes constitute a highly specialized and diverse group of Angiosperms (Devkota, 2005).

The mistletoes have long been credited with one of the most magical, mysterious and feared plants of European folklore. They have found their way into religious rites, mythology, healing arts and superstitions throughout the world (Dengler and Dengler, 1946; Fleming, 1952; Muller, 1939). The aura of legend, symbolism and magic that surrounds *Viscum album* L. of Europe has been the inspiration for countless writings most of which have been summarized and evaluated in a few comprehensive treatments (Bishop, 1951; Kanner, 1939 and Tubeuf, 1923). *Viscum album* has always been of great mystical significance in Northern Europe and has played a role in pagan religious ceremony. Some of this significance persists in its use as a decoration at Christmas time in Europe and the corresponding species in North America is *Phoradendron serotinum* (Raf.) M.C. Johnst. The history and uses of *V. album* and other mistletoes have been the subject of whole books (Becker and Schmoll, 1986) and are also reviewed by Calder (1983) in the invaluable volume on the biology of mistletoes (Calder and Bernhardt, 1983).

The mistletoes are widely recognized as damaging agents to trees. Their victims include garden and shade trees, horticultural, timber yielding, fruit yielding and other forest trees including medicinal plants. The degree of damage varies with the species of mistletoe, its longevity and intensity of parasitism (Gill and Hawksworth, 1961). The parasite competes with its host for water, inorganic nutrients and organic compounds. Some mistletoe may live for decades in association with their host trees and result in little damage (Schulze and Ehleringer, 1984) while others inflict severe damage.

The effects of mistletoes on their hosts may include any combination of the following: reduced vigor and growth rates, poor fruit or seed set, malformation of the woody tissues, sparse foliage, top dying, predisposition to insect and other disease attack and premature death (Gill and Hawksworth, 1961).

Apart from the damaging agents some parasitic plants have provided an invitation to man to make use of them. Even before the discovery of parasitism, the parasitic plants are utilized in a variety of ways throughout the world. The succulent nature of parasitic plants in arid regions has made them a source of food for both man and animals (Kuijt, 1969).

The Loranthaceous plants usually parasitize the aerial parts of autotrophic plants and even on some other Loranthaceous plants (epiparasitism/ hyperparasitism) except a few root parasitic ones (*Atkinsonia*, *Gaiadendron* and *Nuytsia*). They obtain their food, partially or entirely from the host through haustoria. Sometimes, after the parasite dries up and decays, the swollen and ruptured tissues of the host assume a somewhat radiating structure and are termed 'wood flowers' (also called wood roses) or Holzrosen (Tubeuf, 1936). They are known as 'Rosa de Palo' in Mexico and 'Rosa de Madera' in Guatemala (Metcalfe and Chalk, 1950).

Because of this unusual habit and fantastic patterns of 'wood-flowers', they have attracted attention since ages and have been the objects of curiosity. The local people in South Africa harvest the tree branches, remove the mistletoe and sell the remaining wood rose to the tourists as a curio. This has provided a source of income to families living in under developed rural areas where there are few alternative employment possibilities (Dzerefos *et al.*, 1999).

Loranthaceae is one of the largest of all assemblies of parasitic flowering plants and in the stricter sense include about twice as many genera as Viscaceae (Kuijt, 1969).

The members of Loranthaceae are predominantly tropical in distribution with about 1000 species in 74 genera (Barlow *et al.*, 1989). Viscaceae members are widespread in the tropics and extend into the North Temperate Zone and include 400 species in 7 genera (Barlow, 1983).

In India Loranthaceae include nine genera with 35 species and Viscaceae include three genera with 12 species (Razi, 1957). About seven genera comprising 16 species of Loranthaceae and one genus comprising eight species of Viscaceae have been listed from Karnataka State.

The systematics of Loranthaceae is somewhat confusing and the validity of the type genus *Loranthus* is doubtful (Danser, 1933D).

Viscaceae previously treated as a part of mistletoe family Loranthaceae *sensu lato*, consists of hemiparasites, mostly distributed in tropical and warm temperate regions. This family (*sensu stricto*) is represented by seven genera and about 400 species (Barlow and Wiens, 1977). The name Viscaceae was derived from a feature of the seeds-the viscin-that forms a sticky layer on its outer surface; which anchors the seeds to the host branch. The members of Viscaceae cause extensive damage to fruit trees and forest trees.

According to most taxonomists the family Loranthaceae has been divided into two sub families namely Loranthoideae and Viscoideae. Danser (1929) has also accepted the concept of a single family Loranthaceae, including both Viscoid and Loranthoid mistletoes. Maheshwari (1954, 1958) commented that the subfamily Loranthoideae is embryologically distinct from the Viscoideae (Calder, 1983). Later Johri and Bhatnagar (1960, 1972) and Kuijt (1969) have suggested that these two subfamilies should be given the rank of separate families, as they are distinct from each other in many features and each deserves the status of an independent family. Danser (1941) has revised the nomenclature of the genus *Viscum* in India.

Though some of the Parasitic Angiosperms are viewed as important pests of human and agricultural forestry, many are highly valued for food and wood as well as for their medicinal and aesthetic properties. Hence, the parasitic angiosperms pose an added challenge for the conservation biologists and managers who often wish to bolster parasitic populations. Sincere efforts must be needed simultaneously at both the parasites and their hosts.

Although many species of parasitic flowering plants listed in floras based on the early works of Hooker (1872-1897), Cooke (1901-1908) and Gamble and Fischer

(1915-1936), the information on the distribution, host range, degree of abundance and phenology are inadequate. Past research on systematic account of parasitic flowering plants is sporadic and fragmentary. Karnataka state is rich in many parasitic flowering plants. No thorough systematic work on parasitic Angiosperms has been undertaken and executed in any of the states of India. There are no extensive surveys on host ranges, factors responsible for the differential distribution of parasitic weeds on specific hosts etc. Therefore, a systematic study on Loranthaceae and Viscaceae of Karnataka has been investigated and the following are the results.

Observations

Plant exploration trips were conducted to different parts of Karnataka during 2007-2012 covering all the seasons. About 150 collection trips were carried out to collect the specimens belonging to the family Loranthaceae and Viscaceae. A total of 24 species have been recorded from the study area. Loranthaceae comprising seven genera with 16 species and Viscaceae with one genus and 8 species have been collected. We have recorded a new species *Viscum bandipurense* Thriveni *et al.* (2013) from Bandipur National park.

The members of Loranthaceae have large, bright coloured, showy bisexual flowers with haustorial runners. Some species of *Scurrula* and *Taxillus* produce epicortical roots which run along the host stem or bark and at intervals produce haustoria. These haustoria are ectophytic in nature. In Viscaceae the flowers are inconspicuous, very small, unisexual, not bright coloured and the haustorium is endophytic in nature *i.e.*, it enters and becomes embedded in the host tissue to absorb the water and dissolved minerals. The presence of leaves is rare in the members of Viscaceae. Sometimes the leaves absent and if present they are reduced. Usually the Viscaceae members have tuft of pendulous and trichotomous branches on the tree trunks.

Most of the District floras published so far are generally silent with regard to the number of host species, distribution pattern of mistletoes and intensity of parasitism. There is no information on the host range and abundance. In the present study, a complete list of host plants along with their binomials and family names are provided. Nomenclatural citations including the references from available literature, floras, revisions and monographs are provided for each species.

Artificial Keys

Artificial keys are very useful in the identification of plants. They are defined as an orderly arrangement of a series of contrasting characters/statements of which one is to be accepted and other rejected. Each time, a choice is made and one or more taxa are eliminated. These keys ease in a faster preliminary identification of plants. This does not offer the description of the plant concerned. It is artificial in the sense that the pairs of contrasting characters are selected. Indented keys are also called as yoked keys and are most commonly used in floras and manuals. In indented keys, each successive couplet is indented under the one preceding it. This key is better understood and easy to follow, because similar elements are grouped in such a way that they are visible at a glance. Hence, we have prepared the indented keys for the identification of Loranthaceae and Viscaceae up to species level in the present study. These keys are mainly based on the morphological characters observed in the field. This could be used both in the field and in the laboratory for easy identification.

Comparative Account of Mistletoes

Table 18.1: Comparison of Number of Genera and Species of Mistletoes Recorded in Various district Floras of Karnataka State and the Present Study.

Sl.No.	*Name of the District Flora*	*Loranthaceae*		*Viscaceae*	
		Number of Genera	*Number of Species*	*Number of Genera*	*Number of Species*
1.	Flora of Bangalore District	4	5	1	2
2.	Flora of Chikmagalore District	4	6	1	1
3.	Flora of Coorg District	4	7	1	2
4.	Flora of Davangere District	3	4	1	2
5.	Flora of Gulburga District	–	–	–	–
6.	Flora of Hassan District	6	12	1	4
7.	Flora of Karnataka	7	19	1	8
8.	Flora of Kolar District	3	5	1	2
9.	Flora of Mysore District	5	15	1	6
10.	Flora of Shimoga District	6	8	1	1
11.	Flora of Udupi District	5	6	–	–
12.	Present study	7	16	1	8

Taxonomic Treatment

Systematic Position

Most of the taxonomists placed the family Loranthaceae (s.l.) under the order Santalales of dicotyledons. The order Santalales have been treated under various groups in different systems of classifications proposed by prominent taxonomists. For example, Bentham and Hooker (1862-1883) placed the family Loranthaceae (148) under the series 12, Achlamydosporeae of the subclass Monochlamydeae earlier to Santalaceae. Engler and Prantl treated this family under the sub order Santalineae of the sub class Archichlamydeae. Hutchinson (1926-1934) placed the family Loranthaceae under the order Santalales of Lignosae. In both Takhtajan's (1969) and Cronquist's (1968) system of classifications we find the family Loranthaceae under the subclass Rosidae of the class Magnoliopsida and division Magnoliophyta.

The family Loranthaceae (s.l.) have been divided into two subfamilies Loranthoidae and Viscoidae. In recent years, many of the morphologists and taxonomists have elevated these two subfamilies to the level of independent families as it has been dealt in detail in the introductory chapter.

Key to the Families

1. Haustoria ectophytic not endophytic; flowers bisexual, in racemes or spikes; stamens with filaments, and opposite to the petals, anthers free; leaves always present and are variable .... **Loranthaceae**

1. Haustoria endophytic; flowers unisexual, usually in triads; stamens without filaments, anthers adnate and opposite to the petals; leaves absent or when present are reduced .... **Viscaceae**

LORANTHACEAE A.L. Jussieu

Ann. Mus. Natl. Hist. Nat. 12: 292. 1808. *nom. cons.*

Plants parasitic on the branches of trees and shrubs, attached to their hosts through haustoria. Stem branched, sometimes dichotomous, glabrous or with raised tubercles. Leaves alternate or opposite or sub-opposite, whorled or simple, entire; petiolate or sessile; ovate or oblanceolate or cordate; tomentose or pubescent or glabrous. Inflorescence axillary raceme or spike. Flowers bracteate, bracteolate or ebracteolate, complete, bisexual, regular or irregular, epigynous. Calyx short, cylindrical or cup shaped. Corolla fused or lobes free to base. Stamens as many as corolla lobes, epipetalous, opposite and adnate to petals, anthers free and basifixed. Ovary inferior. Style slender or filiform, stigma globose or capitate. Fruit ovoid to oblong or pyriform.

Key to the Genera

1. Flowers subtended by a bract and 2 bracteoles clasping the calyx...... *Macrosolen*
1. Flowers subtended by a single bract and bracteoles absent:
    2. Flowers many, enclosed within a large coloured involucre .... *Tolypanthus*
    2. Flowers not enclosed in an involucre:
        3. Branches dichotomous; corolla lobes elastically coiled at anthesis .... *Helicanthes*
        3. Branches not dichotomous; corolla lobes not elastically coiled:
            4. Flowers regular; corolla fused:
                5. Flowers large, in racemes; petals 5; fruit oblong .... *Dendrophthoe*
                5. Flowers small, in racemes; petals 4; fruit ovoid .... *Helixanthera*
            4. Flowers irregular; corolla split on one side:
                6. Petals 4, tube red inside; fruit pyriform .... *Scurrula*
                6. Petals 4 or 5, tube not red inside; fruit oblong .... *Taxillus*

DENDROPHTHOE Martius

Flora 13: 109. 1830

Shrubs. Stem branched, cylindrical or trigonous. Leaves alternate, opposite or whorled, petiolate. Inflorescence axillary raceme. Each flower subtended by 1 bract, ebracteolate, complete, bisexual, regular, epigynous. Calyx cylindrical. Corolla 5-lobed, lobes spreading or reflexing at the constriction at anthesis. Stamens 5, epipetalous, opposite to petals, anthers free, basifixed, introrse. Ovary inferior, stigma green or bright red, capitate. Fruit oblong.

1. Branchlets terete; leaves alternate or opposite or sub-opposite; corolla creamish yellow or bright orange; stigma dark green .... *D. falcata*
1. Branchlets trigonous; leaves whorled, three at each node; corolla golden yellow; stigma bright red .... *D. trigona*

Dendrophthoe falcata (L.f.) Ettingsh. var. ***falcata***., Denkschr. Kaiserl. Akad. Wiss., Math. Naturwiss. Kl. 32: 52. 187; Wiens in Abeywick., Fl. Ceylon 1: 68. 1973; in Dassan. and Fosberg, Rev. Handb. Fl. Ceylon 6: 134. 1987; Razi, Lloydia 20 (4): 242. 1957; Rama. & Razi, Fl. Bangalore Dist. 153. 1973; Sald. & Nicol., Fl. Hassan Dist. 305. 1976; Rao & Razi, A Syn. Fl. Mysore 460. 1981; Yoga. *et al.*, Fl. Chikmagalur Dist. 285. 1982; Singh, Fl. East. Karnataka 550 (2): 1988; Keshava. & Yoga., Fl. Coorg, 389. 1990; Sald., Fl. Karnataka 2: 75. 1996; Rama. *et al.*, Fl. Shimoga Dist. 511. 2001; Bhat, Fl. Udupi 547. 2003; Manju. *et al.*, Fl. Davanagere Dist. 342. 2004. *Loranthus falcatus* L.f., Suppl. Sp. Pl. 221. 1782. *Loranthus longiflorus* Desr., in Lam., Encycl. Méth., Bot. 3: 598. 1789; Hook. f., Fl. Brit. India 5: 215. 1886; Talb., For. Fl. Bombay 2: 415. 1911; Fischer, Rec. Bot. Surv. Ind. 11: 176. 1926. *Loranthus longiflorus* var. *falcatus* (L.f.) Kurz, For. Fl. 2: 321. 1877; Gamble, Fl. Pres. Madras 2: 1253. 1925; Fischer, Rec. Bot. Surv. Ind. 11: 177. 1926; Cooke, Fl. Pres. Bombay 3: 43. 1906; Wt. Ic. t. 302. *Loranthus longiflorus* var. *amplexifolius* (DC.) Thwaites, Enum. Pl. Zey. 214. 1859; Talb., For. Fl. Bombay 2: 412. 1911; Fischer, Rec. Bot. Surv. Ind. 11: 178. 1926. *Loranthus longiflorus* var. *coccinea* Talb., For. Fl. Bombay 2: 412. 1911.

Large semi parasitic shrubs with haustoriferous runners. Stem branched, cylindrical. Leaves alternate or opposite or sub opposite, petiolate, up to 0.9 cm long, simple, entire, ovate, 10-25 cm x 2-4.5 cm, 5-6 nerved, acute at apex, attenuate at base, glabrous. Inflorescence axillary raceme, peduncle up to 0.4 cm long. Flowers in fascicles of about 25 flowers, pedicel up to 0.3 cm long, complete, bisexual, regular, epigynous. Mature flower up to 4 cm, narrow at the base, gradually widens towards upper part. Shortly above this a dark green band followed by a constriction and a swollen portion. Above this the perianth lobes joined to form a club shaped some what stiff structure. Calyx cylindrical, up to 0.4 cm long. Corolla up to 4 cm long, tube slender, creamish yellow, lobes 5, green, reflexed at anthesis. Stamens 5, epipetalous, filaments adnate and opposite to the perianth lobes, up to 3.8 cm, anthers red, basifixed up to 0.3 cm long, introrse. Ovary inferior. Style long, up to 4 cm long, green, stigma capitate, dark green. Fruit oblong, green, 1.5 cm.

Fl.: March to July.

Distr.: Found in all the districts of Karnataka.

Specimens examined: D.B. Deb 59972, 59973 (MH); M. Mohanan 170281, 170217, 170365 (MH); E. Vajravelu 35278, 68089, 68090, 87067, 74189, 74190, 67758, 67759 (MH); K.M. Sebastine 24681, 24682, 4979, 4980, 14517, 14518, 4393, 4394 (MH); K.

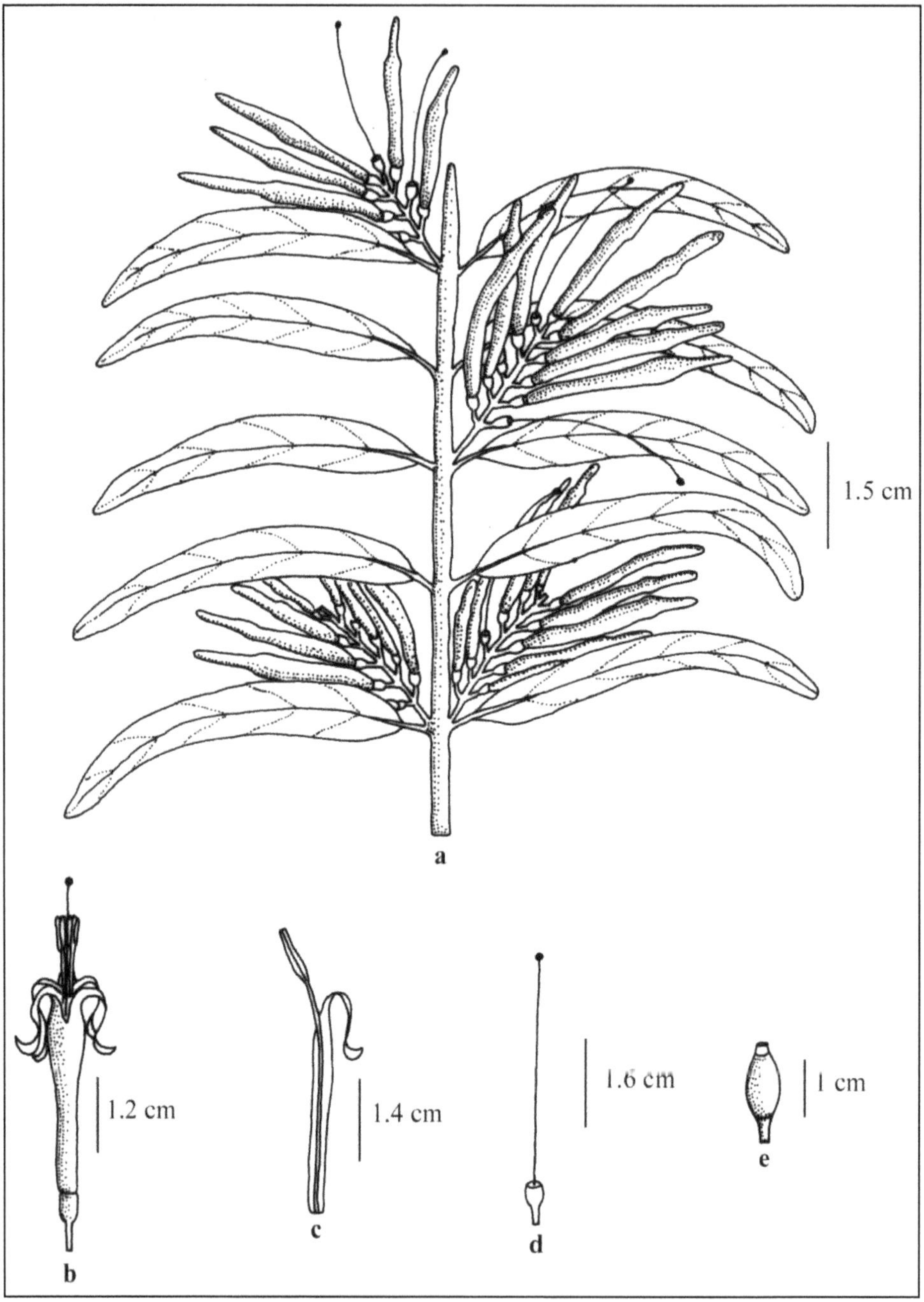

Figure 18.1a: ***Dendrophthoe falcata*** **(L.f.) Ettingsh. var.** ***falcata.***
a: Flowering twig; b: Flower; c: Epipetalous stamen; d: Gynoecium; e: Fruit.

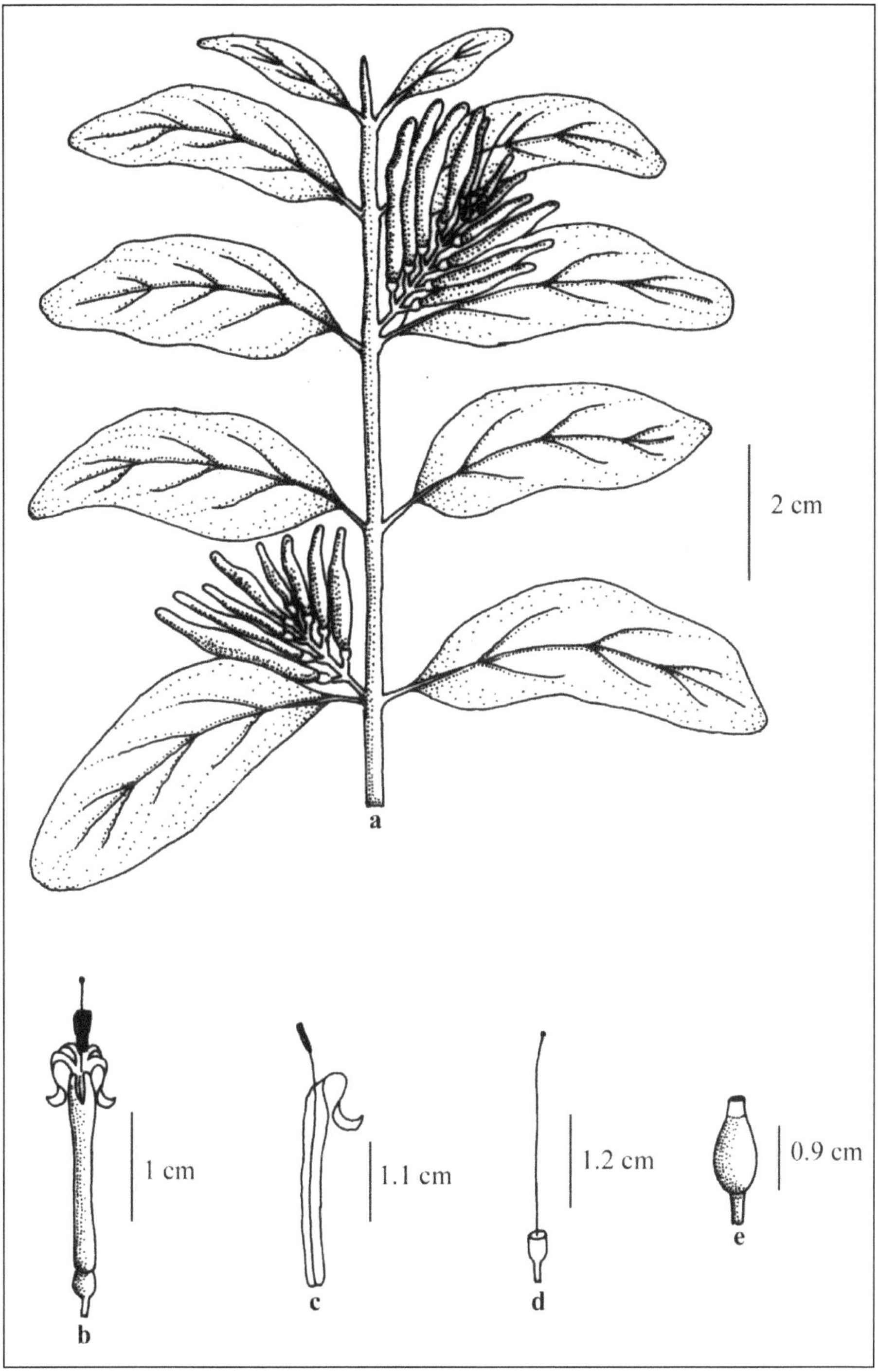

Figure 18.1b: ***Dendrophthoe falcata*** **(L.f.) Ettingsh. var.** ***oblonga.***
a: Flowering twig; b: Flower; c: Epipetalous stamen; d: Gynoecium; e: Fruit.

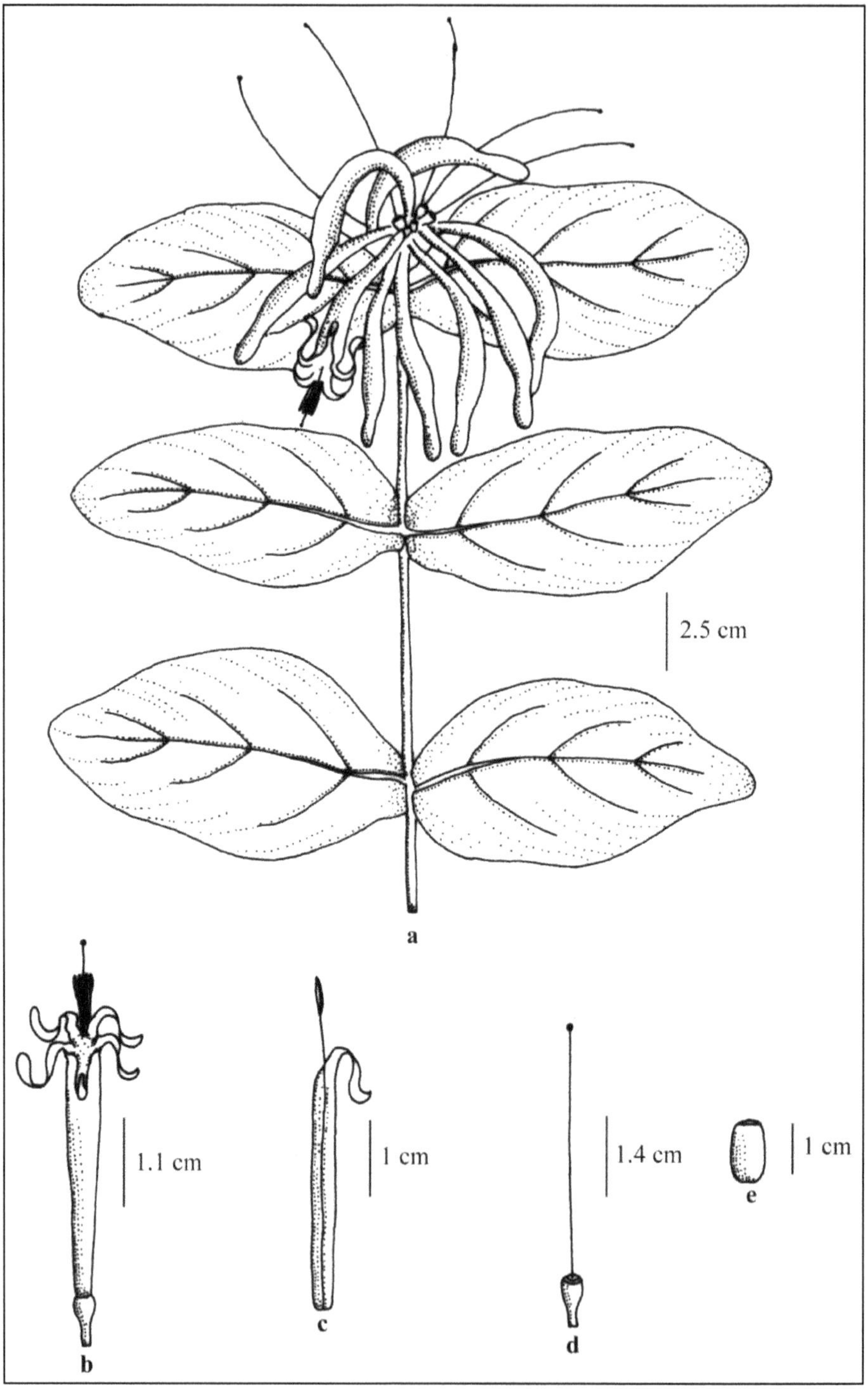

Figure 18.1c: *Dendrophthoe falcata* (L.f.) Ettingsh. var. *amplexifolius*.
a: Flowering twig; b: Flower; c: Epipetalous stamen; d: Gynoecium; e: Fruit.

Plate 18.1a: *Dendrophthoe falcata* (L.f.) Ettingsh. var. *falcata.*
A: Flowering branches; B: Twig with flowers; C: Close up of flowers; D: Young fruits.

Plate 18.1b: *Dendrophthoe falcata* (L.f.) Ettingsh. var. *oblonga.*
A: Flowering twig; B: Inflorescence; C: Flowers; D: Fruits.

Plate 18.1c: ***Dendrophthoe falcala*** **(L.f.) Ettingsh. var.** ***amplexifolius.***
A: Habit; B: Inflorescence; C: Close up of a flower; D: Fruit.

Subramanyam 10515, 10516, 13397, 3673, 3674, 4606, 4605, 4752 (MH); B.V. Shetty 63785, 23445, 23446 (MH); Henry 105522, 91703 (MH); G.V. Subba Rao 71865, 71866, 91151, 70493, 70492 (MH); B.D. Sharma 78795, 78796, 69426, 69427 (MH); K. Ramamurthy 100795, 100796, 46587, 46588 (MH); T. Ravishankar 156089 (MH); M. Chandra Bose 87571 (MH); J. Joseph 27476 (MH); M. Vishwanathan 96155 (MH); Shanavas Khan 7213, 15611, 18363 (CALI); MCT 33, 321.

Host plants: *Acacia auriculiformis* A. Cunn ex Benth, *A. chundra* (Rottl.) Willd., *A. decurrens* Willd., *A. ferruginea* DC., *A. nilotica* (L.) Del., *A. sinuata* (Lour.) Merr., *A. concinna* (Willd.) DC. and *A. suma* Buch.-Ham. (Mimosaceae), *Achras zapota* L. (Sapotaceae), *Adenanthera pavonia* L. (Caesalpiniaceae), *Aegle marmelos* Corr. (Rutaceae), *Albizia amara* Boiv., *A. lebbeck* Benth. and *A. odoratissima* Benth. *Albizia saman* F. Muell. (Mimosaceae), *Aphanomixis polystachya* (Wall.) Parker (Meliaceae), *Anogeissus latifolia* Wall. (Combretaceae), *Annona squamosa* L. (Annonaceae), *Artocarpus heterophyllus* Lam. (Moraceae), *Azadirachta indica* A. Juss. (Meliaceae), *Barringtonia acutangula* (L.) Gaertn. (Lecythidaceae), *Bauhinia purpurea* L., *B. racemosa* Lam. and *B. variegata* L. (Caesalpiniaceae), *Bombax ceiba* L. (Bombacaceae), *Bridelia scandens* (Roxb.) Willd. (Euphorbiaceae), *Boswellia serrata* Roxb. ex Coleb. (Burseraceae), *Calotropis gigantea* R. Br. (Asclepiadaceae), *Canthium dicoccum* (Gaertn.) Teys. & Binn. (Rubiaceae), *Careya arborea* Roxb. (Lecythidaceae), *Cassia fistula* L., *C. montana* Heyne, *C. siamea* Lam., *C. spectabilis* (Caesalpiniaceae), *Casuarina equisetifolia* Forst. (Casuarinaceae), *Catunaregam spinosa* (Thunb.) Tirv. (Rubiaceae), *Toona ciliata* Rolmer (Meliaceae), *Ceiba pentandra* (L.) Gaertn. (Bombacaceae), *Chukrasia velutina* M. Roem. (Meliaceae), *Citrus medica* L. (Rutaceae), *Commiphora caudata* Engl. (Burseraceae), *Dalbergia latifolia* Roxb., *D. paniculata* Wight, *D. sissoo* Roxb. (Papilionaceae), *Delonix regia* Raf. (Caesalpiniaceae), *Dolichandrone* sp. (Bignoniaceae), *Elaeodendron glaucum* Pers. (Celastraceae), *Ervatamia coronaria* Stapf (Apocyanaceae), *Eucalyptus globulus* Labill. (Myrtaceae), *Limonia acidissima* L. (Rutaceae), *Ficus carica* L., *F. hispida* L.f., *F. religiosa* L. (Moraceae), *F. racemosa* L., *Gossypium herbaceum* L. (Malvaceae), *Grevillea robusta* A. Cunn. (Proteaceae), *G. tiliifolia* Vahl (Tiliaceae), *Hardwickia binata* Roxb. (Caesalpiniaceae), *Hibiscus rosa-sinensis* L. (Malvaceae), *Holoptela integrifolia* Pl. (Ulmaceae), *Jacaranda mimosifolia* D. Don (Bignoniaceae), *Kydia calycina* Roxb. (Malvaceae), *Lagerstroemia indica* L., *L. lanceolata* Wall. (Lythraceae), *Macaranga peltata* Mull.Arg (Euphorbiaceae), *Mangifera indica* L. (Anacardiaceae), *Melia dubia* Cav. (Meliaceae), *Millingtonia hortensis* L.f. (Bignoniaceae), *Mitragyna parvifolia* Korth. (Rubiaceae), *Moringa oleifera* Lam. (Moringaceae), *Morus alba* L. (Moraceae), *Muntingia calabura* L. (Tiliaceae), *Murraya koenigii* Spr., *Naringi crenulata* (Roxb.) Nicol. (Rutaceae), *Nerium oleander* L. (Apocyanaceae), *Nyctanthes arbor-tristis* L. (Nyctaginaceae), *Persea macrantha* (Nees) Kosterm. (Lauraceae), *Peltophorum pterocarpum* (DC.) K. Heyne, *Phyllanthus acidus* (Euphorbiaceae), *Pithecellobium dulce* (Roxb.) Benth. (Mimosaceae), *Polyalthia longifolia* L. (Annonaceae), *Pongamia pinnata* (L.) Pierre (Papilionaceae), *Psidium guajava* L. (Myrtaceae), *Punica granatum* Vent. (Punicaceae), *Radermachera xylocarpa* K. Schum. (Bignoniaceae), *Saraca asoca* L. (Caesalpiniaceae), *Salix tetrasperma* Roxb. (Salicaceae), *Stereospermum colais* Mabb., *S. suaveolens* DC. (Bignoniaceae), *Swietenia macrophylla* King. (Meliaceae), *Tabebuia argentea* (Bur. & Schum.) Britton (Bignoniaceae), *Tectona grandis* L.f. (Verbenaceae), *Terminalia arjuna* Wt. & Arn., *T.*

bellirica (Gaertn.) Roxb., *T. catappa* L., *T. chebula* (Gaertn.) Roxb., *T. crenulata* Roth., *T. paniculata* Roth. and *T. tomentosa* Wt. & Arn. (Combretaceae), *Cascabelathe vitia* (L.) Lippold. (Apocynaceae), *Trema orientalis* Bl. (Ulmaceae), *Vitex altissima* L.f. (Verbenaceae), *Ziziphus jujuba* Lam. (Rhamnaceae).

Note: Very common parasite and the foregoing lists reveal that it has a very wide host range. It is a polymorphic species widely distributed throughout Karnataka. Three morphological varieties have been observed during the collection. Variety *falcata* is having linear falcate leaves and flowers are bright orange yellow in colour. Variety *amplexifolious* has large leaves with clasping leaf base which surrounds the stem.

Key to Varieties

1. Leaves linear and falcate; flowers orange yellow .... var. *falcata*
1. Leaves oblong; flowers creamish yellow .... var. *oblonga*
1. Leaves oblong with clasping leaf base; flowers pinkish .... var. *amplexifolius*

Dendrophthoe trigona (Wight & Arn.) Danser ex Santapau, Rec. Bot. Surv. India 16 (1): 163. 1953; Razi, Lloydia 20 (4): 243. 1957; Sald. & Nicol. Fl. Hassan Dist. 307. 1976; Rao and Razi, A Syn. Fl. Mysore 460. 1981; Yoga. *et al.*, Fl. Chikmagalur Dist., 286. 1982; Singh, Fl. East. Karnataka 551 (2): 1988; Keshava. & Yoga., Fl. Coorg, 389. 1990; Sald. Fl. Karnataka 2: 76. 1996. *Loranthus trigonus* Wight & Arn., Prodr. 386. 1834; Hook. f., Fl. Brit. India 5: 219. 1886; Talb., For. Fl. Bombay 2: 416. 1911; Gamble, Fl. Pres. Madras 2: 1253. 1925; Cooke, Fl. Pres. Bombay 3: 43. 1906.

Robust semi parasitic shrub. Stem branched, trigonous, glabrous. Leaves 3 in a whorl, petiolate, up to 1 cm, simple, entire, broadly elliptic, 12-17 x 6-10 cm, lateral nerves 4 or 5 pairs, cuneate or rounded and decurrent at base, obtuse or rounded at apex, glabrous. Inflorescence axillary racemes in fascicles with about 15 flowers. Flowers complete, bisexual, regular, epigynous, bract connate into a shallow, oblique cup, laterally acuminate and separate from the calyx. Calyx small, up to 0.3 cm long. Corolla slender up to 2 cm long, tube orange yellow, green towards upper part, lobes 5, tip reflexed at anthesis. Stamens 5, epipetalous, anthers basifixed up to 0.3 cm long. Style short, green, up to 2 cm long, stigma bright red, capitate. Fruit ovoid, up to 1.5 cm long, green, crowned with calyx-limb.

Fl.: September to November

Distr.: Found in all the districts of Karnataka.

Specimens examined: E. Vajravelu 83842, 83850 (MH); B.D. Sharma 77767 (MH); B.D. Nainthani 44664 (MH); Shanavas Khan 63500, 23325, 42180 (CALI); MCT 326, 320.

Host plants: *Manilkara zapota* (L.) P. Royen (Sapotaceae), *Aegle marmelos* Corr. (Rutaceae), *Artocarpus heterophyllus* Lam. (Moraceae), *Calotropis gigantea* R. Br. (Asclepiadaceae), *Dalbergia latifolia* Roxb. (Papilionaceae), *Eugenia* sp. L. (Myrtaceae), *Ficus benghalensis* L., *F. exasperata* Vahl., *F. virens* Aiton., *F. drupacea* var. *pubescens* (Roth.) Thunb., *F. racemosa* Roxb. and *F. religiosa* L. (Moraceae), *Mangifera indica* L.

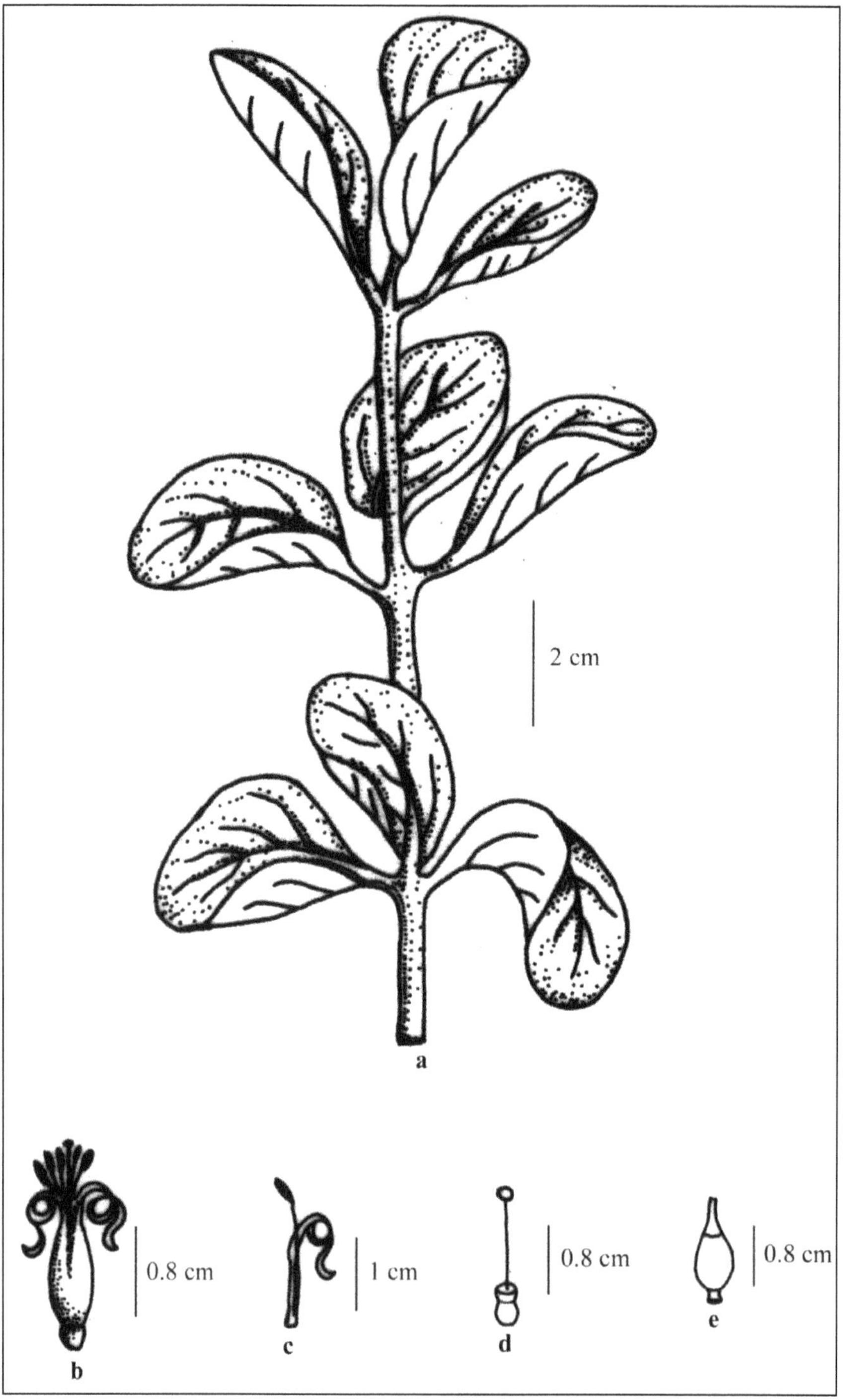

Figure 18.2: ***Dendrophthoe trigona*** **(Wt. & Arn.) Dans. ex Santap.**
a: Twig; b: Flower; c: Epipetalous stamen; d: Gynoecium; e: Fruit.

Plate 18.2: *Dendrophthoe trigona* (Wt. & Arn.) Dans. ex Santap.

A: Branches; B: A twig showing three leaves in each whorl; C: Close up of flowers; D: Fruits.

(Anacardiaceae), *Murraya koenigii* Spr. (Rutaceae), *Salix tetrasperma* Roxb. (Salicaceae), *Swietenia macrophylla* King. (Meliaceae), *Syzygium caryophyllatum* (L.) Alston (Myrtaceae).

Note: Usually very common on *Ficus* species with robust branching. Sometimes the leaves mimic the host leaves where it becomes very difficult to distinguish parasite from the host.

HELICANTHES Danser

Verh. Kon. Ned. Akad. Wetensch., Afd. Tweede Sect. 29 (6): 55. 1933.

Shrubs. Stem dichotomously branched. Nodes swollen. Leaves opposite, sessile. Inflorescence axillary, fascicles. Flowers bracteate, ebracteolate, complete, bisexual, actinomorphic, epigynous. The lower one third portion of the corolla tube becomes broader and tapers towards apex as a pointed structure. Calyx small. Corolla 5-lobed; lobes linear, elastically coiled at anthesis. Stamens 5, epipetalous, opposite to petals, anthers basifixed. Ovary inferior. Stigma dark red. Fruit ovoid, green, up to 0.7 cm.

Helicanthes elastica (Desr.) Danser, Verh. Kon. Ned. Akad. Wetensch., Afd. Tweede Sect. 29 (6): 55. 1933; Razi, Lloydia 20 (4): 245. 1957; Sald. Fl. Karnataka 2: 77. 1996; Bhat, Fl. Udupi 548. 2003. *Loranthus elasticus* Desr., in Lamk., Encycl. Méth. Bot. 3: 599. 1789; Wight, Ic. t. 343. 1840; Hook. f., Fl. Brit. India 5: 216. 1886; Talb., For. Fl. Bombay 2: 414. 1911; Roxb. Fl. Ind. Carey and Wall. (Ed.) 2. 217. 1824; Gamble, Fl. Pres. Madras 2: 1254. 1925; Fischer, Rec. Bot. Surv. Ind. 11: 178. 1926; Cooke, Fl. Pres. Bombay 3: 41. 1906.

Large semi parasitic, woody shrubs. Stem dichotomously branched, cylindrical, nodes swollen. Leaves opposite, sessile, simple, entire, ovate or elliptic, 7-12 cm x 3-4.5 cm, coriaceous, obtuse at apex, glabrous. Flowers in axillary fascicles, with about 15 flowers, sessile, complete, bisexual, regular, epigynous. Mature flower up to 4 cm long, the lower one-third portion becomes broader and tapers towards apex. Calyx small, truncate. Corolla up to 4 cm long, tube white with green veins, lobes 5, linear, light green, elastically coil inwards like a spring at anthesis. Stamens 5, epipetalous, opposite to the petals, filaments reddish, adnate to petals for about one third distance, anthers linear, basifixed up to 0.6 cm long, introrse. Ovary inferior. Style long, up to 4 cm long, reddish. Stigma capitate, dark red. Fruit ovoid, green when young, reddish with age, up to 0.7 cm.

Fl.: October to November

Distr.: Kodagu, Uttar Kannada, Dakshina Kannada, Udupi, Shimoga.

Specimens examined: K. Ramamurthy 29476, 29477 (MH); M.V. Vishwanathan 96733, 96734 (MH); K.M. Sebastine 554 (MH); K. Subramanyam 763, 11263, 11264, 25816 (MH); T. Ravishankar 157149, 157150 (MH); B.D. Sharma 77851, 77852 (MH); A.N. Henry 100426, 100427 (MH); Shanavas Khan 19133, 13059, 11043, 5272 (CALI); MCT 345, 356.

Host plants: *Anacardium occidentale* L. (Anacardiaceae), *Aporusa lindleyana* Baill. (Euphorbiaceae), *Artocarpus heterophyllus* Lam. (Moraceae), *Casuarina equisetifolia* Forst.

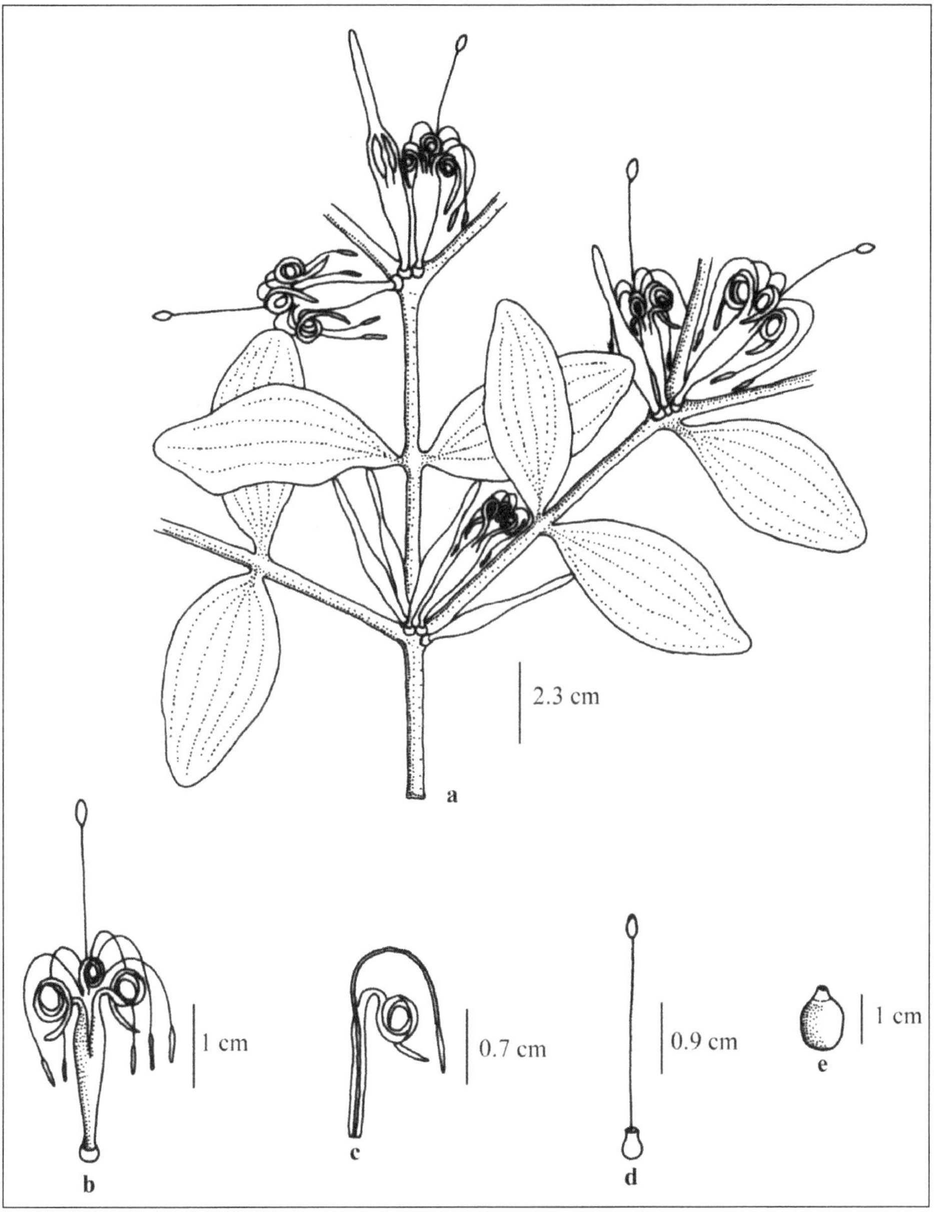

Figure 18.3: *Helicanthes elastica* (Desr.) Dans.
a: Flowering twig; b: Flower; c: Epipetalous stamen; d: Gynoecium; e: Fruit.

Plate 18.3: *Helicanlhes elastica* (Desr.) Dans.

A: Flowering branch; B: Inflorescence; C: Flower showing elastically coiled corolla lobes (arrow: spring like); C: Fruits.

(Casuarinaceae), *Ceiba pentandra* (L.) Gaertn. (Bombacaceae), *Barringtonia acutangula* (L.) Gaertn. (Lecythidaceae), *Evodia lunu-ankenda* Merr. (Rutaceae), *Ficus racemosa* L. (Moraceae), *Lagerstroemia reginae* Roxb. (Lythraceae), *Mallotus philippensis* (Lam.) Muell. Arg. (Euphorbiaceae), *Mangifera indica* L. (Anacardiaceae), *Muntingia calabura* L. (Tiliaceae), *Psidium guajava* L. (Myrtaceae), *Swietenia macrophylla* King. (Meliaceae), *Syzygium cumini* (L.) Skeels (Myrtaceae).

Note: Very common on Mango trees in Dakshin Kannada and Udupi.

HELIXANTHERA Loureiro

Fl. Cochinch. 1: 142. 1790.

Shrubs. Stem branched, glabrous. Leaves opposite or sub-opposite, petiolate or sessile. Inflorescence raceme or spike. Individual flower subtended by a bract. Flowers sessile, bracteate, ebracteolate, complete, bisexual, regular, epigynous. Calyx cylindrical. Corolla tetramerous, sometimes angular. Stamens 4, epipetalous, dorsifixed. Ovary inferior. Style short, stigma capitate. Fruit ovoid or globose.

1. Flowers in spike .... *H. hookeriana*
1. Flowers not in spike:
    2. Leaves simple, base auricled .... *H. obtusata*
    2. Leaves simple, base not auricled:
        3. Corolla tube angular .... *H. intermedia*
        3. Corolla tube not angular .... *H. wallichiana*

Helixanthera hookeriana (Wight & Arn.) Danser, Bull. Jard. Bot. Buitenz. Ser. 3. 10: 317. 1929; Razi, Lloydia 20 (4): 241. 1957; Rao and Razi, A Syn. Fl. Mysore 460. 1981; Sald. Fl. Karnataka 2: 77. 1996. *Loranthus hookerianus* Wight & Arn., Prodr. 381. 1834; Hook. f., Fl. Brit. India 5: 207. 1886; Gamble, Fl. Pres. Madras 2: 1251. 1925; Fischer, Rec. Bot. Surv. Ind. 11: 173. 1926.

Large semiparasitc shrubs. Stem branched with minute tubercles. Leaves opposite, petiolate, petiole up to 0.5 cm long, simple, entire, lanceolate, 6.5-8 cm x 1.5-2.5 cm, obtuse at apex, narrowed at base, glabrous. Inflorescence spike up to 5 cm long. Mature flower subtended by a bract, complete, bisexual, regular, epigynous. Calyx small, cylindrical. Corolla dark red, up to 1 cm long, lobes 4, reflexed at anthesis. Stamens 4, epipetalous, filaments opposite and adnate to the petals, anthers red, basifixed, 0.3 cm long, introrse. Ovary superior. Style short up to 1 cm long, stigma capitate. Fruit ovoid.

Fl.: April to May.

Distr.: B.R. Hills.

Specimens examined: Shanavas Khan 7585, 22348 (CALI).

Host plants: Common on *Mallotus philippensis* (Lam.) Muell. (Euphorbiaceae).

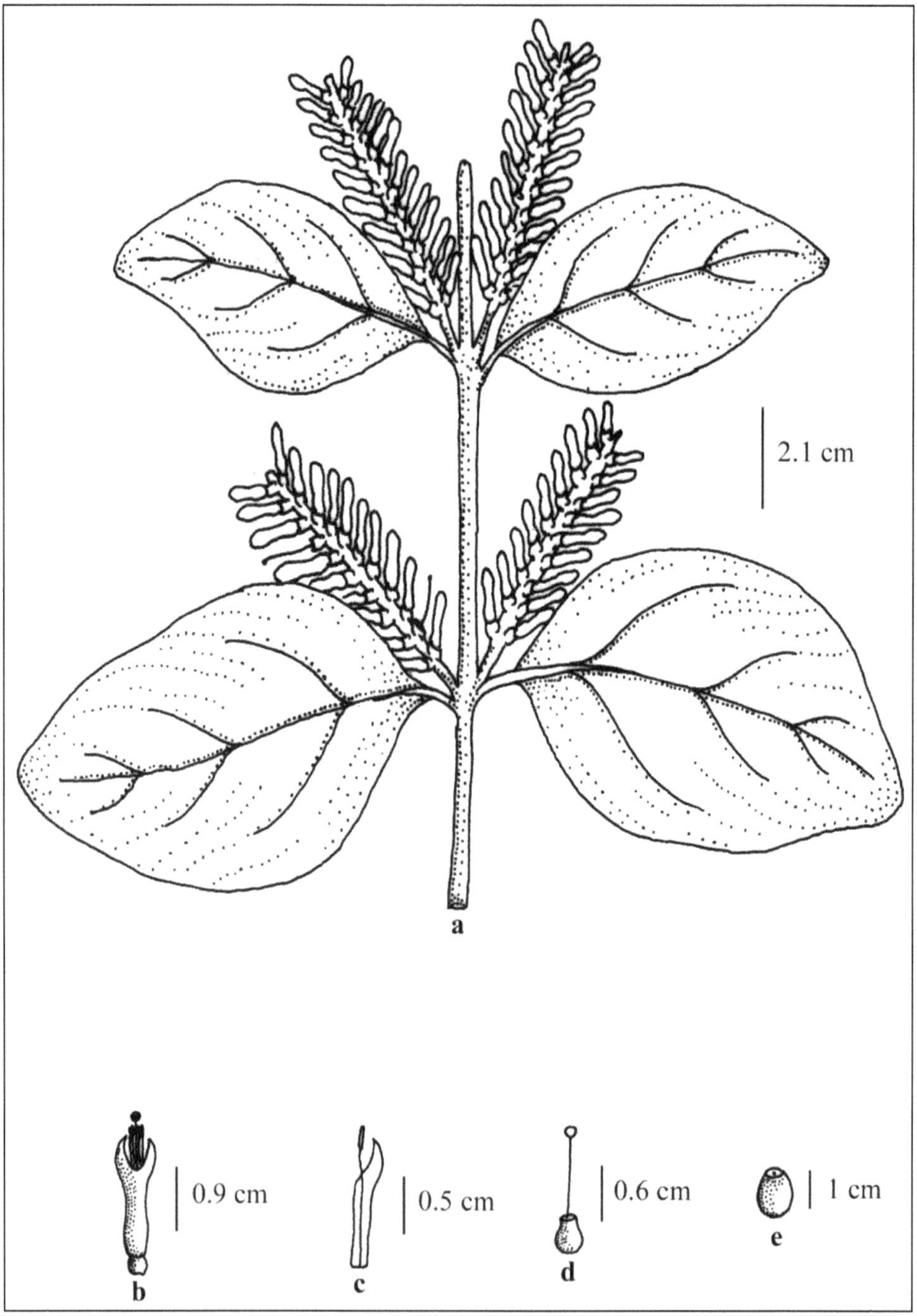

Figure 18.4: ***Helixanthera hookeriana*** **(Wt. & Arn.) Dans.**
a: Flowering twig; b: Flower; c: Epipetalous stamen; d: Gynoecium; e: Fruit.

Plate 18.4: ***Helixanthera hookeriana*** **(Wt. & Arn.) Dans.**
A: Flowering twig; B and C: Inflorescence; D: Twig with fruits.

Inlcuded on the authority Kammathy *et al.* (1967).

Helixanthera intermedia (Wight) Danser, Bull. Jard. Bot. Buitenz. Ser. 3. 10: 317. 1929; Razi, Lloydia 20 (4): 242. 1957; Sald. & Nicol. Fl. Hassan Dist. 307. 1976; Rao & Razi, A Syn. Fl. Mysore 461. 1981; Yoga. *et al.*, Fl. Chikmagalur Dist., 287. 1982; Sald. Fl. Karnataka 2: 77. 1996; Rama. *et al.*, Fl. Shimoga Dist. 514. 2001. *Loranthus intermedius* Wight, Calcutta J. Nat. Hist. 6: 361. 1846; Hook. f., Fl. Brit. India 5: 205. 1886; Fischer, Rec. Bot. Surv. Ind. 11: 173. 1926; Gamble, Fl. Pres. Madras 2: 1251. 1925.

Large semi parasitic shrubs. Stem branched, glabrous. Leaves opposite, petiolate, petiole up to 1.5 cm long, simple, entire, ovate, 11-15.5 cm x 5.5-8 cm, acute at apex, obtuse at base, glabrous. Inflorescence solitary racemes at the nodes of old branches. Mature flower up to 1.7 cm long, subtended by a bract, actinomorphic, bisexual, epigynous. Calyx short, cylindrical. Corolla red, 1.2 cm, lobes 4, tube angular. Stamens 4, epipetalous, filament adnate and opposite to petals, anthers basifixed, 2 mm long, introrse. Style short, 1.2 cm long, stigma capitate. Fruit

Fl.: April to May

Distr.: B.R. Hills, Shimoga, Dakshina Kannada, Hassan, Kodagu.

Occurrence and specimens examined: K.M. Sebastine 36140, 18377 (MH); V.S. Ramachandran 123524, 53822 (MH); M. Mohan 117384 (MH); J.L. Ellis 48230 (MH); C.N. Mohanan and B. Ramanujan 72115, 149996 (MH); Shanavas Khan 27903, 13213, 37622, 13095 (CALI); MCT 137, 255, 208.

Host plants: *Aphanamixis polystachya* (Wall.) Parker (Meliaceae), *Aporusa lindleyana* Baill. (Euphorbiaceae), *Artocarpus heterophyllus* Lam. (Moraceae), *Casuarina equisetifolia* Forst. (Casuarinaceae), *Celtis tetrandra* Roxb. (Ulmaceae), *Chukrasia velutina* M. Roem. (Meliaceae), *Citrus aurantium* L. (Rutaceae), *Dalbergia sissoo* Roxb. (Papilionaceae), *Evodia lunu-ankenda* Merr. (Rutaceae), *Ficus racemosa* L. (Moraceae), *Glochidion malabaricum* Bedd. (Euphorbiaceae), *Mangifera indica* (Anacardiaceae), *Neolitsea zeylanicum* Merr. (Lauraceae), *Terminalia tomentosa* Wight & Arn. (Combretaceae).

Helixanthera obtusata (Schult.) Danser, Bull. Jard. Bot. Buitenz. Ser. 3. 10: 317. 1929; Razi, Lloydia 20 (4): 242. 1957; Rao & Razi, A Syn. Fl. Mysore 461. 1981; Sald. Fl. Karnataka 2: 72. 1996. *Loranthus obtusatus* Schult., Syst. 7: 653. 1830; Hook. f., Fl. Brit. India 5: 205. 1886; Talb., For. Fl. Bombay 2: 406. 1911; Gamble, Fl. Pres. Madras 2: 1251. 1925; Fischer, Rec. Bot. Surv. Ind. 11: 173. 1926; Cooke, Fl. Pres. Bombay 3: 39, 1967.

Large semi parasitic shrubs with haustoriferous runners. Stem branched, glabrous. Leaves opposite, sessile, simple, entire, ovate, 8-10.5 cm x 4-6 cm, obtuse at apex, auricled at base, glabrous. Inflorescence solitary racemes with short peduncle at the nodes of old branchlets. Flowers pedicellate, bracteate, ebrateolate, complete, bisexual, regular, epigynous. Mature flower up to 2.5 cm long, pedicel 0.3 cm long. Calyx cylindrical, up to 3 mm long. Corolla red, up to 1.7 cm long, lobes 4, tube angular. Stamens 4, epipetalous, filaments opposite and adnate to the petals, anthers red, basifixed, 0.2 cm long, introrse. Ovary inferior. Style short up to 1.7 cm long,

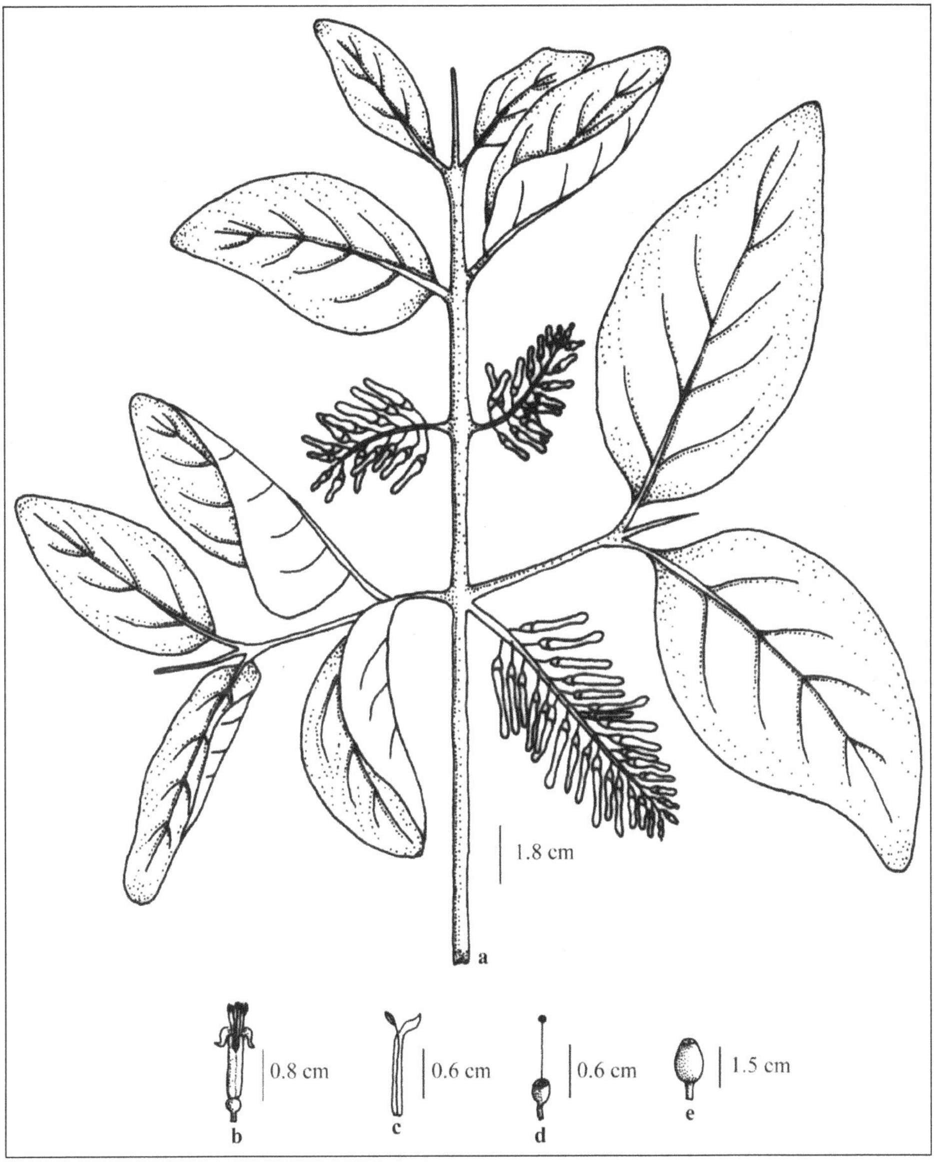

Figure 18.5: *Helixanthera intermedia* (Wt.) Dans.

a: Flowering twig; b: Flower; c: Epipetalous stamen; d: Gynoecium; e: Fruit.

Plate 18.5: ***Helixanthera intermedia*** **(Wt.) Dans.**

A: Flowering branches; B: Young inflorescence; C: Close up of flowers; D: Fruits.

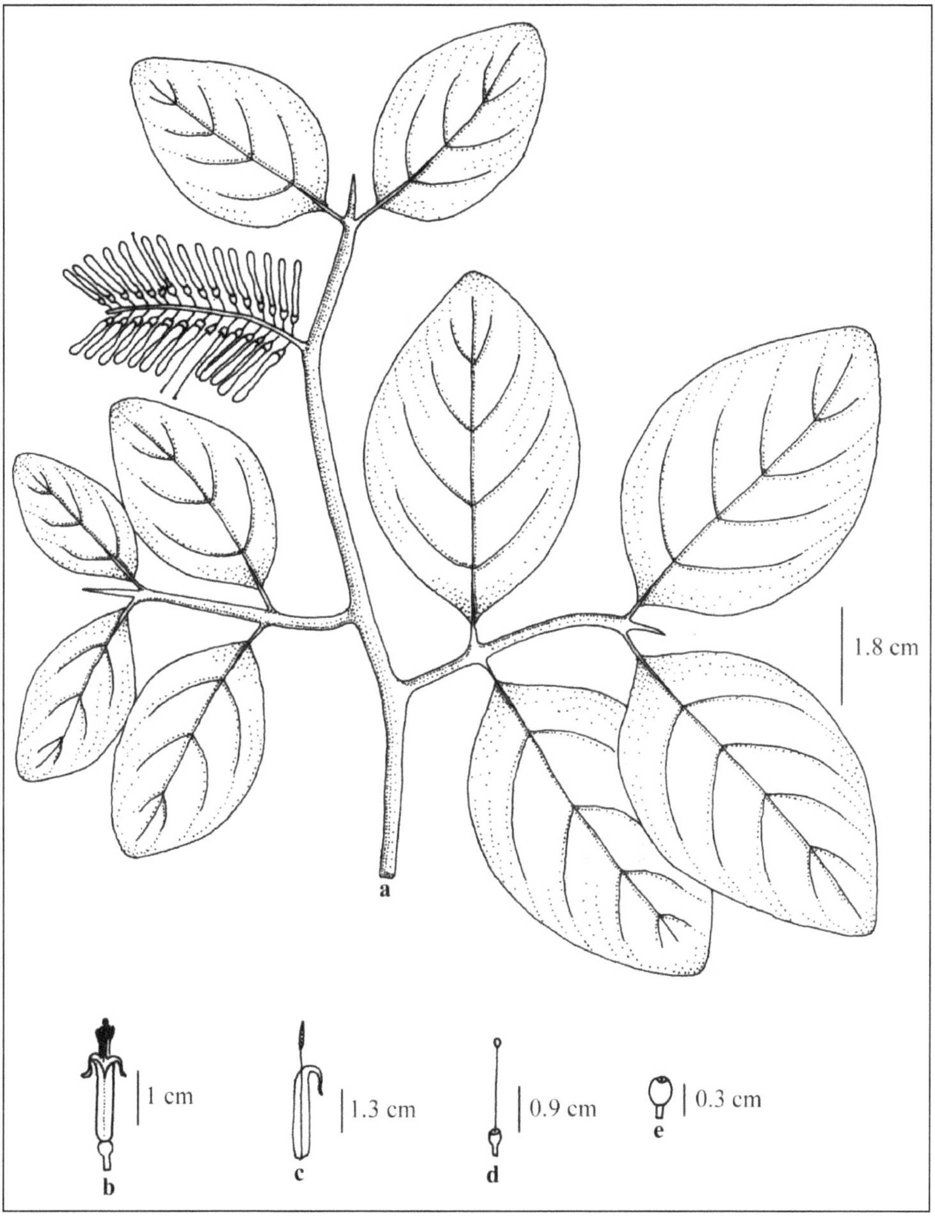

Figure 18.6. ***Helixanthera obtusata*** **(Schult.) Dans.**
a: Flowering twig; b: Flower; c: Epipetalous stamen; d: Gynoecium; e: Fruit.

stigma capitate. Fruit globose.

Fl.: April to May.

Distr.: B.R. Hills, Shimoga, Dakshina Kannada, Hassan, Kodagu.

Specimens examined: Bhargavan 87309 (MH), Shanavas Khan 42142, 37611 (CALI).

Host plants: Aporusa lindleyana Baill. (Euphorbiaceae), *Casuarina equisetifolia* Forst. (Casuarinaceae), *Citrus aurantium* L. (Rutaceae).

Included on the authority of Saldanha (1976).

Helixanthera wallichiana (Schult.) Danser, Bull. Jard. Bot. Buitenz. Ser. 3. 10: 319. 1929; Razi, Lloydia 20 (4): 242. 1957; Sald. & Nicol. Fl. Hassan Dist. 307. 1976; Rao & Razi, A Syn. Fl. Mysore 461. 1981; Yoga. *et al.*, Fl. Chikmagalur Dist., 287. 1982; Keshava. & Yoga., Fl. Coorg, 390. 1990; Sald. Fl. Karnataka 2: 72. 1996; Rama. *et al.*, Fl. Shimoga Dist. 515. 2001; Bhat, Fl. Udupi 548. 2003. *Loranthus wallichianus* Schult., Syst. 7: 100. 1829; Hook. f., Fl. Brit. India 5: 205. 1886; Wight Ic. t. 143. 1846; Talb., For. Fl. Bombay 2: 405. 1909; Gamble, Fl. Pres. Madras 2: 1251. 1925; Fischer, Rec. Bot. Surv. Ind. 11: 173. 1926; Cooke, Fl. Pres. Bombay 3: 39. 1967.

Large semi parasitic shrubs with haustoriferous runners. Stem branched, elongated, glabrous. Leaves opposite, sessile, simple, entire, ovate, 6-13 cm x 3-6 cm, obtuse at apex, attenuate at base, glabrous. Inflorescence axillary racemes with about 15 flowers. Flowers pedicellate, bracteate, ebracteolate, complete, bisexual, regular, epigynous. Mature flower up to 0.5 cm, pedicel 0.3 cm. Calyx cylindrical, 0.1 cm. Corolla scarlet red, 0.4 cm, lobes 4. Stamens 4, epipetalous, filaments opposite and adnate to the perianth lobes, anthers basifixed, —introrse. Ovary inferior. Style short, 0.4 cm, stigma capitate. Fruit globose and red.

Fl.: March to April.

Distr.: B.R. Hills, Shimoga, Dakshina Kannada, Hassan, Kodagu.

Specimens examined: Mohanan 69257 (MH), J. Joseph 41991 (MH), Ramachandran 61636 (MH), Ramamurthy 49635 (MH), Vajravelu 32164 (MH), 49793 (CALI), Shanavas Khan 5422 (CALI), N.C. Nair 56638 (CALI), Balakrishnan 40328 (CALI), Ramarao 1511 (CALI), MCT 152, 270.

Host plants: *Bischofia javanica* Bl. (Euphorbiaceae), *Citrus aurantium* L. and *C. medica* L. (Rutaceae), *Coffea arabica* L. (Rubiaceae), *Memecylon talbotianum* Brandis and *M. umbellatum* Burm.f. (Melastomaceae), *Neolitsea zeylanicum* (Nees.) Merr. (Lauraceae).

MACROSOLEN (Blume) Reichenbach

Syst. 7: 1731. 1830.

Large semi parasitic shrubs. Stem branched with elongated internodes, terete. Leaves opposite, petiolate. Inflorescence axillary raceme, rarely umbels. Individual flower subtended by 1 bract and 2 bracteoles closely clasping the base of the flower, complete, bisexual, regular and epigynous. Calyx cyndrical. Corolla 6-lobed; lobes reflexed near the constriction at anthesis. Stamens 6, epipetalous, anthers basifixed,

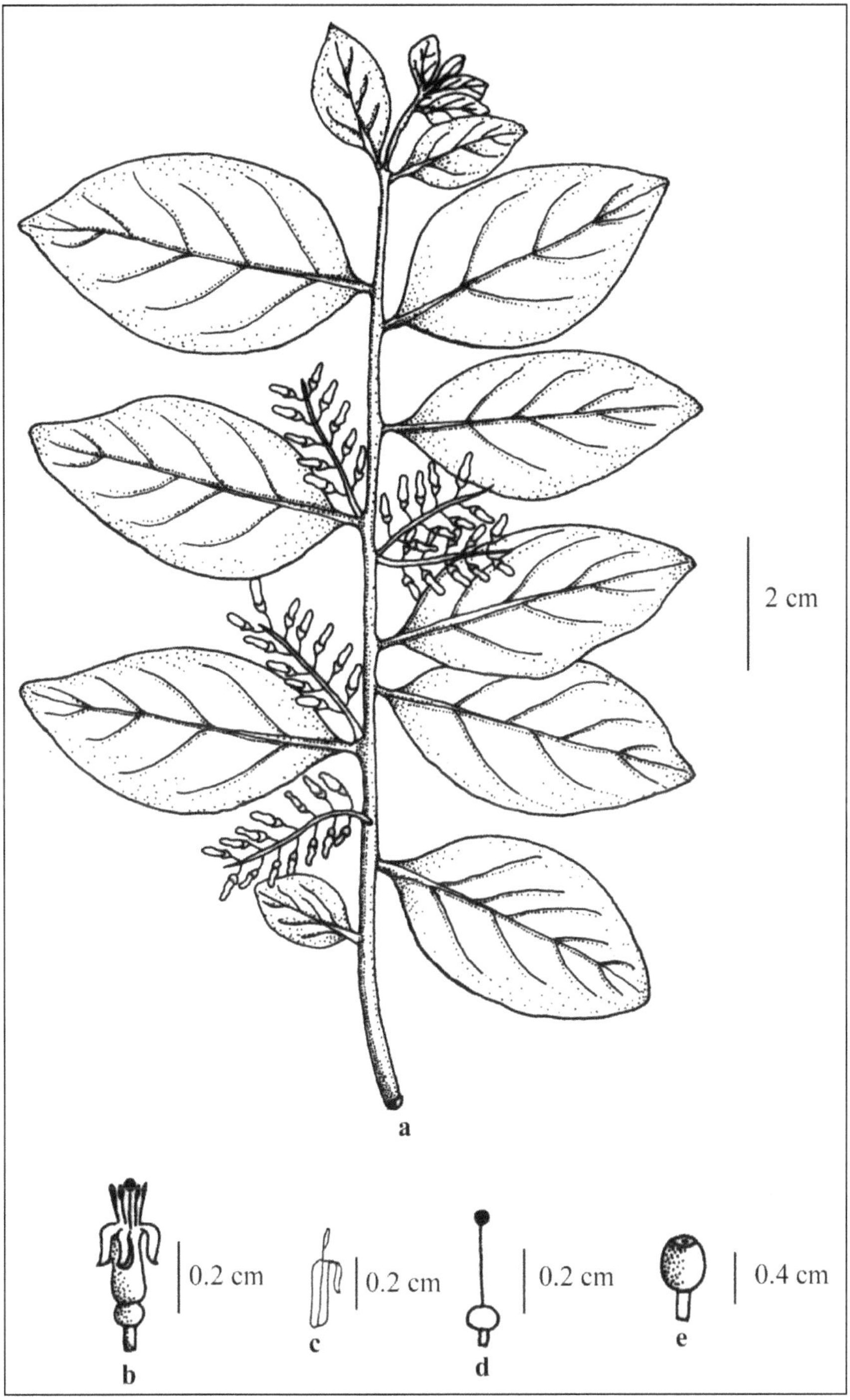

Figure 18.7. ***Helixanthera wallichiana*** **(Schult.) Dans.**
a: Flowering twig; b: Flower; c: Epipetalous stamen; d: Gynoecium; e: Fruit.

Plate 18.6: ***Helixanthera wallichiana*** **(Schult.) Dans.**

A: Flowering branches; B: A twig with young inflorescence; C: Branch with flowers; D: Twig with fruits.

introrse. Ovary inferior. Stigma globose, exerted. Fruit globose or oblong.

1. Corolla less than 3 cm long, tube swollen in the middle and straight .... *M. capitellatus*
1. Corolla more than 5 cm long, tube slender below and slightly swollen towards the tip and curved .... *M. parasiticus*

Macrosolen capitellatus (Wt. & Arn.) Danser, Blumea 2: 36. 1936; Wiens in Abeywick., Fl. Ceylon 1: 64. 1973; in Dassan. & Fosberg, Rev. Hadndb. Fl. Ceylon 6: 126. 1987; Razi, Lloydia 20 (4): 241. 1957; Sald. Fl. Karnataka 2: 78. 1996; Danser, Bull. Jard. Bot. Buitenz. Ser. 3, 10: 313. 1929; Sald. & Nicol. Fl. Hassan Dist. 308. 1976; Singh, Fl. East. Karnataka 551 (2): 1988; Rama. *et al.*, Fl. Shimoga Dist. 515. 2001; Bhat, Fl. Udupi 548. 2003. *Loranthus capitellatus* Wt. & Arn., Prodr., 382. 1834; Wight, Ic. t. 304. 1840; Hook. f., Fl. Brit. India 5: 221. 1886; Talb., For. Fl. Bombay 2: 417. 1911; Fischer, Rec. Bot. Surv. Ind. 11: 179. 1926; Cooke, Fl. Pres. Bombay 3: 44. 1906. *Elytranthe capitellata* (Wt. & Arn.) Engl., Pflanzenf. 3 (1): 189. 1889; Gamble, Fl. Pres. Madras 2: 1255. 1925.

Large semi parasitic shrubs with haustoriferous runners. Stem branched, cylindrical, glabrous, internodes elongated, terete. Leaves opposite, petiolate up to 1.5 cm, simple, entire, elliptic, 8-12.5 cm x 2-4.5 cm, 3-5 nerved, acute at apex, attenuate at base, glabrous. Inflorescence axillary raceme with short peduncled clusters with about 3-6 together in a cluster. Mature flower pink, up to 3 cm, subtended by 1 bract and 2 bracteoles which coalesce closely clasping the base of the flower actinomorphic, bisexual, epigynous. Calyx cylindrical, up to 0.5 cm long. Corolla up to 2.5 cm long, constricted at the base and dilated at the middle, narrowed towards the tip, tube pink with green limb, lobes 6, reflexed at anthesis, green, spathulate. Stamens 6, epipetalous, filaments opposite and adnate for about one third the distance, anthers basifixed, up to 0.2 cm long, introrse. Ovary inferior. Style short, up to 2 cm long. Stigma globose, slightly 2 lobed, exerted. Fruit globose.

Fl.: March to April.

Distr.: Mandya, Mysore, Bangalore, Dharwad, Shimoga, Chitradurga, Udupi, Hassan,

Specimens examined: J. Joseph 24997, 24998, 45762, 45759, 84920, 89921 (MH); M.V. Vishwanathan 96658 (MH); V. Narayanaswamy 79996, 96659, 97878, 97379, 97428 (MH); M. Mohanan 116077, 116076 (MH), 66642 (CALI); E. Vajravelu 101946, 101947, 48884, 48885, 86694, 80695 (MH); V.S. Ramachandran 123767, 123768 (MH); A.N. Henry and M. Chandra Bose 42094, 42095 (MH); G.V. Subbarao 74427, 74428 (MH); N.C. Rathna Krishnan 73413 (MH); Shanavas Khan5443, 37623, 22363 (CALI), MCT 344, 356.

Host plants: *Anogeissus latifolia* Wall. (Combretaceae), *Artocarpus heterophyllus Lam.* (Moraceae), *Ehretia laevis* Roxb. (Boraginaceae), *Cassine glauca* (Rottb.) Kuntze, *Ficus religiosa* L. (Moraceae), *F. drupacea* var. *pubescens* (Roth.) Thunb. (Moraceae), *Flacourtia montana* Grah. (Flacourtiaceae), *Mangifera indica* L. (Anacardiaceae), *Mitragyna parvifolia* Korth. (Rubiaceae), *Terminalia paniculata* Roth. and *Terminalia tomentosa* Wt.

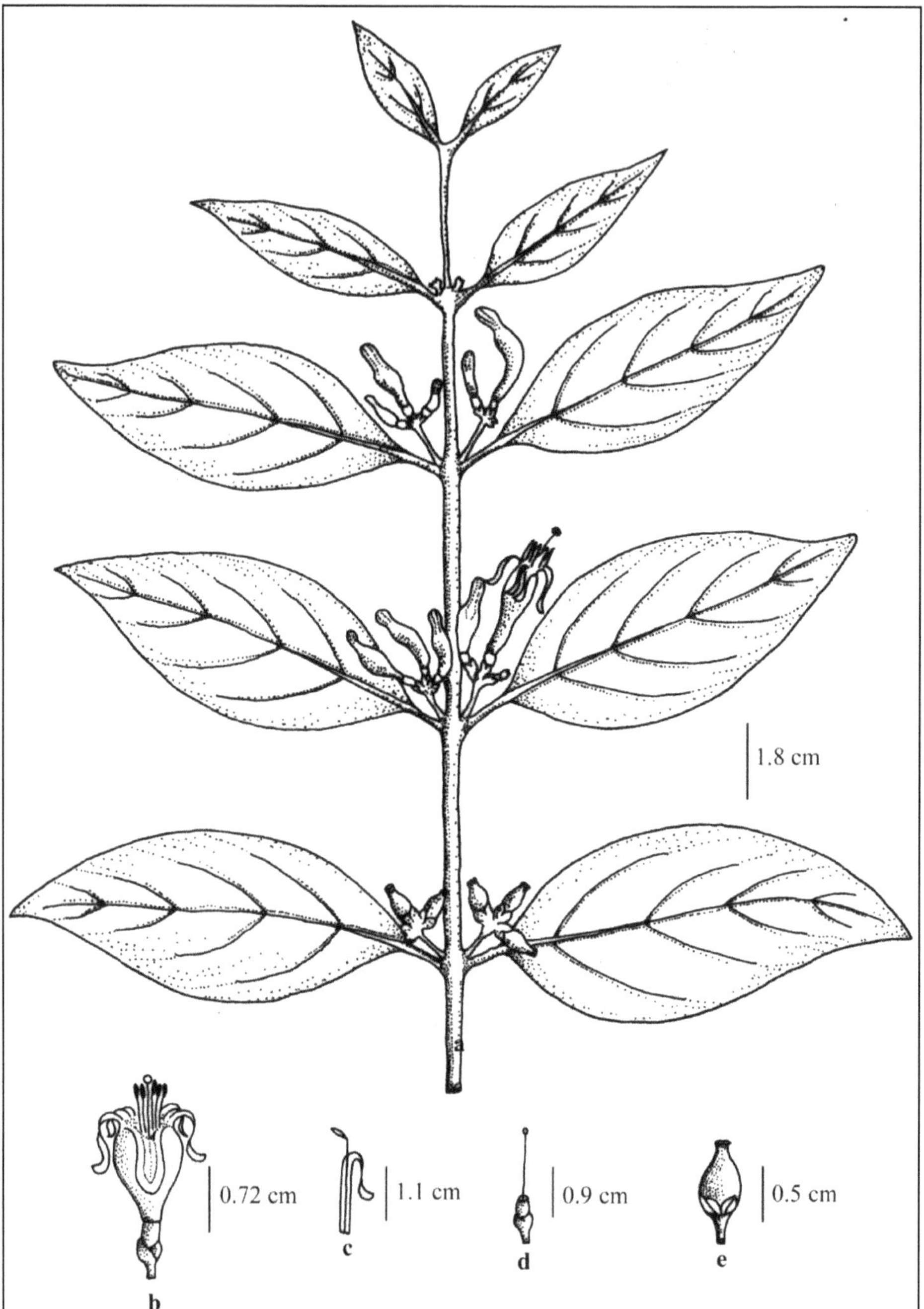

Figure 18.8: *Macrosolen capitellatus* (Wt. & Arn.) Dans.
a: Flowering twig; b: Flower; c: Epipetalous stamen; d: Gynoecium; e: Fruit.

Plate 18.7: ***Macrosolen capitellatus*** **(Wt. & Arn.) Dans.**
A: Flowering branches; B: Flowering twig; C: Close up of flowers; D: Fruits.

& Arn. (Combretaceae), *Schefflera venulosa* (Wt. & Arn.) Harms (Araliaceae).

Macrosolen parasiticus (L.) Danser, Blumea 2: 36. 1936; Wiens in Abeywick., Fl. Ceylon 1: 65. 1973; in Dassan. and Fosberg, Fl. Ceylon 6: 128. 1987; Sald. & Nicol. Fl. Hassan 308. 1976; Yoga. *et al.*, Fl. Chikmagalur Dist., 286. 1982; Sald. Fl. Karnataka 2: 79. 1996; Rama. *et al.*, Fl. Shimoga Dist. 516. 2001. *Loranthus loniceroides* L., Sp. Pl. ed. 2, 473. 1762; Talb., For. Fl. Bombay 2: 417. 1911; Fischer, Rec. Bot. Surv. Ind. 11: 179. 1926. *Lonicera parasitica* L., Sp. Pl. 175. 1753; Cooke, Fl. Pres. Bombay 3: 45. 1906; Wight Ic. t. 203. 1840; Hook. f., Fl. Brit. India 5: 221. 1886. *Elytranthe loniceroides* (L.) G. Don, Gen. Syst. 3: 427. 1834; Gamble, Fl. Pres. Madras 2: 1255. 1925; Danser, Bull. Jard. Bot. Buitenz. Ser 3, 10: 315. 1929. *Elytranthe parasitica* (L.) Danser, Bull. Jard. Bot. Buitenz. Ser 3, 10: 315. 1929; Razi, Lloydia 20 (4): 241. 1957; Rama. & Razi, Fl. Bangalore Dist. 1973; Rao & Razi, A Syn. Fl. Mysore 460. 1981.

Large semi parasitic shrubs with haustoriferous runners. Stem branched, cylindrical, glabrous, internodes elongated. Leaves opposite, petiolate up to 1.5 cm, simple, entire, elliptic, 10 -14 cm x 3-6.5 cm, 3-5 nerved, acute at apex, attenuate at base, glabrous. Inflorescence axillary raceme, in short peduncled clusters. Flowers pedicellate, bracteate, bracteolate, complete, bisexual, actinomorphic, epigynous, subtended by 1 bract and 2 bracteoles which coalesce at base, complete, bisexual, regular and epigynous. Calyx cylindrical up to 0.5 cm long. Corolla up to 5.5 cm long, tube pink up to half way, limb green, lobes 6 reflexed at anthesis, green. Stamens 6, epipetalous, filaments opposite and adnate up to one third distance, anthers basifixed, up to 0.4 cm long, introrse. Ovary inferior. Style long, up to 5.7cm long. Stigma globose, exerted. Fruit oblong.

Fl.: August to September.

Distr.: Mandya, Mysore, Bangalore, Dharwad, Shimoga, Chitradurga, Udupi, Hassan.

Specimens examined: Shivakumar 45702, 45703 (MH); R. Maruthan 60452, 60453 (MH); K. Subramanyan 5839, 5840, 3843, 3844, 2245, 2246 (MH); K.M. Sebastine 8163, 8164, 3923, 3924, 11534, 11535, 11555, 11556 (MH); M.V. Vishwanathan 97067, 97068 (MH); M. Chandrabose 55374, 55375 (MH); D.B. Deb 61030, 61031 (MH); K. Vivekananthan 79513, 79514 (MH); E. Vajravelu 80774, 80775, 83177, 83178 (MH); N. Mohanan 9895 (CALI), Shanavas Khan7276, 7278 (CALI); MCT 275, 279.

Host plants: *Acacia decurrens* Willd. and *Acacia* sp. (Mimosaceae), *Ailanthus* sp. (Simarubaceae), *Anacardium occidentale* L. (Anacardiaceae), *Anogeissus latifolia* Wall. (Combretaceae), *Artocarpus heterophyllus* Lam. (Moraceae), *Bauhinia racemosa* Lam. (Caesalpiniaceae), *Dalbergia lanceolaria* L.f. and *D. latifolia* Roxb. (Papilionaceae), *Ficus drupacea* var. *pubescens* (Roth.) Corner (Moraceae), *Lagerstroemia lanceolata* Wall. (Lythraceae), *Mangifera indica* L. (Anacardiaceae), *Mimusops elengi* L. (Sapotaceae), *Syzygium caryophyllatum* Gaertn. (Myrtaceae), *Terminalia paniculata* Roth and *T. tomentosa* Wt. & Arn. (Combretaceae).

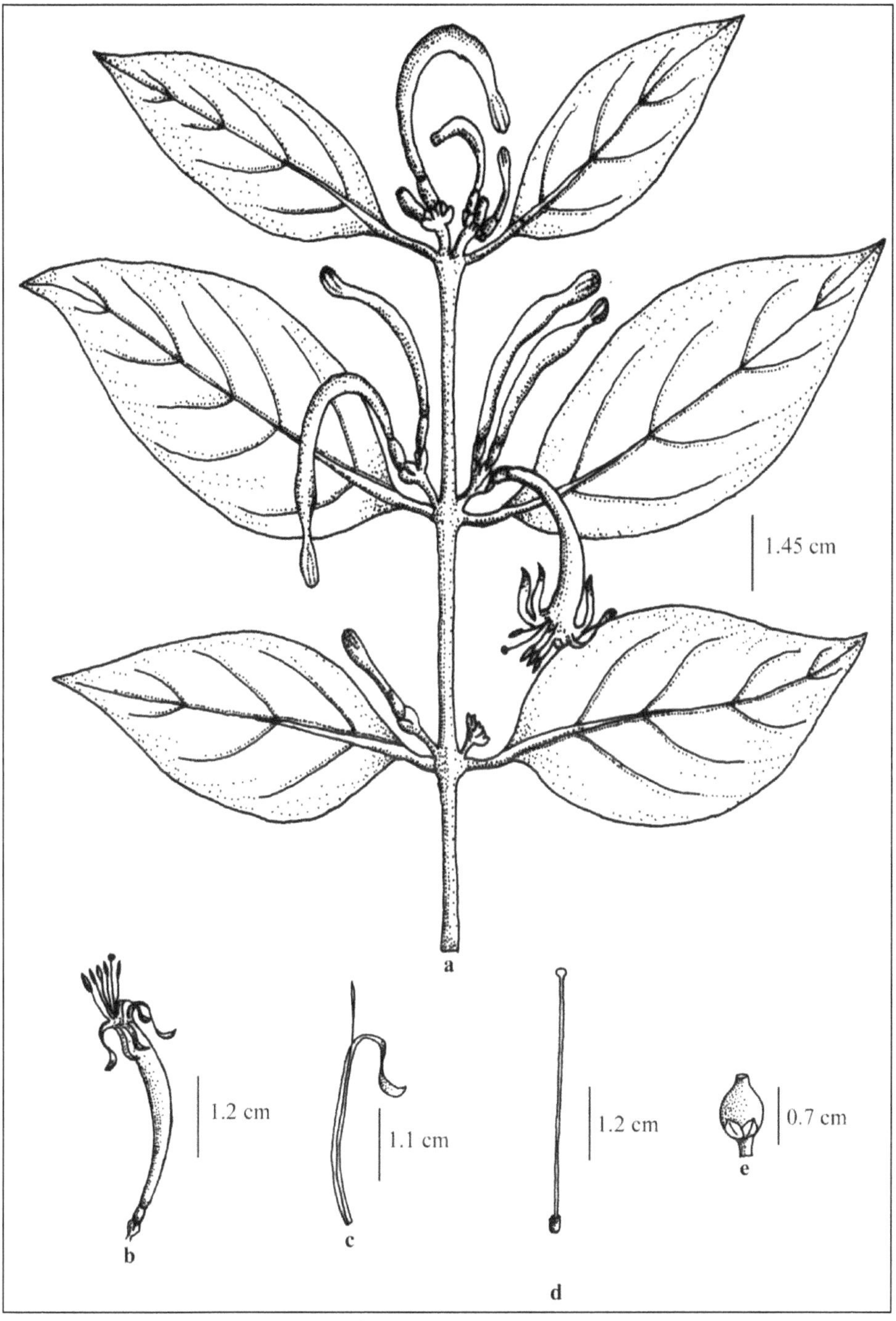

Figure 18.9: *Macrosolen parasiticus* (L.) Dans.
a: Flowering twig; b: Flower; c: Epipetalous stamen; d: Gynoecium; e: Fruit.

Plate 18.8: ***Macrosolen parasiticus*** **(L.) Dans.**

A: Flowering branches; B: A twig with inflorescence;
C: Flower showing anthers (a) and stigma (s); D: Fruits.

Note: Very common on Mango trees in Mysore, Mandya, Bangalore, Hassan.

SCURRULA Linnaeus

Sp. Pl. ed. 1. 110. 1753.

Shrubs. Stem branched. Leaves opposite, tomentose, glabrous with age. Inflorescence axillary racemes in fascicles. Individual flowers subtended by a single bract, complete, bisexual, irregular, epigynous. Calyx short. Corolla lobes 4, split on one side. Stamens 4, epipetalous, filaments opposite and adnate to petal lobes, anthers linear. Ovary inferior. Stigma capitate. Fruit pyriform.

1. Leaves cordate, white tomentose beneath .... *S. cordifolia*
1. Leaves ovate, rusty tomentose beneath .... *S. parasitica*

Scurrula cordifolia (Wall.) G. Don, Gen. Syst. 3: 423. 1834; Wiens in Abeywick., Fl. Ceylon 1: 73, 1973; in Dassan & Forsberg, Rev. Handb. Fl. Ceylon 6: 147. 1987; Razi, Lloydia 20 (4): 243. 1957; Rama. & Razi, Fl. Bangalore 154. 1973; Sald. & Nicol. Fl. Hassan Dist. 309. 1976; Sald. Fl. Karnataka 2: 79. 1996; Bhat, Fl. Udupi 549. 2003. *Loranthus cordifolius* Wall. in Roxb., Fl. Indica ed. Carey & Wall. 2: 222. 1824; Hook. f., Fl. Brit. India 5: 209. 1886; Gamble, Fl. Pres. Madras 2: 1252. 1925; Fischer, Rec. Bot. Surv. Ind. 11: 174. 1926.

Large semi parasitic shrubs, with haustoriferous runners. Stem branched, cylindrical. Leaves opposite, petiolate up to 1.5 cm long, simple, entire, cordate, 6-8 cm x 3-5.5 cm, 4-5 nerved, obtuse at apex, cordate at base, white tomentose beneath, glabrous with age. Inflorescense axillary raceme with 6 or 7 flowered fascicles, pedicel 0.2 cm, whitish tomentose. Flowers bracteate, ebracteolate, complete, bisexual, irregular, epigynous. Calyx short, up to 0.2 cm. Corolla up to 1.5 cm, outer tube grey, red inside, lobes 4, split on one side. Stamens 4, epipetalous, opposite and adnate to the petals, anthers linear, up to 0.2 cm, introrse. Ovary inferiror. Style short up to 1.5 cm long. Stigma capitate. Fruit pyriform.

Fl.: April to June.

Distr.: Mysore, Chamarajanagara, Bandipur National park, Bangalore, Shimoga, Hassan, Dakshina Kannada, Uttara Kannada, Kodagu.

Specimens examined: K. Subramanyam 13565, 13566, 6929, 1958 (MH); M. Mohanan 170339, 170340, 170371, 170372, 100895, (MH); J. Joseph 24072, 30578, 15461, 22240, 11308 (MH); E. Vajravelu 102973, 102974, 51991, 122456, 122457, 60534, (MH); J.S. Gamble 45221, 45228, 45230 (MH); B.B. Shetty 23563, 23564, 11965, 4099 (MH); K.M. Sebastine 4100, 2125 (MH); B.D. Sharma 68715, 68716, 35687; S. Karthikeyan 56157, 56158, 26896; D.B. Deb 61783, 61784, 31445 (MH); B.H. Beddome 45225, 45227, (MH); K. Ramamurthy 131594, 131597, 75578 (MH); C.A. Barber 45229, 6875 (MH); Shanavas Khan15623 (CALI); MCT 329, 331.

Host plants: *Butea monosperma* (Lam.) Taub. (Papilionaceae), *Cordia dichotoma* Forst. f. (Boraginaceae), *Dalbergia paniculata* Roxb. (Papilionaceae), *Lagerstroemia reginae*

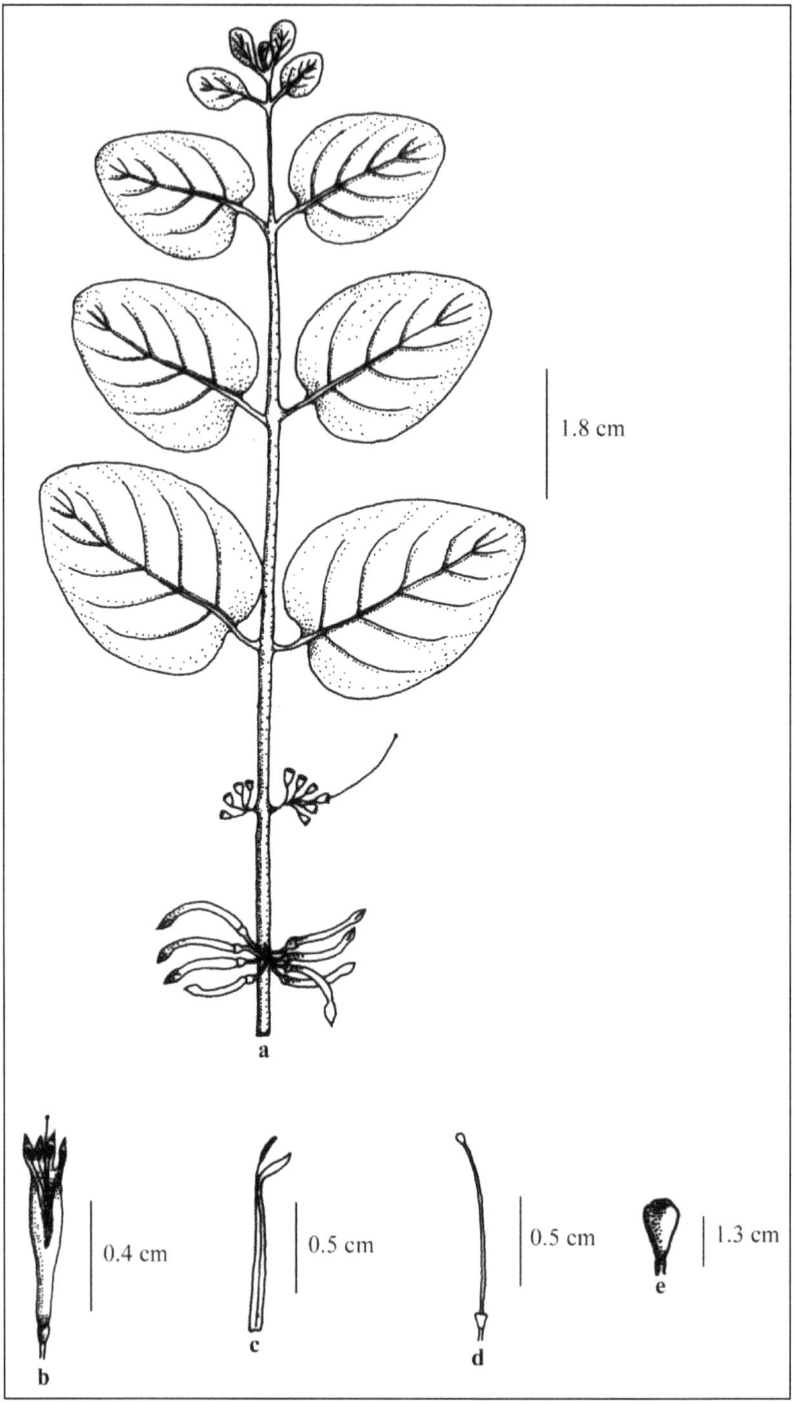

Figure 18.10: *Scurrula cordifolia* (Wall.) G. Don
a: Flowering twig; b: Flower; c: Epipetalous stamen; d: Gynoecium; e: Fruit.

Plate 18.9: *Scurrula cordifolia* (Wall.) G. Don
A: Flowering twig; B: Close up of flowers; C: Fruits.

Roxb. (Lythraceae).

Scurrula parasitica L., Sp. Pl. 110. 1753; Wiens in Abeywick., Fl. Ceylon 1: 73. 1973; in Dassan and Fosberg, Fl. Ceylon 6: 146. 1987; Razi, Lloydia 20 (4): 243. 1957; Sald. & Nicol. Fl. Hassan Dist. 309. 1976; Rao & Razi, A Syn. Fl. Mysore 461. 1981; Keshava. & Yoga., Fl. Coorg, 390. 1990; Sald. Fl. Karnataka 2: 79. 1996; Rama. *et al.*, Fl. Shimoga Dist. 517. 2001; Manju. *et al.*, Fl. Davanagere Dist. 342. 2004. *Loranthus scurrula* L. Sp. Pl. ed. 2, 472. 1762; Hook. f., Fl. Brit. India 5: 208. 1886; Talb., For. Fl. Bombay 2: 407. 1909; Fischer, Rec. Bot. Surv. Ind. 11: 173. 1926; Cooke, Fl. Bombay 3: 40. 1967; Roxb., Fl. Indica. Ed. Carey & Wall. 2: 232. 1824. *Loranthus buddleioides* Desr. in Lam., Encycl. 3: 600. 1792; Roxb., Pl. Corom. t. 140. 1800; Gamble, Fl. Pres. Madras 2: 1251. 1925.

Large semi parasitic shrubs. Stem branched, internodes short, glabrous. Leaves opposite, petiolate up to 2 cm, simple, entire, ovate, 5-8.5x3-5 cm, 3-6 nerved, obtuse at apex, attenuate at base, rusty tomentose beneath when young, glabrous when mature. Inflorescence axillary raceme, flowers 6 or 7 together in fascicles. Flower pedicellate up to 0.3 cm, bracteates, ebracteolate, complete, bisexual, irregular, epigynous, whitish tomentose. Calyx short, up to 0.3 cm. Corolla up to 2.5 cm, outer tube grey, red inside, lobes 4, split on one side. Stamens 4, epipetalous, opposite and adnate to the petals, exerted, anthers linear, up to 0.2 cm long, introrse. Ovary inferior. Style long, slender, up to 2.8 cm long. Stigma capitate. Fruit pyriform, up to 0.8 cm long.

Fl.: September to October

Distr.: Mandya, Mysore, Bangalore, Shimoga, Hassan, Kodagu.

Occurrence and specimens examined: K. Ramamurthy 97252, 49365 (MH); K. Rajasekar 45181, 6983 (MH); C.A. Barber 45180 (MH); V.S. Ramachandran 114945, 114946, 52325, 58692, 136002, 136003 (MH); P. Bhargavan 92097, 157800, 157799, 129464, (MH), P. Bhargavan 65714 (CALI); C.N. Mohanan and B. Ramanujan 72112, 149992, 149993 (MH); K. Ramamurthy 72978, 124672, 45665, 132372, 132373, 80863, 124671, 72978, (MH); K. Ramamurthy 75578 (CALI); K. Vivekananthan 88650, 45665, 88649, 44027, 44026, 60477, 22379, 60476, 29379 (MH); A.G. Pandurangan 149969, 66478, 149968 (MH); V.T. Antony 1031, 143500 (MH); N.C. Rathna Krishnan 73038, 73039, 3749, (MH); C.A. Barber 45211, 45213, 45191, 45192, (MH); J.L. Ellies 83381, 83382, 43249 (MH); K.M. Sebastine 14325, 7312, 5211, 5213, 2701, 14326, 7312, 4378, 8487, 8488, 18824, 9597 (MH); B.D. Sharma 68471, 68472, 35562, (MH); E. Vajravelu 67658, 67659, 35019, 71059, 36807, 89932, 89933, 46200, 50673, 50674, 50208, 50209, 26019, 26280, 48784, 94383, 94384,; K. Subramanyam 9641, 9642, 4959 (MH); A.N. Henry and M. Chandra Bose 42005, 19256, 45181, 9480, 129465, 65714 (MH); N.C. Nair 126087, 126088, 64661 (MH); V.G. Nair 113652, 113653 (MH); C.N. Mohan 112971 (MH); Anil Kumar 147050; J. Joseph 17827, 34640, 34641, 44536, 85952 (MH); Shanavas Khan26850 (CALI); MCT 179, 336, 339.

Host plants: *Acacia decurrens* Willd. and *Albizia* sp. (Mimosaceae), *Anacardium occidentale* L. (Anacardiaceae), *Artocarpus heterophyllus* Lam. (Moraceae), *Buahinia racemosa* Lam. (Caesalpiniaceae), *Careya arborea* Roxb. (Lecythidaceae), *Citrus medica* L. (Rutaceae) *Cordia dichotoma* Forst. f. (Boraginaceae), *Dalbergia paniculata* Roxb. (Papilionaceae), *Ficus drupacea* var. *pubescens* (Roth.) Corner (Moraceae), *Gardenia*

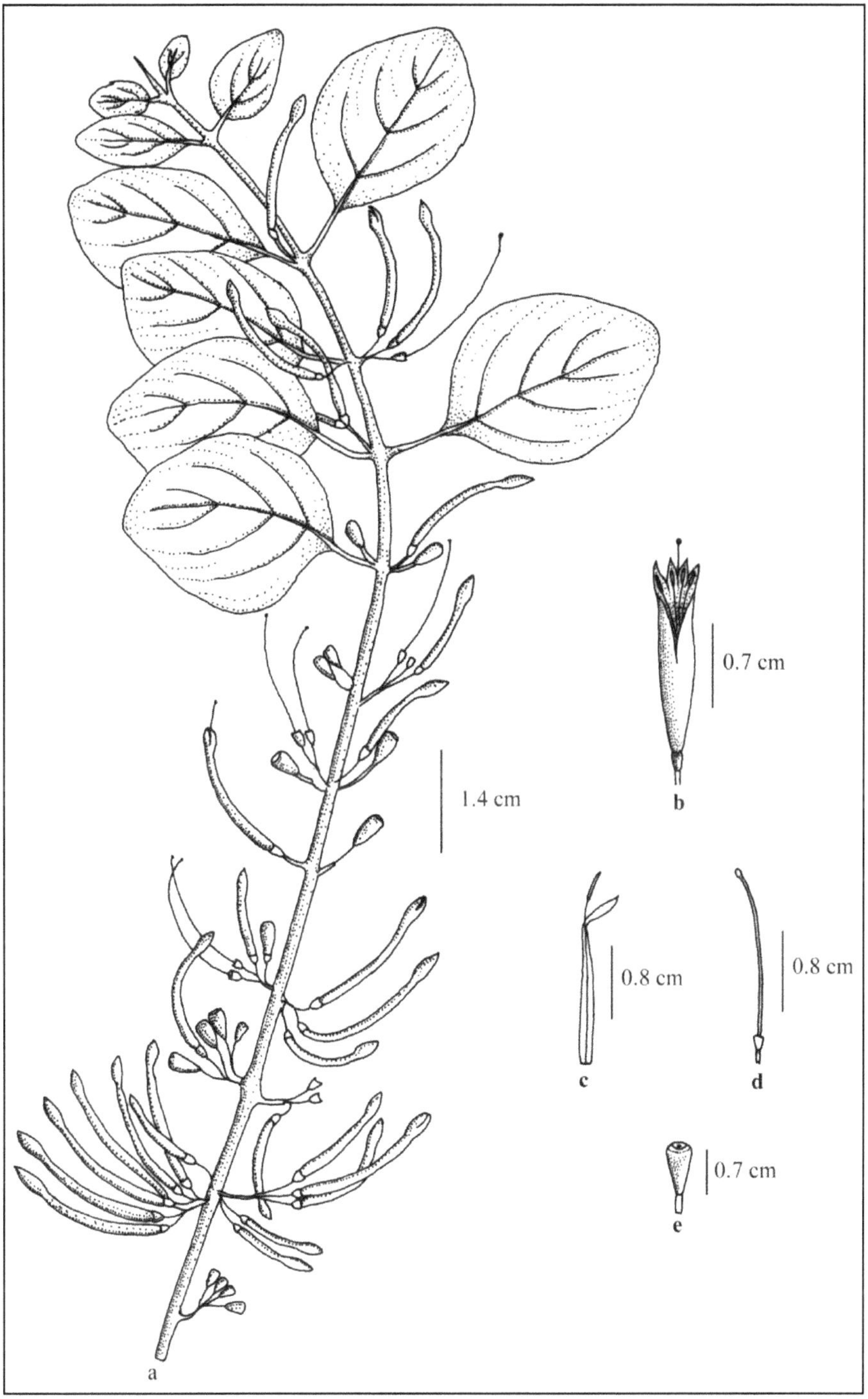

Figure 18.11: *Scurrula parasitica* L.

a: Flowering twig; b: Flower; c: Epipetalous stamen; d: Gynoecium; e: Fruit.

Plate 18.10: ***Scurrula parasitica*** **L.**

A: Flowering twig; B: Inflorescence; C: Close up of flowers; D: Fruits.

gummifera L.f. (Rubiaceae), *Garuga pinnata* Roxb. (Burseraceae), *Gmelina arborea* Roxb. (Verbenaceae), *Grewia* sp. (Tiliaceae), *Holoptelea integrifolia* Pl. (Ulmaceae), *Litsea floribunda* Gamb. (Lauraceae), *Morinda citrifolia* L. (Rubiaceae), *Phyllanthus indofishceri* Bennet (Euphorbiaceae), *Pithecellobium dulce* (Roxb.) Benth. and *P. bigeminum* Mart. (Mimosaceae), *Psidium guajava* L. (Myrtaceae), *Scleropyrum wallichianum* Arn. (Santalaceae), *Tectona grandis* L.f. (Verbenaceae), *Terminalia bellirica* (Gaertn.) Roxb. (Combretaceae), *Vitex altissima* L.f. and *V. negundo* L. (Verbenaceae), *Zanthoxylum rhetsa* (Roxb.) DC. (Rutaceae).

TAXILLUS Van Tieghem

Bull. Soc. Bot. France 42: 256. 1895.

Shrubs. Stem branched. Leaves alternate or opposite, sessile or shortly petioled, glabrous or or highly tomentose. Inflorescence axillary racemes, few flowered or in cluster. Flowers bracteate, ebracteolate, complete, bisexual, irregular, epigynous. Calyx short. Corolla lobes 5, rusty tomentose, pubescent or glabrous, tube split on one side. Stamen 5, epipetalous, anthers basifixed. Fruit oblong.

1. Leaves glabrous .... *T. cuneatus*
1. Leaves variously pubescent:
    2. Leaves orbicular; bract longer than calyx .... *T. bracteatus*
    2. Leaves obovate; bract shorter than calyx:
        3. Leaves ashy pubescent above rusty tomentose below .... *T. tomentosus*
        3. Leaves grey pubescent glabrous with age .... *T. recurvus*

OR

1. Leaves orbicular, bract longer than calyx .... *T. bracteatus*
1. Leaves obovate; bract shorter than calyx:
    2. Leaves glabrous .... *T. cuneatus*
    2. Leaves variously pubescent:
        3. Flowers usually 3-5 together in a cluster; corolla lobes 5 ...... *T. tomentosus*
        3. Flowers always 2 in a cluster; corolla lobes 4 .... *T. recurvus*

Taxillus bracteatus (Van Tiegh.) Danser Bull. Soc. Bot. France 42: 256. 1895; Rama. & Razi Fl. Bangalore 154. 1973; Sald. & Nicol. Fl. Hassan Dist. 309. 1976; Singh, Fl. East. Karnataka 552 (2): 1988; Keshava. & Yoga., Fl. Coorg, 391. 1990; Manju. *et al.*, Fl. Davanagere Dist. 343. 2004. *Loranthus bracteatus* Heyne in Roxb., Fl. Ind. Ed. Carey and Wall. 2: 220. 1824; Wight, Ic. t. 378. 1840; Hook. f., Fl. Brit. India 5: 213. 1886; Gamble, Fl. Pres. Madras 2: 1252. 1925; Fischer, Rec. Bot. Surv. Ind. 11: 174. 1926.

Large semi parasitic shrubs with haustoriferous runners. Stem branched, white pubescent glabrous with age. Leaves alternate, petiolate up to 0.5 cm long, simple, entire, orbicular, 2.5-3.5 cm x 2.1-3.1 cm, obtuse at apex and base, white tomentose. Inflorescence axillary racemes with umbellate flowers, white tomentose. Mature flower

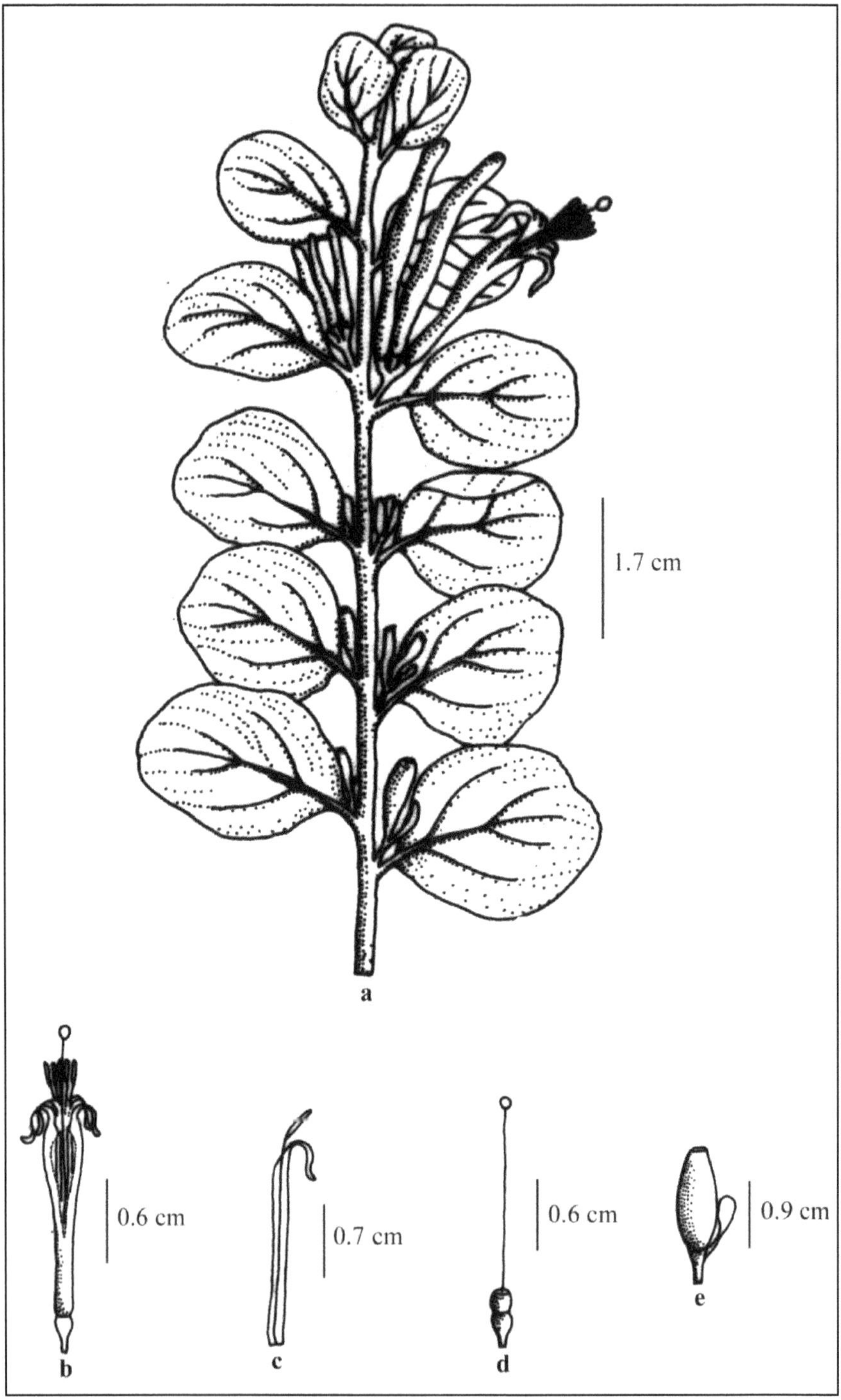

Figure 18.12: *Taxillus bracteatus* (Van Tiegh.) Dans.
a: Flowering twig; b: Flower; c: Epipetalous stamen; d: Gynoecium; e: Fruit.

Plate 18.11 : *Taxillus bracteatus* (Van Tiegh.) Dans.
A: Flowering branches; B: Inflorescence; C. Flowers; D: Fruits.

up to 3 cm long, subtended by a lateral bract, up to 1.1 cm, complete, bisexual, irregular, epigynous. Calyx up to 0.5 cm, irregularly lobed, shorter than bract. Corolla up to 2.5 cm, greyish white, lobes 5, reflexed at anthesis. Stamens 5, epipetalous, anthers linear, 0.2 cm, basifixed. Style 2.2 cm, stigma capitate. Fruit oblong or ellipsoid, up to 2 cm.

Fl.: July to August.

Distr.: Chitradurga, Chamarajanagar, Kodagu, Shimoga.

Specimens examined: E. Vajravelu 77197, 77198, 32576, 83187, 83188, 71182 (MH); K. Subramnayam 2053, 1073 (MH); K.M. Sebastine 8161, 4214 (MH); J.L. Ellis 66974, 66975, 34649, 1006 (MH); MCT 242, 244, 264.

Host plants: *Grewia tiliifolia* Vahl., *Grewia orbiculata* Rottler (Tiliaceae); *Ixora pavetta* Andr. (Rubiaceae).

Taxillus cuneatus (Roth) Danser, Bull. Jard. Bot. Buitenz. Ser. 3, 10: 1929; Razi, Lloydia 20 (4): 244. 1957; Sald. & Nicol. Fl. Hassan Dist. 309. 1976; Rao & Razi, A Syn. Fl. Mysore 461. 1981; Keshava. & Yoga., Fl. Coorg 391. 1990; Sald. Fl. Karnataka 2: 80. 1996; Rama. *et al.*, Fl. Shimoga Dist. 517. 2001. *Loranthus cuneatus* Heyne in Roth, Nov. Pl. Sp. 193. 1821; Hook. f., Fl. Brit. India 5: 214. 1886; Talb., For. Fl. Bombay 2: 410. 1911; Gamble, Fl. Pres. Madras 2: 1253. 1925; Cooke, Fl. Pres. Bombay 3: 42. 1967.

Large semi parasitic shrubs with haustoriferous runners. Stem branched, glabrous. Leaves alternate, sessile, simple, entire, obovate, 3-4.5 cm x 1.4-2.2 cm, 3 nerved, obtuse at apex, cuneate at the base, glabrous. Inflorescence axillary raceme. Flowers 3 or 5 together in fascicles, bracteates, ebracteolate, complete, bisexual, epigynous. Calyx short, irregularly lobed, up to 0.3 cm long. Corolla up to 2 cm long, greenish outside, yellow inside, lobes 5. Stamens 5, epipetalous, filaments red, anthers linear, basifixed, up to 0.2 cm long. Style short, up to 1.8 cm long. Stigma red, capitate. Fruit oblong, up to 1.5 cm long.

Fl.: August to September.

Distr.: Mysore, Chamarajanagar, Kodagu, Shimoga.

Specimens examined: B.D. Nainthani 45356, 45357, 45375, 45376, 45380, 45391, 45380 (MH); D.B. Deb 58563, 30851 (MH); K. Ramamurthy 46353, 23390 (MH); B.D. Sharma 39918, 77879, 77880 (MH); B.V. Shetty 54387, 54388, 28000, 63766, 32989 (MH); M. Chandra Bose 98468, 98469, 54268 (MH); J. Joseph 26131, 26132, 13249, 101279 (MH); Vishwanathan 96726, 96727, 96700, 96701 (MH); Shanavas Khan5880, 5480, 26851 (CALI); MCT 212, 277, 295.

Host plants: *Anogeissus latifolia* Wall. (Combretaceae), *Bauhinia racemosa* Lam. (Caesalpiniaceae), *Grewia orientalis* L. (Tiliaceae), *Ixora coccinea* L., *Canthium dicocum* (Gaertn.) Tlivs & Brom. (Rubiaceae), *Premna tomentosa* Willd. (Verbenaceae), *Spondias pinnata* (L.f.) Kurz. (Anacardiaceae).

Taxillus recurvus (DC.) Van Tiegh. Bull. Soc. Bot. France 42: 256. 1895; Razi, Lloydia 20 (4): 244. 1957; Rao & Razi, A Syn. Fl. Mysore 462. 1981; Sald., Fl. Karnataka 2: 81. 1996. *Loranthus recurvus* Wall. in DC., Prodr. 4: 299. 1830; Hook. f., Fl. Brit. India 5: 213. 1886; Wt. Ic. t. 305. 1840; Gamble, Fl. Pres. Madras 2: 1252. 1925;

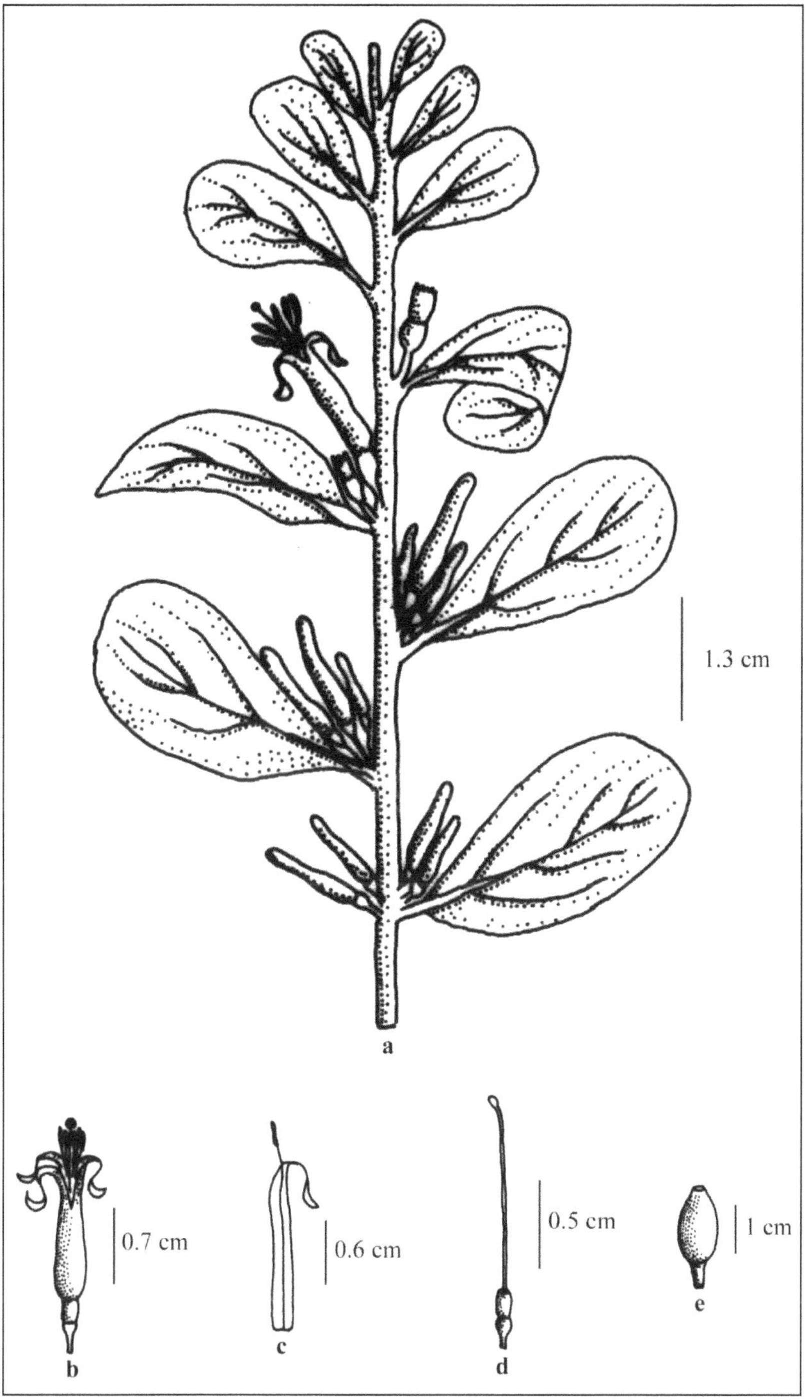

Figure 18.13: ***Taxillus cuneatus*** **(Roth.) Dans.**
a: Flowering twig; b: Flower; c: Epipetalous stamen; d: Gynoecium; e: Fruit.

Plate 18.12: ***Taxillus cuneatus*** **(Roth.) Dans.**

A: Flowering branches; B: Inflorescence;
C: Flowers (inset-close up of flowers); D: Fruits.

Fischer, Rec. Bot. Surv. Ind. 11: 175. 1926.

Large semiparasitic shrubs with haustoriferous runners. Stem branched, young branches grey pubescent, older ones glabrous with age. Leaves alternate, sessile, simple, entire, obovate, 4-7.8 cm x 1.3-4 cm, obtuse at apex, cuneate at base, younger ones grey pubescent glabrous with age, sometimes clothed with white tomenta beneath. Inflorescence axillary, 2 flowered fascicles, peduncle very short, tomentose. Flowers bracteate, ebracteolate, complete, bisexual, irregular, epigynous. Calyx long, irregularly lobed, up to 0.7 cm. Corolla up to 1.5 cm long, greenish yellow, not much curved and gibbously inflated below the middle, lobes 4, reflexed at anthesis. Stamens 4, epipetalous, opposite and adnate to the petals, anthers linear, up to 0.2 cm long, basifixed, introrse. Ovary inferior. Style short, up to 1.5 cm long. Stigma capitate. Fruit oblong up to 1 cm.

Fl.: March to April.

Distr.: B.R. Hills, Bandipur, Shimoga.

Specimens examined: Vajravelu 77198, 83187 (MH); K. Subramanyam 2053 (MH), K.M. Sebastine 8161 (MH), J.L. Ellis 66974 (MH). Shanavas Khan5491 (CALI).

Inlcuded on the authority of Barnes (1944).

Taxillus tomentosus (Roth) Van Tiegh. Bull. Soc. Bot. France 42: 256. 1895; Razi, Lloydia 20 (4): 244. 1957; Sald. & Nicol. Fl. Hassan Dist. 310. 1976; Rao & Razi, A Syn. Fl. Mysore 462. 1981; Keshava. & Yoga., Fl. Coorg 391, 1990; Sald. Fl. Karnataka 2: 81. 1996; Manju. *et al.*, Fl. Davanagere Dist. 343. 2004. *Loranthus tomentosus* Heyne in Roth, Nov. Pl. Sp. 191. 1821; Hook. f., Fl. Brit. India 5: 212. 1886; Gamble, Fl. Pres. Madras 2: 1252. 1925; Fischer, Rec. Bot. Surv. Ind. 11: 174. 1926.

Large semi parasitic shrubs with haustoriferous runners. Stem branched, pubescent. Leaves alternate, sometimes sub-opposite, shortly petiolate, petiole up to 1 cm long, simple, entire, obovate, 4.5-8.5 cm x 1.5-3.5 cm long, 3 nerved, obtuse at apex, attenuate at base, ashy pubescent above, rusty tomentose below. Inflorescence axillary, 4 flowered fascicles, peduncle short, rusty tomentose. Flowers bracteate, ebracteolate, complete, bisexual, irregular, epigynous. Calyx short, irregularly lobed, up to 0.4 cm long. Corolla up to 1.3 cm long, yellowish outside, greenish inside, lobes 4, reflexed at anthesis. Stamens 4, epipetalous, opposite and adnate to the petals, anthers linear, up to 0.2 cm long, basifixed, introrse. Ovary inferior. Style short, up to 1.7 cm long. Stigma capitate. Fruit oblong up to 1 cm.

Fl.: March to April.

Distr.: Mysore, Kodagu, Chamarajanagar, Hassan, Shimoga, Bangalore.

Specimens examined: J.S. Gamble 45246, 45247, 45251, 45252, 45255, 45256, 45261, 45262, 45263, 45271, 45282 (MH); E. Vajravelu 55658, 55659, 63927 (MH); N.C. Nair 69621 (MH); J.L Ellis 52234, 52235, 25758 (MH); V.S. Ramachandran 128791, 58777 (MH); M. Mohanan 115359 (MH); P. Bhargavan 129067 (MH). Shanavas Khan7219

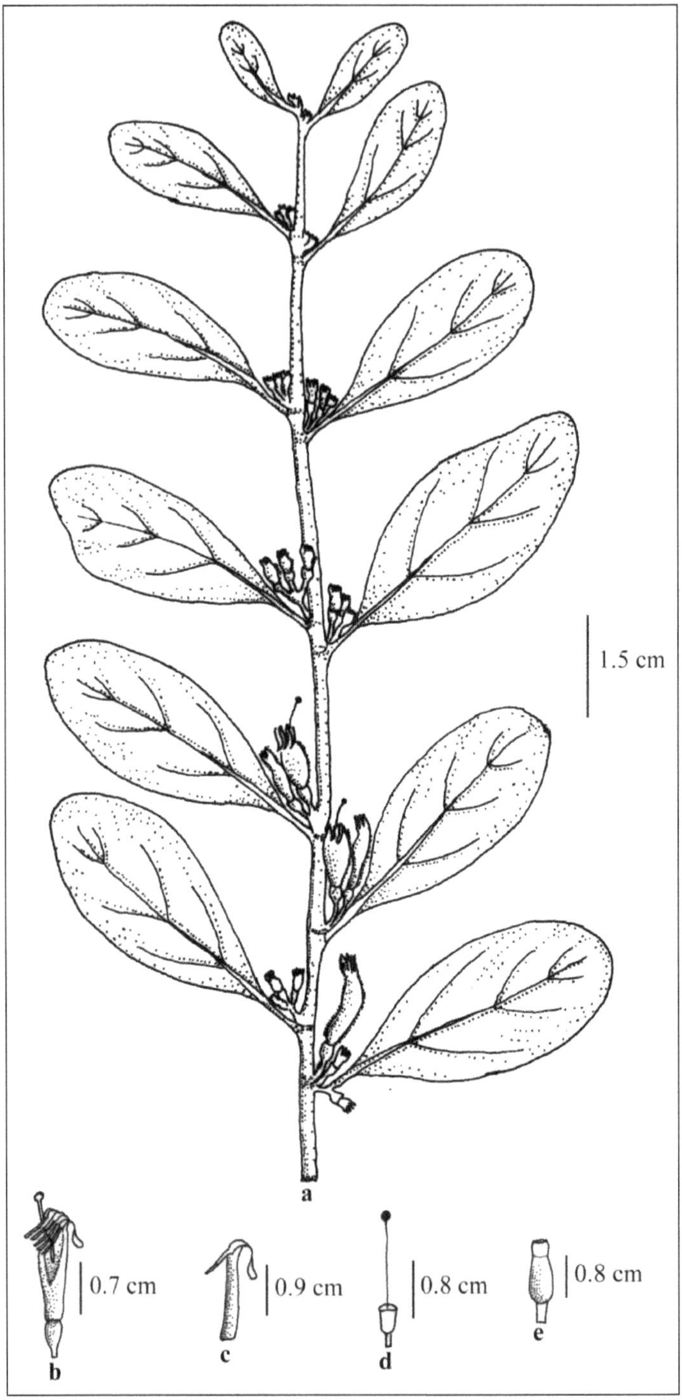

Figure 18.14: *Taxillus tomentosus* (Roth) Van Tiegh.
a. Flowering twig; b. Flower; c. Epipetalous stamen; d: Gynoecium; e: Fruit.

Plate 18.13: *Taxillus tomentosus* (Roth) Van Tiegh.

A: Flowering twig; B: Inflorescence; C: Close up of flower; D: Fruits.

(CALI), Vajravelu 62921 (CALI), K. Ramamurthy 72799 (CALI). MCT 11, 156, 209.

Host plants: *Aphanamixis polystachya* (Wall.) Parker (Meliaceae), *Anogeissus latifolia* Wall. (Combretaceae), *Casuarina equisetifolia* Forst. (Casuarinaceae), *Chuckrasia tabularis* A. Juss. (Meliaceae), *Dalbergia paniculata* Roxb. (Papilionaceae), *Emblica officinalis* Gaertn. (Euphorbiaceae), *Muntingia calabura* L. (Tiliaceae), *Spondias acuminata* Roxb. (Anacardiaceae), *Terminalia tomentosa* Wt. & Arn. (Combretaceae).

TOLYPANTHUS Blume

Syst. 7, 2, 1: 173 1830.

Shrubs. Stem branched, cylindrical, covered with lenticels. Leaves opposite. Inflorescence in fascicles, subtended by large involucral bract, often on old naked branches, bright red, 5 or 6 lobed, companulate. Individual flowers sessile, present within the involucre usually 5 or 6 in a group, bracteate, ebracteolate, complete, bisexual, regular, epigynous. Calyx short. Corolla 5-lobed, reflexed at anthesis. Stamens 5, epipetalous, anthers opposite and adnate to the petals, basifixed. Ovary inferior. Stigma capitate. Fruit ovoid.

Tolypanthus lagenifer (Wight) Van Tiegh., Bull. Soc. Bot. France 42: 249. 1895; Razi, Lloydia 20 (4): 244. 1957; Sald. & Nicol. Fl. Hassan Dist. 310. 1976; Sald. Fl. Karnataka 2: 82. 1996; Rama. *et al.*, Fl. Shimoga Dist. 518. 2001; Bhat, Fl. Udupi 549. 2003. *Loranthus lageniferus* Wight, Icon. Pl. Ind. Orient t. 306. 1840; Hook. f., Fl. Brit. India 5: 218. 1886; Talb., For. Fl. Bombay 2: 415. 1911; Gamble, Fl. Pres. Madras 2: 1254. 1925; Cooke, Fl. Pres. Bombay 3: 44. 1906.

Large semi parasitic shrubs. Stem branched, cylindrical, covered with raised lenticels. Leaves opposite, petiolate, up to 1 cm long, simple, entire, elliptic, 10-15.5 cm x 4-6.5 cm, 3-5 nerved, acute at apex, obtuse at base, glabrous. Involucres shortly peduncled, usually on the old naked branches, companulate, bright red, up to 5 cm long, 5 or 6 lobed, containing 5 or 6 sessile flowers. Flowers bracteate, ebracteolate, complete, bisexual, regular, epigynous. Calyx short, up to 0.2 cm long. Corolla up to 5 cm long, tube red, constricted below and dilated up wards, lobes 5, with reddish tip. Stamens 5, epipetalous, filaments absent, anthers basified, about 0.2 cm, introrse. Ovary inferior. Style red, filiform, up to 5.5 cm long. Stigma capitate. Fruit ovoid, up to 1 cm.

Fl.: October to November

Distr.: Shimoga, Dakshina Kannada, Udupi, Kodagu.

Occurrence and specimens examined: R. Ansari 121445, 121446 (MH); E. Vajravelu 94643, 94644 (MH); V.S. Ramachandran 123783, 123784 (MH); J.S. Ellis 52270 (MH). Bhargavan 48914 (CALI), Shanavas Khan44220, 13211 (CALI). MCT 175, 280.

Host plants: *Hopea ponga* (Dennst.) Mabberly (Dipterocarpaceae), *Garcinia xanthochymus* Hk.f. (Clusiaceae).

Note: The branches of this parasite appear purplish when young. In spite of several visits we were not able to collect the fruits of this taxon. The plant flowers during every season but fruit setting is rare.

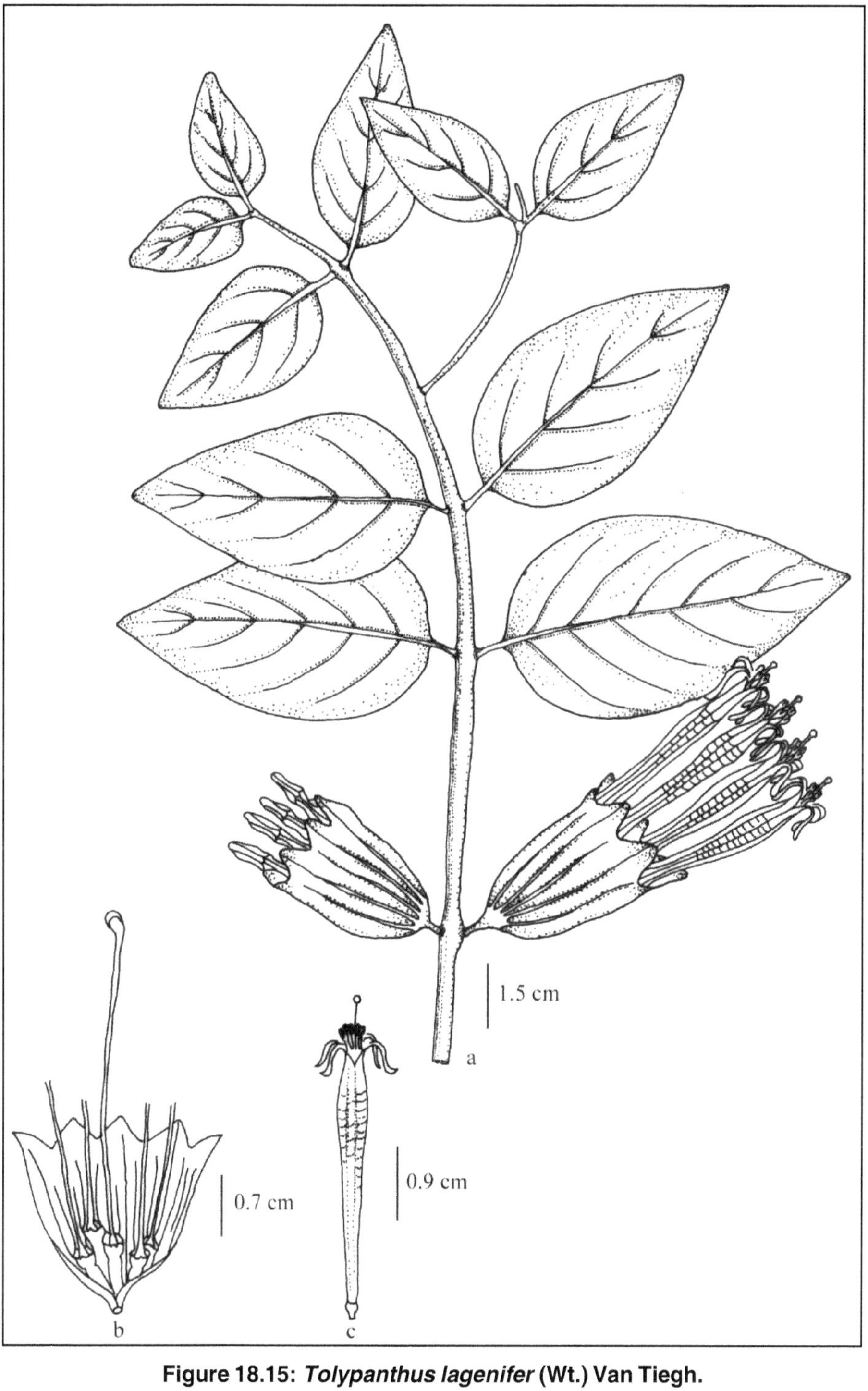

Figure 18.15: ***Tolypanthus lagenifer*** **(Wt.) Van Tiegh.**
a. Flowering twig; b. Inflorescence; c. Flower.

Plate 18.14. *Tolypanthus lagenifer* (Wt.) Van Tiegh.
A: *Tolypanthus lagenifer* on *Hopea ponga;* B: Flowering twig; C: Close up of flower.

VISCACEAE Batsch

Tab. Affin. Regni. Veg. 240. 1802.

Plants parasitic on the branches of trees and shrubs, usually pendulous on the tree trunks. Attached to their host through endophytic haustoria. Stem green, much branched, usually di- or tri-chotomous, branchlets terete, nodes swollen, internodes cylindrical or angular or sometimes flattened. Leaves present, sometimes absent. Flowers unisexual, usually in triads at the nodes or in cymes, bracteate, ebracteolate, incomplete, epigynous. Perianth lobes valvate. Staminate flowers with stamens opposite and adnate to the tepals, filaments absent, anther sacs many. Pistillate flowers with deciduous perianth lobes. Ovary inferior. Fruit ovate or oblong or globose.

VISCUM Linnaeus

Sp. Pl. 1023. 1753.

Shrubs. Stem branched, di- or tri-chotomous, internodes rounded or flattened or angular, terete. Leaves opposite or some times absent. Inflorescence axillary cyme, usually in sessile fascicles. Each flower subtended by 2 bracteoles or sometimes absent, incomplete, unisexual, epigynous. Male: perianth lobes 3 or 4; stamens adnate to the perianth lobes, filaments absent; pollen sacs many, open irregularly. Female: perianth lobes 3 or 4, usually deciduous after anthesis. Ovary inferior, unilocular. Fruit globose or oblong or ovoid.

Key to the Species

1. Leaves present; not reduced:
    2. Leaves opposite, conduplicately folded or both margin rolled inwards on the adaxial side .... *V. capitellatum*
    2. Leaves opposite, not conduplicately folded:
        3. Leaves spathulate, similar to table tennis racket .... *V. bandipurense*
        3. Leaves not spathulate:
            4. Leaves dimorphic, larger at some node; smaller at the base of all ramifications .... *V. mysorense*
            4. Leaves not dimorphic:
                5. Leaves obovate, trinerved, shortly petioled, distinct veins .... *V. orientale*
                5. Leaves obovate-suborbicular sometimes cuneate, sessile, indistinct veins .... *V. trilobatum*
1. Leaves absent:
    6. Internodes flattened, twisted near the base .... *V. nepalense*
    6. Internodes not flattened:
        7. Internodes 4-angled .... *V. angulatum*
        7. Internodes cylindrical .... *V. ramosissimum*

Viscum angulatum Heyne ex DC., Prodr. 4: 283. 1830; Hook. f., Fl. Brit. India 5: 225. 1886; Talb., For. Fl. Bombay 2: 422. 1911; Gamble, Fl. Pres. Madras 2: 1259. 1925; Fischer, Rec. Bot. Surv. Ind. 11: 181. 1926; Danser, Blumea 4 (2): 292. 1941; Razi, Lloydia 20 (4): 245. 1957; R.S. Rao, J. Ind. Bot. Soc. 36 (2): 140. 1957; Cooke, Fl. Pres. Bombay 3: 47. 1906; Sald. & Nicol., Fl. Hassan. 311. 1976; Rao & Razi, A Syn. Fl. Mysore 462. 1981; Yoga. *et al.*, Fl. Chikmagalur Dist., 288. 1982; Sald., Fl. Karnataka 2: 82. 1996; Rama. *et al.*, Fl. Shimoga Dist. 519. 2001; Sanj. & Bala., Rheedea 16 (2): 88. 2006. *Viscum ramosissimum* Wall. ex Wt., Ic. t. 1017. 1845.

Large semi parasitic shrubs, usually hanging. Stem slender, angular, usually 4-angled, (or 4-sided), branches numerous, internodes terete, distinctly tetrangular. Leaves absent. Inflorescence sessile, axillary actinomorphic, usually in triads, central female and laterals male. Male flowers small up to 3 mm, actinomorphic, perianth lobes 3 or 4, free, ovate up to 2 mm. Stamens 3 or 4, opposite and adnate to the perianth lobes, pistillode absent. Female flower present in the center of the triad, up to 5 mm, actinomorphic, epigynous, perianth lobes 4, free, ovate, acute at tip, staminodes absent. Ovary inferior. Style absent, stigma nipple shaped. Fruit sub globose.

Fl.: August to September

Distr.: Madikeri, Shimoga, Chikmagalur.

Specimens examined: J. Joseph 26135, 26136, 45879, 45880, 45881, 45882, 45883, 45887, 45889, 45965, 45966 (MH); D.B. Deb 58616, 58617, 60030, 60031, 60932, 60933 (MH); K. Ramamurthy 46601, 46602 (MH); K. Subramanyam 11285, 11286, 10803, 10804, 5899, 5900, 20720, 20721, 3777, 3778 (MH); E. Vajravelu 130430, 130431, 67594, 67595, 74771, 74772, 71440, 71441, 84302, 84303, 108349, 108350, 48780, 48781, 75422, 75423, 76703, 46704 (MH); M.V. Vishwanathan 96770, 96771, 152812 (MH); V. Narayanaswamy 79147 (MH); J.D. Hooker and TT 70828 (MH); B.D. Sharma 68168, 68169 (MH); B.V. Shetty 66492, 66493, 72580, 72581 (MH); K.M. Sebastine 14343, 14344, 11089, 11090, 4181, 4182 (MH); J.L. Ellis 72740, 72741, 35592, 35593, 48181, 48182 (MH); G.V. Subbarao 72130, 72131 (MH); V.S. Ramachandran 123521, 123522, 125403, 125404 (MH); S.R. Srinivasan 114401, 114402, 109386, 109387 (MH); R.A. Beddome 45878 (MH); B.D. Naithani 44859, 44860 (MH). Shanavas Khan 25456 (CALI), Vajravelu 60403 (CALI); MCT 358, 359.

Host plants: *Schrebera swietenioides* Roxb. (Oleaceae), *Terminalia tomentosa* Wt. & Arn. (Combretaceae), *Pterocarpus marsupium* Roxb. (Papilionaceae).

Note: Very distinct with its characteristic quadrangular internodes and hence it is delimited from *V. articulatum* and its varieties.

Viscum bandipurense Thriveni *et al.*, in J. Syst. Evol. 51 (2): 224. 2013.

Small semiparasitic shrub, growing as hyperparasite on *Dendrophthoe trigona* which in turn parasitic on *Ficus benghalensis.* Haustoria endophytic. Stem branched, usually di- or -trichotomous, glabrous, nodes swollen, internodes elongated, terete. Leaves opposite-decussate, petiolate, up to 0.5cm; simple, entire, spathulate (resembling a table tennis racket), 3–4.4 cm x 2–3.5 cm; obtuse, attenuate at base; glabrous, 3–5 nerved. The mature leaves show a shallow depression in the middle of

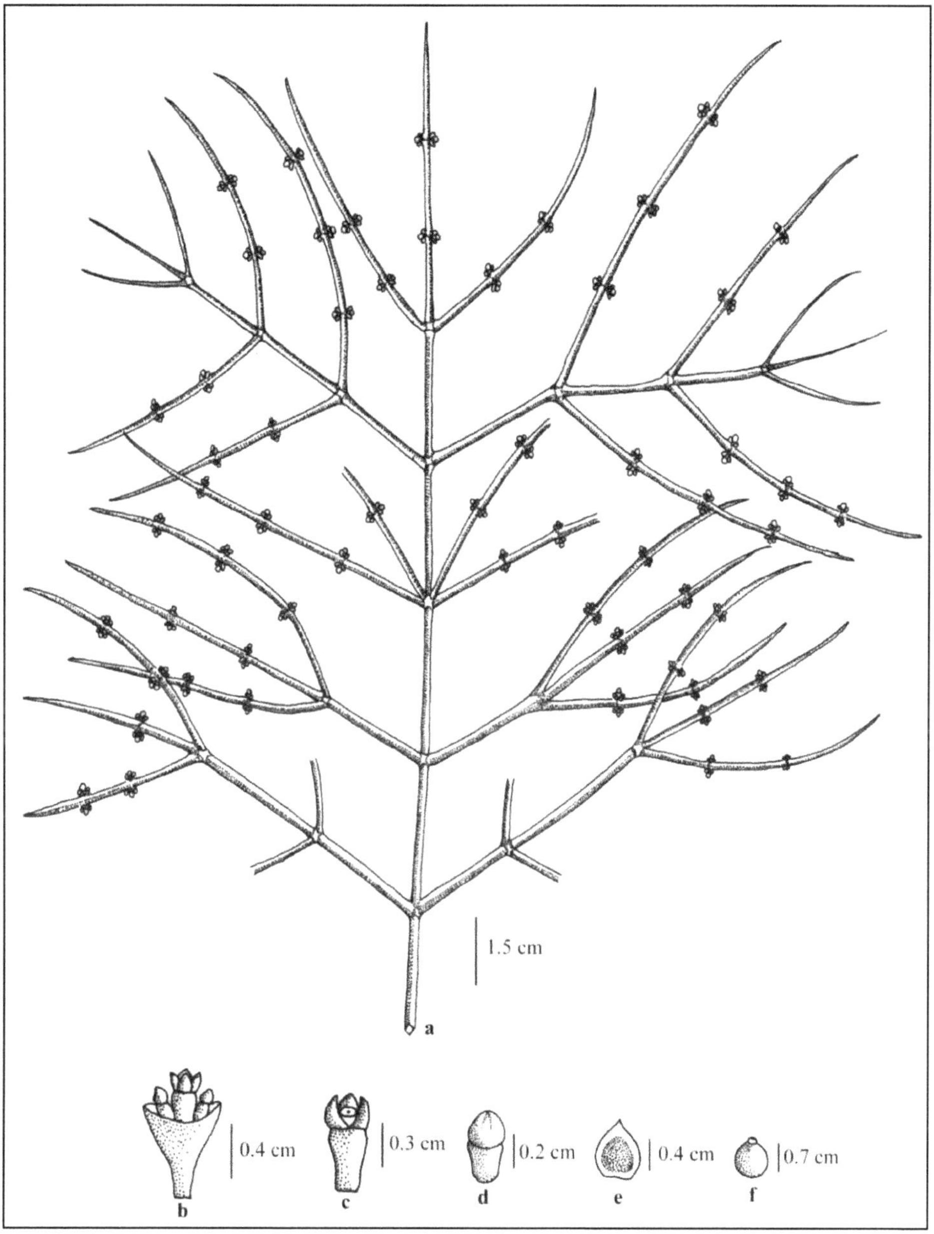

Figure 18.16: *Viscum angulatum* Heyne ex DC.

a: Flowering twig; b: Inflorescence; c: Female flower; d: Male flower; e: Male perianth lobe with anther; f: Fruit.

Plate 18.15: *Viscum angulatum* Heyne ex DC.

A: Flowering branches; B: Inflorescence (quadrangular stem-arrow); C: Young fruits; D: Mature fruit.

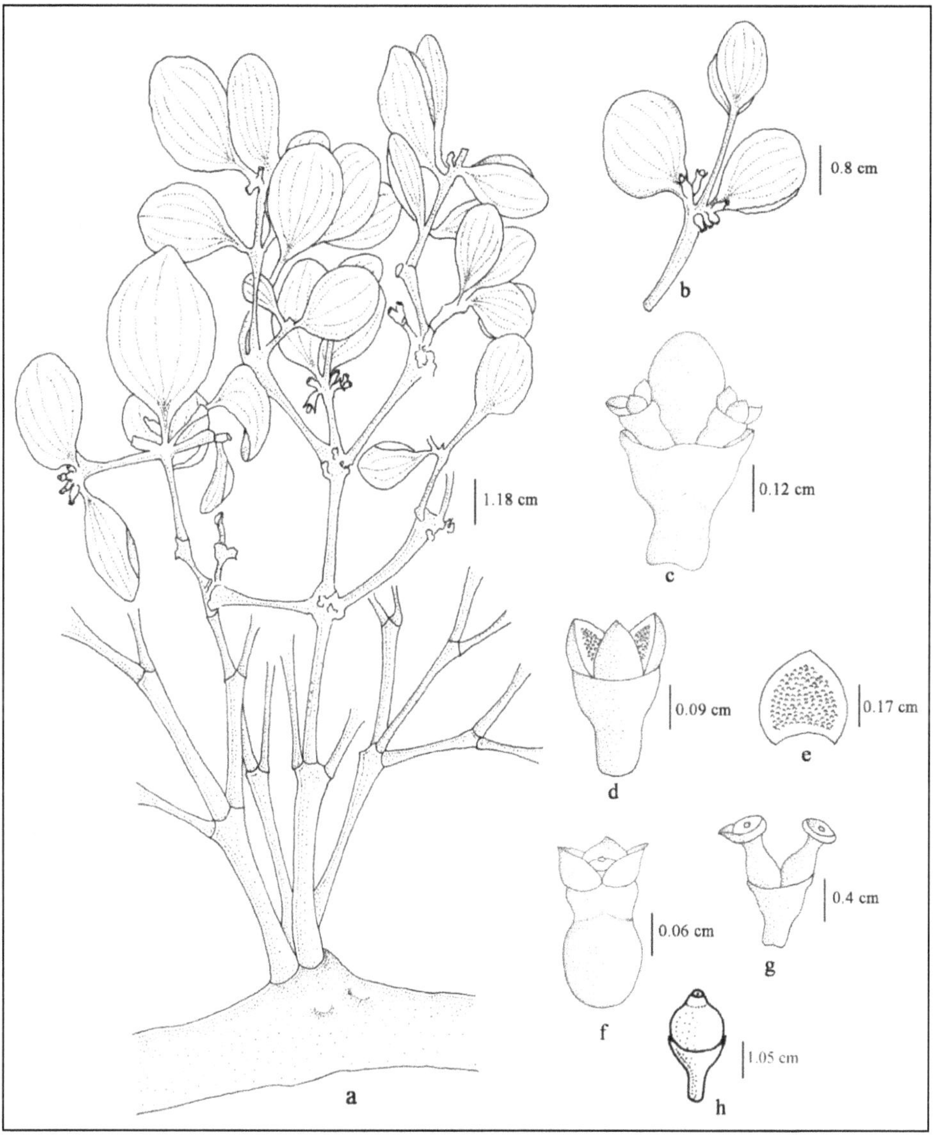

Figure 18.17: *Viscum bandipurense* Thriveni *et al.*

a: Habit; b: Twig; c: Inflorescence; d: Male flower; e: Male perianth lobe with pollen sac; f: Female flower; g: Young fruits; h: Mature fruit.

Plate 18.16: *Viscum bandipurense* Thriveni *et al.*

A: Habit; B: Young shoots on *Dendrophthoe trigona* (arrow); C: Young inflorescence; D: Close up of flowers; E: Young fruit; F: Mature fruit.

the lamina. The leaves fleshy when young and become somewhat thick (2mm) and hard at maturity. Inflorescence axillary cyme, up to 0.5–1cm long. Flowers small, pedicellate, fascicled, unisexual, monoecious; actinomorphic and enclosed in a boat shaped involucre, up to 0.5 cm; usually in triads in the axils of leaves; central flower of the triad (male) and lateral ones (female). Many a time male flowers do not develop, in which case inflorescence bears only two female flowers. Male flower present in the centre of the triad (sometimes absent), up to 0.5 cm; actinomorphic. Tepals 3 or 4, green, triangular, valvate, up to 0.3 cm. The male flowers mature earlier than female flowers. Stamens 3 or 4, opposite to the tepals, without filament, adnate to the perianth lobes; pollen sacs several of various shapes, dehisce introrsely. Pollen grains spherical, 42 µm; exine smooth, tricolporate. Female flowers usually 2 in a triad, one on either side of the central male flower, 0.5 cm. Tepals 3, green, triangular, very small (obscure), (transform into a rim like structure around the sub-terminal region of the stigma), valvate, caducous. Ovary inferior, but becomes apparently superior at maturity since the tepals fall off early leaving no trace of their presence. Style absent; stigma very small, dot like structure due to the failure of pollination and fertilization. Only a few female flowers develop into fruit where stigma appears capitate. Fruit a berry, greenish when young, white at maturity, globose, up to 1 cm, with thick viscous pulp.

Fl.: August-September

Distr.: Bandipur National Park, Bandipur and also found in Nagarhole and Madikeri.

MCT 181 (CAL); 182, 187, 350, 355 (MGB).

Host plants: Hyperparasite on *Dendrophthoe trigona* which is parasitic on *Ficus benghalensis* L., *Pterocarpus marsupium* Roxb., *Ficus racemosa* L.

Viscum capitellatum Smith in Rees, Cycl. 37: 18. 1817; Hook. f., Fl. Brit. India 5: 225. 1886; Talb., For. Fl. Bombay 2: 421. 1911; Gamble, Fl. Pres. Madras 2: 1258. 1925; Fischer, Rec. Bot. Surv. Ind. 11: 181. 1926; Danser, Blumea 4 (2): 309. 1941; Razi, Lloydia 20 (4): 245. 1957; R.S. Rao, J. Ind. Bot. Soc. 36 (2): 166. 1957; Cooke, Fl. Pres. Bombay 3: 47. 1906; Rao & Razi, A Syn. Fl. Mysore 462. 1981; Sald., Fl. Karnataka 2: 82. 1996; Sanj. & Bala., Rheedea 16 (2): 92. 2006.

Small stout semi parasitic shrubs, usually as hyperparasite on *Dendrophthoe falcata*, with endophytic haustoria. Stem branched, older branches whorled at nodes, nodes swollen, internodes short, rounded, terete, glabrous. Leaves opposite, sessile, conduplicately folded, entire, obovate, up to 2.5 cm long, obtuse at apex, decurrent at base, glabrous. Inflorescence axillary cyme, usually in triads at nodes. Flowers small, sessile, unisexual, monoecious, actinomorphic, epigynous, central flower of the triad is male and lateral ones female. Male flower sessile, present in the centre of the triad, up to 0.4 cm, actinomorphic, epigynous, perianth lobes 3, green, triangular, valvate, up to 0.3 cm. Stamens 3, opposite and adnate to the perianth lobes, filaments absent, pollen sacs many, dehisce introrsely. Female flower sessile, usually 2 in a triad one on either side of central male flower, up to 0.4 cm long, actinomorphic, perianth lobes 3, green, triangular, valvate up to 0.2 cm long. Ovary inferior. Style short. Stigma cone shaped. Fruit globose, green when young, white at maturity, up to 0.5 cm.

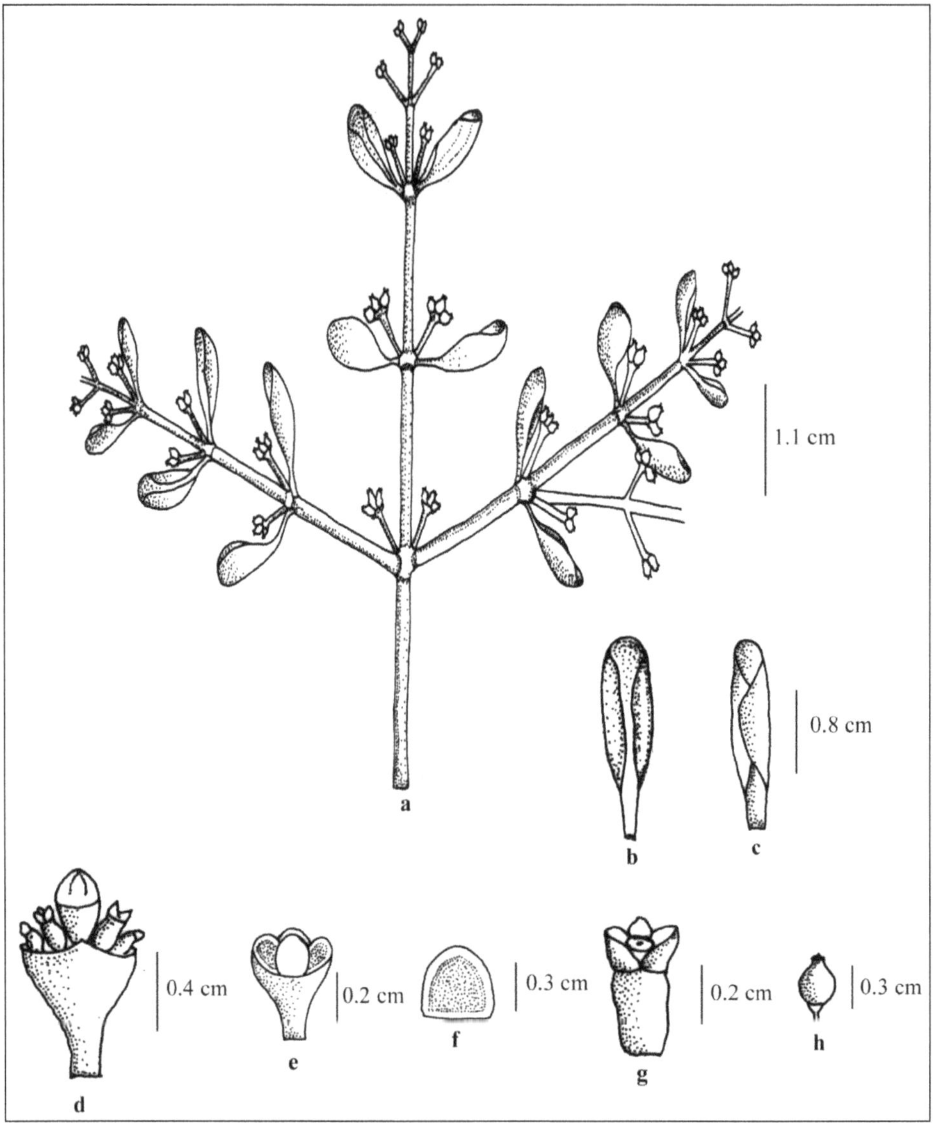

Figure 18.18: ***Viscum capitellatum*** **Sin.**

a: Flowering twig; b and c: Conduplicately folded leaf; d: Inflorescence; e: Male flower; f: Male perianth lobe with anthers; g: Female flower; h: Fruit.

Plate 18.17: *Viscum capitellatum* Sm.

A: Branches; B: Flowering twig (inset male flowers); C: Branch with young fruits; D: Mature fruits.

Fl.: September to October.

Distr.: Shimoga, B.R. Hills, Bandipur, Dashina Kannada, Dharwad, Kodagu, Chitrdurga.

Specimens examined: E. Vajravelu 75940, 75941, 45333, 45334, 87081, 87082, 53256, 53257 (MH); S. Raghupathy 147482, 147844 (MH); J.L. Ellis 23271, 23272 (MH). Shanavas Khan25448 (CALI); Sivarajan 1115 (CALI). MCT 322.

Host plants: *Azadirachta indica* A. Juss. (Meliaceae), *Terminalia paniculata* Roxb. (Combretaceae), *Terminalia tomentosa* Wt. & Arn. (Combretaceae), *Mitragyna parvifolia* Korth. (Rubiaceae).

Note: Found exclusively as hyperparasite on *Dendrophthoe falcata* (Loranthaceae) which was parasitic on *Nyctanthes arbor-tristis*, *Azadirachta indica* and *Syzigium cumini*. Found also growing on *Helixanthera intermedia* (Loranthaceae) which in turn parasitic on *Terminalia paniculata*.

Viscum mysorense Gamble in Bull. Misc. Inform. 329. 1925; Gamble, Fl. Pres. Madras 2: 1259. 1925; Danser, Blumea 4 (2): 279. 1941; R.S. Rao, J. Indian Bot. Soc. 36 (2): 125. 1957; Sald. & Nicol., Fl. Hassan Dist. 311. 1976; Sald., Fl. Karnataka 2: 83. 1996; Manju. *et al.*, Fl. Davanagere Dist. 344. 2004; Sanj. & N.P.Balakr., Rheedea 16 (2): 99. 2006.

All parts with a golden-yellow hue. Only stem available, slender, over 50 cm long, at nearly all the nodes di- or tri-chotomous, its basal portion terete, 5-6 cm long, up to 3 mm in diameter, longitudinally wrinkled, hardly striped, slightly thickened at the nodes, the young internodes usually 2.5-4 cm long, distinctly longitudinally striped with shallow grooves, nearly terete or slightly flattened near the base, 1.5–2 mm broad. Leaves normally developed only on a part of the nodes, the largest obtusely lanceolate to spathulate, up to 4 cm long by 10 mm broad, often smaller, rounded beneath at apex, tapering into a short petiole that is rounded beneath, flat or slightly canaliculated above, the lamina rather thick-coriaceous, with 3 longitudinal nerves, that are some what more distinct above than beneath and connected by indistinct veins. Leaves scale-like on most of the nodes, nearly 0.5 mm long, acute; also 2 scales (prophylls) at the base of all ramifications. Inflorescence rarely terminal, usually axillary or at both sides of the axillary ones, sessile or shortly pedunculate 1-3 flowered cymes; peduncle flattened, up to 1 mm long and broad, bearing at its apex 2 opposite acute bracts forming together a naviculate cup up to 2 mm long, each bearing one sessile flower in their axil devoid of a bracteal cup and usually male, rarely female, nearly 1 mm long and compressed between the bract and the middle flower; female (middle flower), rarely sessile and without bracteal cup, usually very shortly pedicellate and surrounded by a cup composed of two small bracts alternating with those of the lower pair. Fruit unknown.

(Description from the type specimen in the Kew Herbarium).

Fl.: Not known.

Distr.: Known only from the type locality Arasikere.

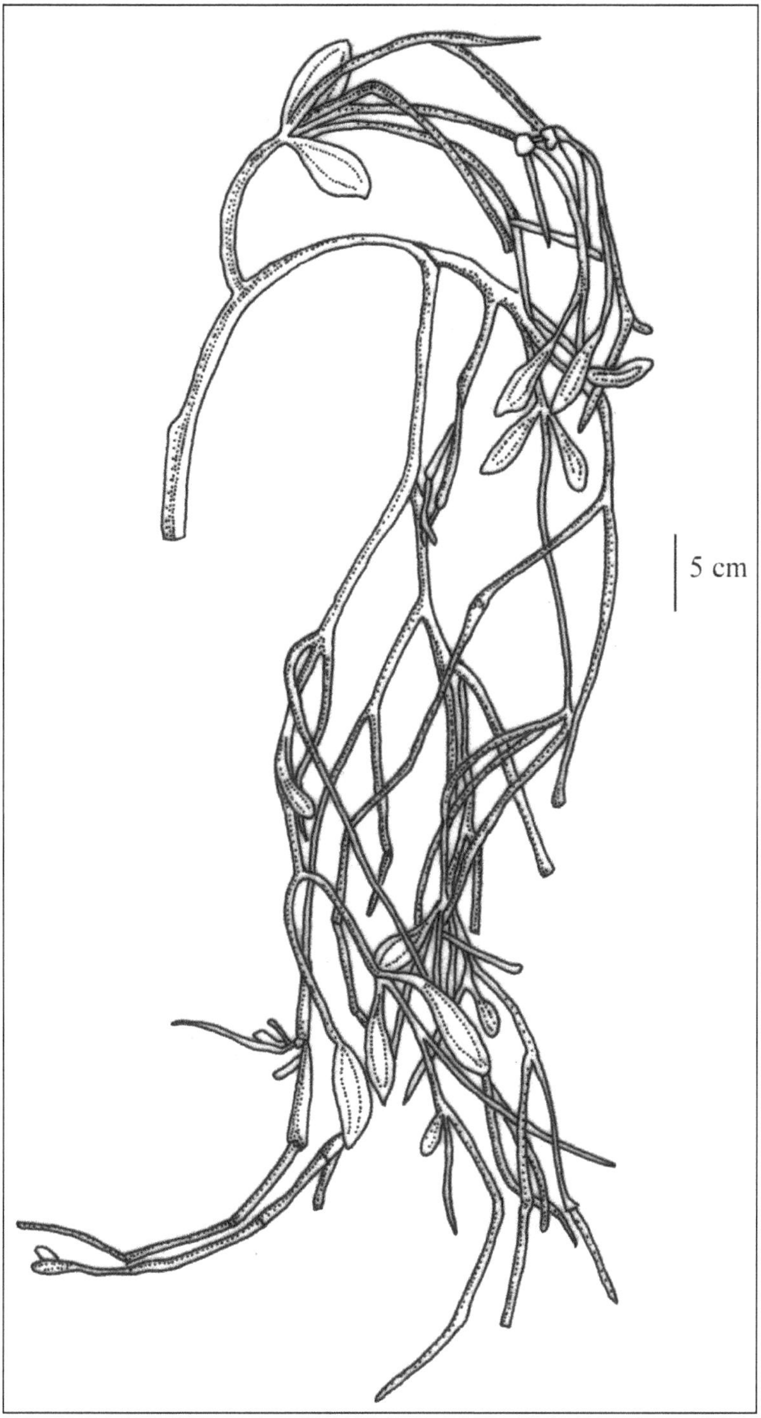

Figure 18.19: ***Viscum mysorense*** **Gamble,**
(drawn from the photo of type specimen cited in Danser, 1941).

Occurrence and specimens examined: Nobody has collected the specimen from the type locality after Meebold 1908.

Host plants: not mentioned in the original description.

Note: Danser (1941) opined that *V. mysorense* is a remarkable and very distinct species. According to the structure and distribution of the inflorescence, it belongs to the leafless form of *Viscum articulatum*, but the internodes are not flattened. Hence he said that it is intermediate between leafy and leafless forms of *Viscum wrayi* Gamble but it is entirely different from this species and its allies by the structure of its inflorescences. Even Seshagiri Rao (1957) and Sanjai and Balakrishnan (2006) has revised the Indian genus *Viscum* were also not able to collect the type specimen. There are only two specimens of which the type specimen is deposited at Kew herbarium (Kew) and another at Berlin-Dahlem. Except these two no another specimen is available.

In spite of several field visits to the type locality, the specimen could not be collected. No one has collected this species from the type locality after Meebold (1908). Although, Manjunatha (2002) mentioned in his thesis that he has collected the *V. mysorense* from Kallapura, but the description given in it are not of *V. mysorense*. Therefore this appears to be a part of misidentification.

Viscum nepalense Spreng., Syst. Veg. 4: 47. 1827; Danser, Blumea 4: 283. 1941; Sald. & Nicol. Fl. Hassan Dist. 311. 1976; Singh, Fl. East. Karnataka 553 (2): 1988; Sald., Fl. Karnataka 2: 83. 1996; *Viscum articulatum* Burm. f., Fl. Ind. 311. 1768; Hook. f., Fl. Brit. India 5: 226. 1886; Wiens in Dassan. & Fosberg, Rev. Handb. Fl. Ceylon 6: 134. 1987; Talb., For. Fl. Bombay 2: 423. 1911; Gamble, Fl. Pres. Madras 2: 1259. 1925; Fischer, Rec. Bot. Surv. Ind. 11: 181. 1926; Danser, Blumea 4 (2): 280. 1941; Razi, Lloydia 20 (4): 245. 1957; R.S. Rao, J. Ind. Bot. Soc. 36 (2): 126. 1957; Cooke, Fl. Pres. Bombay 2: 47. 1906; Rao & Razi, A Syn. Fl. Mysore 462. 1981. Sanj. & Bala., Rheedea 16 (2): 90. 2006.

Small semi parasitic shrub, evergreen, drooping, up to 1 m. Stem branched, usually dichotomous, sometimes more than 2 branches at each node, flattened, at right angle to the preceding one, gradually broadening from the base to tip, twisted near the base, longitudinally prominently grooved. Leaves absent. Inflorescence sessile. Flowers unisexual, monoecious, actinomorphic, usually in triads, central flower of the triad is female and lateral ones male. Male flower sessile, usually 2 in a triad on either side of the central female, up to 0.3 cm long, actinomorphic, epigynous, perianth lobes 3 or 4, green, triangular, valvate, up to 0.2 cm long. Stamens 3 or 4, opposite and adnate to the perianth lobes, filaments absent, pollen sacs many dehisce introrsely. Female flower sessile, present in the centre of the triad, up to 0.4 cm long, actinomorphic. Perianth lobes 3 or 4, green, triangular, valvate up to 0.2 cm long. Ovary inferior. Style absent. Stigma nipple shaped. Fruit globose, green when young, yellowish at maturity, up to 0.4 cm.

Fl.: September to October

Distr.: B.R. Hills, Bandipur, Shimoga, Chitradurga, Dharwad, Hassan, Belgaum.

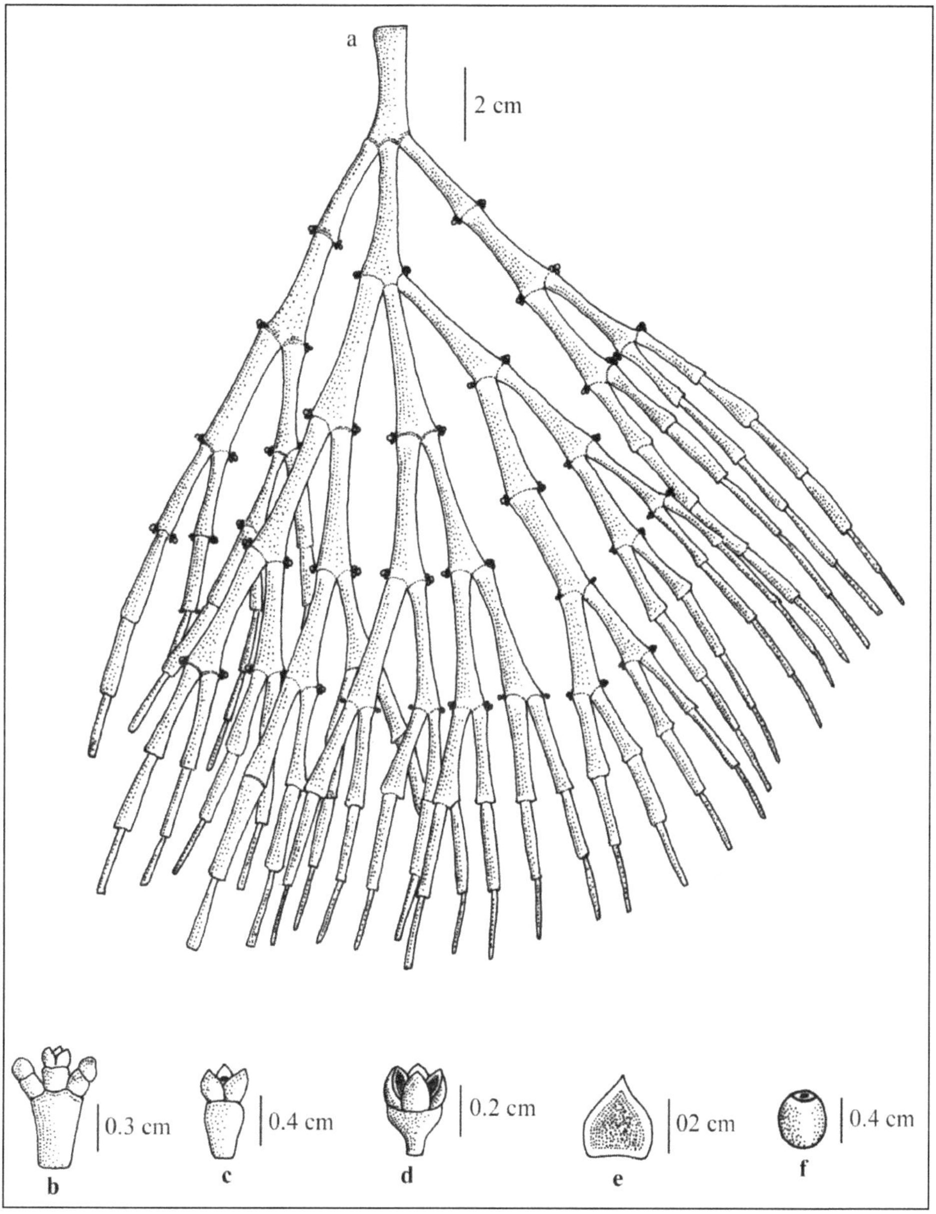

Figure 18.20: *Viscum nepalense* Spreng.

a: Flowering twig; b: Inflorescence; c: Female flower; d: Male flower; e: Male perianth lobe with stamen; f: Fruit.

Plate 18.18: ***Viscum nepalense*** **Spreng.**

A: Twig; B: Flowering twig (inset-male flower); C: Branches with young fruit; D: Mature fruit.

Specimens examined: J.S. Gamble 70840, 45923, 45918, 45919, 45920 (MH); J.D. Hooker 70834 (MH); G. King 70835 (MH); George Watt. 70811 (MH); J.D. Hooker and J.J. 70833, 70836, 70839 (MH); K.M. Sebastine 30552, 27220, 27221, 30551, 13198, 24741, 4647, 4648, 2297, 2298, 5315, 5316, 8475, 8476, 7209, 7210 (MH); J. Joseph 24421, 24422, 24392, 24932, 24937 (MH); K. Subrmanyam 14021, 14022, 13403, 13404, 13575, 13575, 9653, 9654 (MH); N.P. Balakrishnan and A.N. Henry 23361, 23362 (MH); G.V. Subbarao 38036, 38037, 52913, 47463, 47464 (MH); M. Mohanan 170367, 170368, 110208, 110209, 61532, 61533 (MH); A. Sreenivasa Rao 148789 (MH); D. Ranga Charyulu 148025 (MH); G.V. Subbarao 91496, 91497, 70124, 70125 (MH); T. Ravishankar 156061, 156062, 155918 (MH); M.R. Rajendra Prasad 148544, 148570 (MH); T. Pullaiah and N. Yeshoda 126898, 126899 (MH); K.C. Jacob 45904, 45905, 45906, 81196, 45908, 45909, 45910, (MH); J.L. Ellis 33466, 33467, 49635, 49636 (MH); P.V. Sreekumar and T. Ravishankar 15469, 154690 (MH); A.N. Henry 31413, 31414, 107081, 107082 (MH); K. Subramanyam 17601, 17602, 10839, 10840, 4753, 4754, 2557, 2558, 12689, 12690 (MH); R. Gopalan 144966, 144967 (MH); S.R. Sreenivasan 160007, 160008 (MH); E. Vajravelu 130690, 130691, 134997, 134998, 102967, 102968, 56034, 56035, 75810, 75811 (MH); M.V. Vishwanathan 96587, 96588, 45853, 45896, 45897 (MH); V. Narayanaswamy 78245 (MH); A.R. Ansari 101499 (MH); D.B. Deb 65765, 65766 (MH); S. Karthikeyan 56066, 56067 (MH); D.V. Shetty 54287, 54288 (MH); N.C. Nair 120151, 120152 (MH); J. Joseph 34646, 34677 (MH); B.D. Naithani 40476, 44682, 44683 (MH). N. Mohanan 8172 (CALI), Shanavas Khan7507 (CALI); MCT 132, 326, 337, 338.

Host plants: *Cassia fistula* L. (Caesalpiniaceae), *Cordia* sp. (Boraginaceae), *Dalbergia paniculata* Roxb. and *D. latifolia* W. (Papilionaceae), *Diospyros montana* Roxb. (Ebenaceae), *Dolichondrone atrovirens* Schum. (Bignoniaceae), *Ehretia laevis* Roxb. (Boraginaceae), *Erythroxylon monogynum* Roxb. (Ehythroxylaceae), *Gmelina arborea* Roxb. (Verbenaceae), *Grewia orientalis* L. and *G. tiliifolia* Vahl (Tiliaceae), *Mitragyna parvifolia* Korth. (Rubiaceae), *Catunaregum spinosa* (Thunb.) Tirv. (Rubiaceae), *Rhus mysorensis* Heyne (Anacardiaceae), *Terminalia paniculata* Roxb. and *T. tomentosa* Wt. & Arn. (Combretaceae), *Zizyphus oenoplia* Lam. (Rhamnaceae).

Viscum orientale Willd., Sp. Pl. 4: 737. 1806; Hook. f., Fl. Brit. India 5: 224. 1886; Gamble, Fl. Pres. Madras 2: 1258. 1925; Talb. For. Fl. Bombay 2: 420. 1909; Fischer, Rec. Bot. Surv. Ind. 11: 180. 1926; Danser, Blumea 4 (2): 299. 1941; Razi, Lloydia 20 (4): 246. 1957; R.S. Rao, J. Ind. Bot. Soc. 36 (2): 154. 1957; Cooke, Fl. Pres. Bombay 3: 46. 1906; Rao & Razi, A Syn. Fl. Mysore 462. 1981; Keshava. & Yoga., Fl. Coorg, 392. 1990; Sald., Fl. Karnataka 2: 85. 1996; Sanj. & Bala., Rheedea 16 (2): 100. 2006.

Small semi parasitic shrub, with endophytic haustoria and bushy in appearance. Stem branched usually dichotomous, cylindrical, nodes swollen, internodes terete. Leaves opposite, subsessile, simple, entire, obovate, 4-6 cm x 1.5-2.2 cm, 3-nerved, obtuse at apex, tapering into a short petiole at base, glabrous. Inflorescence axillary cyme, borne at nodes. Flowers small, sessile, unisexual, monoecious, actinomorphic, epigynous, usually in triads rarely up to 5 flowers, surrounded by a boat shaped bracteal cup. The central flower of the triad is female, larger than male and laterals small. Male flower short, compressed, up to 0.3 cm, actinomorphic perianth lobes 4,

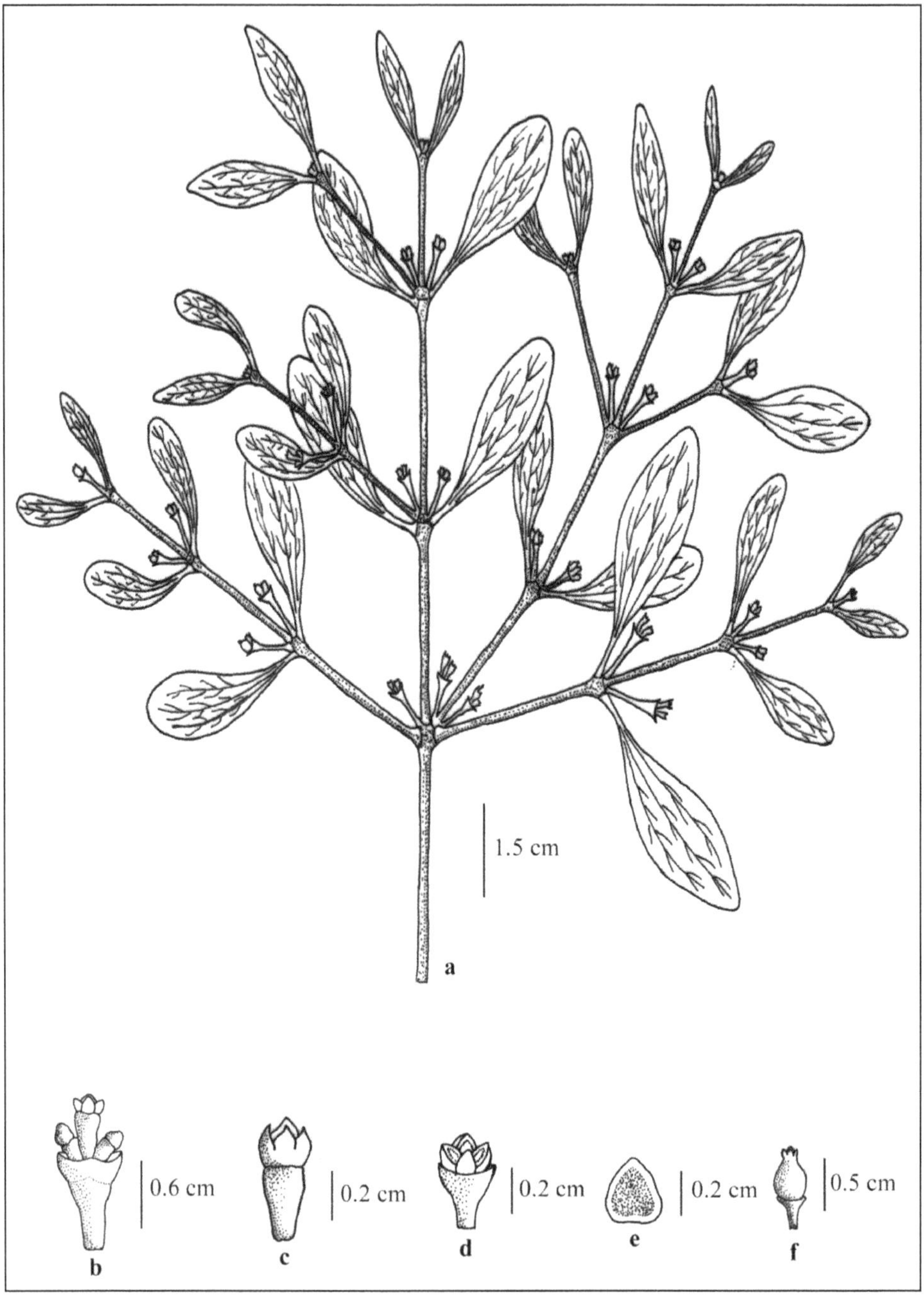

Figure 18.21: *Viscum orientale* Willd.

a: Flowering twig; b: Inflorescence; c: Female flower; d: Male flower; e: Male perianth lobe with anther; f: Fruit.

Plate 18.19: ***Viscum orientale*** **Willd.**

A: Flowering branches; B: Inflorescence;
C: Close up of male (m) and female (f) flowers; D: Mature fruits.

green, triangular, valvate up to 0.2 cm long. Stamens 4, opposite and adnate to the perianth lobes, filaments absent, pollen sacs many, dehisce introrsely. Female flower sessile usually two in a triad, one on either side of the central male flower up to 0.4 cm long, actinomorphic, epigynous, perianth lobes 4, green, triangular, valvate, up to 0.2 cm long. Ovary inferior. Style absent, stigma capitate. Fruit ovoid, green, up to 0.6 cm long.

Fl.: March to April

Distr.: found in all the districts.

Specimens examined: K. Subramanyam 45837, 13949, 13950, 11953, 912, 10423, 11183, 11184 (MH); V. Narayanaswamy 82817, 78208 (MH); K.M. Sebastine 28867, 28868, 45813, 45860, 45828, 15649, 15650, 13257, 25785, 25786 (MH); K. Ramamurthy 145128, 138867 (MH); M.V. Vishwanathan 96836, 96837 (MH); M. Mohanan 170369, 170370, 45830, 45831, 45796, 45835 (MH); J.L. Ellis 32328, 32329 (MH); J.S. Gamble 45825, 45826, 45832, 45838, 45839, 1887 (MH); E. Vjaravelu 130550, 130551 (MH); T. Ravishankar 155943, 155944 (MH); G.V. Subbarao 52799, 47417, 37349 (MH). Shanavas Khan7220, 23380, 11012 (CALI); MCT 315, 316.

Host plants: *Albizia amara* Boiv. (Mimosaceae), *Flacourtia indica* (Burm.) Merr. (Flacourtiaceae), *Holoptela integrifolia* Pl. (Ulmaceae), *Pongamia pinnata* (L.) Pierre (Papilionaceae), *Punica granatum* Vent. (Punicaceae), *Dalbergia paniculata* Roxb. (Papilionaceae), *Macaranga peltata* M. Arg. (Euphorbiaceae).

Note: Very common on *Pongamia pinnata.*

Viscum ramosissimum Roxb. ex DC. Prodr. 4: 278. 1830; Hook. f., Fl. Brit. India 5: 225. 1886; Gamble, Fl. Pres. Madras 2: 1258. 1925; Fischer, Rec. Bot. Surv. Ind. 11: 181. 1926; Danser, Blumea 4 (2): 294. 1941; Razi, Lloydia 20 (4): 246. 1957; R.S. Rao, J. Indian Bot. Soc. 36 (2): 142. 1957; Rao and Razi, A Syn. Fl. Mysore 462. 1981; Keshava. & Yoga., Fl. Coorg, 392. 1990; Sanj. & Bala., Rheedea 16 (2): 104. 2006.

Large semi parasitic shrubs, pendulous, very rarely erect on the branches as well as on main trunks of the host. Stem slender, cylindrical or rounded, branches numerous, nodes swollen, internodes terete, the basal internodes are longer than the upper ones. Leaves absent. Inflorescence sessile, axillary cymose at the nodes, rarely terminal. Flowers small, sessile, unisexual, monoecious, actinomorphic, epigynous, usually in triads, central female and laterals male. Male flowers small, compressed up to 3 mm long, actinomorphic, perianth lobes 4, free, ovate, valvate, up to 2 mm long. Stamens 4, opposite and adnate to the perianth lobes, filaments absent, pollen sacs many, dehisce introrsely, pistillode absent. Female flower oblong, larger than male, present in the center of the triad, 0.4 cm long, actinomorphic, epigynous, perianth lobes 4 free, conical, staminodes absent. Ovary inferior. Style absent. Stigma capitate. Fruits small, globose, up to 0.5 cm long.

Fl.: July to August.

Distr.: Bandipur, B.R. hills, Hassan, Chikmagalur, Shimoga, Muthtathi, Mandya, Gundlupet.

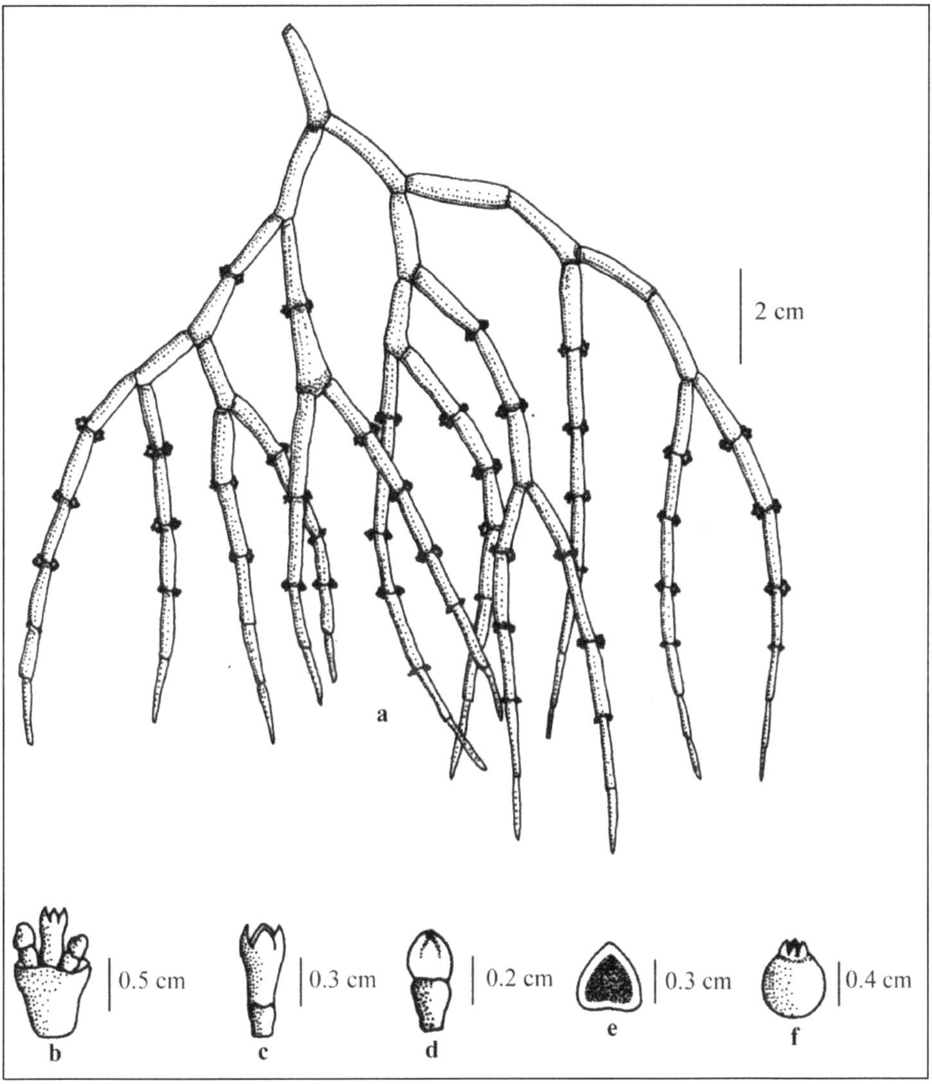

Figure 18.22: *Viscum ramosissimum* Roxb. ex DC.

a: Flowering twig; b: Inflorescence; c: Female flower; d: Male flower; e: Male perianth lobe with anther; f: Fruit.

Plate 18.20: ***Viscum ramosissimum*** **Roxb. ex DC.**

A: Robust branch; B: Twig with inflorescence; C: Close up of flowers; D: Fruits.

Specimens examined: K. Ramamurthy 27818, 27819 (MH); K.M. Sebastine 5317, 5318, 4799, 4800 (MH); M.V. Vishwanathan 96748, 96749 (MH); K.C. Jacob 81272, 31194, 31195 (MH); K. Subramanayam 10511, 10512 (MH); A. Henry 100198, 100199, 11957, 11958 (MH); B.V. Shetty 63760, 63761, 54383, 54384 (MH); B.D. Sharma 78027, 78028 (MH); E. Vajravelu 134921, 134922, 43160, 58806, 45871, 45872, 45873, 45875, 45876, 45877, 45885 (MH). Shanavas Khan7224, 7558 (CALI). Vajravelu 27753 (CALI). MCT 57, 63, 68, 78.

Host plants: *Anogeissus latifolia* Wall. (Combretaceae), *Bauhinia racemosa* Lam. (Caesalpiniaceae), *Erythroxylon monogynum* Roxb. (Erythroxylaceae), *Ficus benghalensis* L. (Moraceae), *Grewia orientalis* L. (Tiliaceae), *Holoptelea integrifolia* Pl. (Ulmaceae), *Phyllanthus indofischeri* (Euphorbiaceae), *Schrebera swietenioides* Roxb. (Oleaceae), *Ziziphus jujuba* Lam. (Rhamnaceae), *Ziziphus oenoplea* Lam. (Rhamnaceae).

Note: *Viscum ramosissimum* which is parasitic on *Ficus* spp. is very robust on the tree trunks, usually pendulous in nature. It resembles *V. angulatum* Heyne ex DC. in the structure of inflorescence and fruit. But differs in the shape of internodes which are terete and sometimes longitudinally grooved where as in case of *V. angulatum* Heyne ex DC. the internodes are quadrangular in nature. During the collection we have observed two morphologically varied specimens one with slender and the other with thick, terete branchlets.

Much controversy has been centered around the distinction of *V. articulatum* Bl. It mainly differs from *V. angulatum* Heyne ex DC. and *V. ramosissimum* by its decussately flattened internodes. Danser (1941) points out that *V. articulatum* Bl. and *V. nepalense* Spreng. share similar characters except that the former is usually with strongly flattened internodes and small, globose, white or light yellowish fruits and is usually parasitic on Loranthceae. But other leafless forms of British-Indian *Viscum* spp. are sometimes having distinctly broader and larger and dark coloured fruits. These fruits are more oblong. They rarely grow on Loranthaceae members. Gamble distinguished *V. nepalense* Spreng. from *V. articulatum* Bl. based on the breadth of internodes. He (Gamble) merges all the specimens under *V. nepalense* Spreng. while doing so he mentions much more southern localities.

Later Danser distinguishes *V. nepalense* from *V. liquidambaricolum* based on the shape of the fruit which is globose in *V. nepalense* and more oblong in *V. liquidambaricolum*. To this adds that *V. nepalense* has usually broader internodes, but this difference is not depended on. Therefore, it is very doubtful that whether *V. nepalense* and *V. liquidambaricolum* are distinct are not.

Viscum trilobatum Talb., For. Fl. Bombay Pres. & Sind. 2: 419. 1911; Danser, Blumea 4 (2): 308. 1941; R.S. Rao, J. Ind. Bot. Soc. 36 (2): 163. 1957; Sanj. and Bala., Rheedea 16 (2): 106. 2006.

Small semi parasitic shrub growing as hyperparasite on *Macrosolen capitellatus* with endophytic haustoria. Stem branched, glabrous, nodes swollen, internodes terete. Leaves opposite, sessile, simple, unequal sides, obovate-suborbicular, 2-3.5 cm x 1.5-2.8 cm, triangular at apex, obtuse at base. Inflorescence short peduncled axillary cyme at nodes, rarely terminal, peduncle bearing a boat shaped bracteal cup enclosing 3 flowers. Flowers small, sessile, unisexual, monoecious, actinomorphic, epigynous.

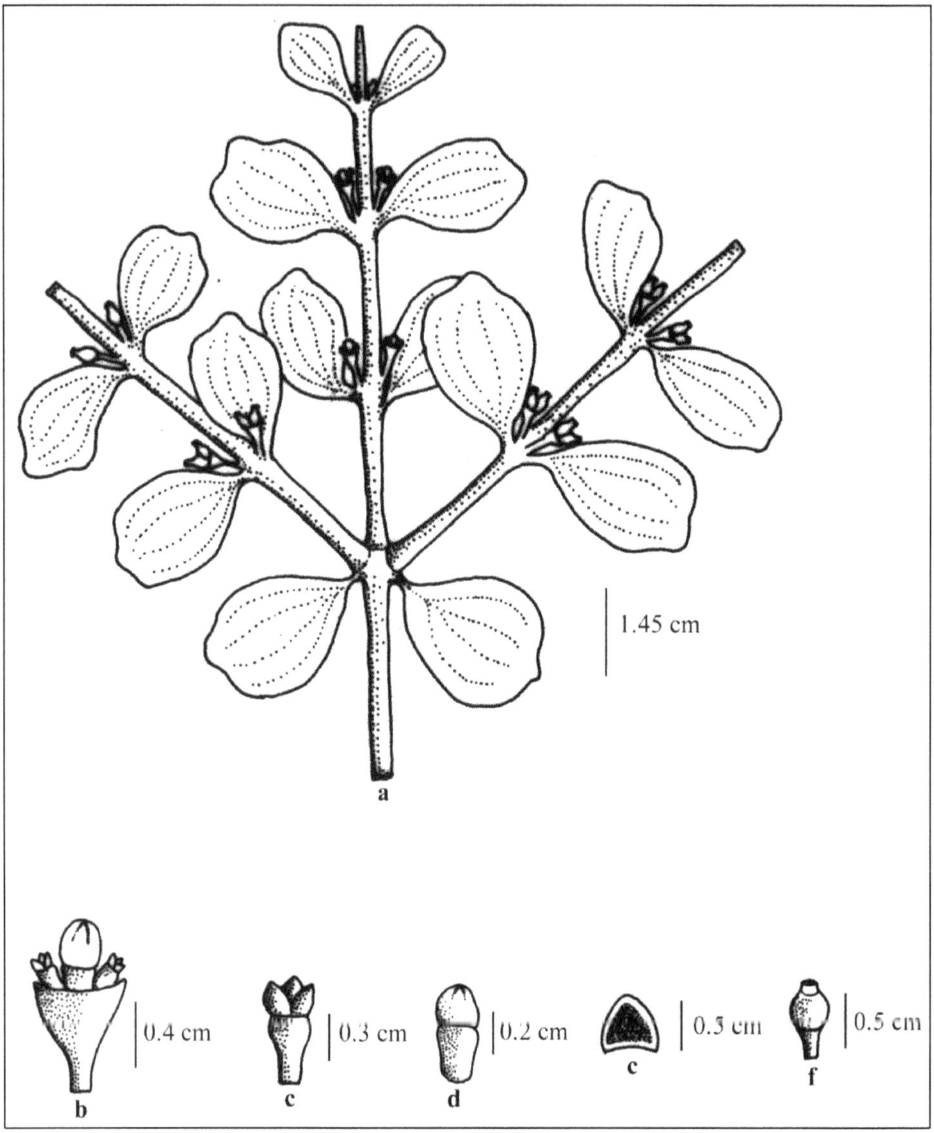

Figure 18.23: ***Viscum trilobatum*** **Talb.**

a: Flowering twig; b: Inflorescence; c: Female flower; d: Male flower; e: Male perianth lobe with anther; f: Fruit.

Plate 18.21: *Viscum trilobatum* Talb.

A: Branches in flowering; B: Trilobate nature of leaf; C: Young fruits; D: Mature fruit.

Central flowers of the triad are male and laterals female. Male flower sessile present in the center of the triad, up to 0.3 cm long, actinomorphic, epigynous, perianth lobes 3, free, valvate, green, triangular. Stamens 3, opposite and adnate to the perianth lobe, pistillode absent. Female flower sessile, present on either side of the central male, up to 0.4 cm long, actinomorphic, epigynous, perianth lobes 4, free, valvate, green, triangular, staminodes absent. Ovary inferior. Style absent, stigma capitate. Fruit roundish or ovoid, up to 0.6 cm long.

Fl.: February to March.

Distr.: *Belgaum*, Dakshinakannada, Chikmagalur.

Specimens examined: B.D. Sharma 77765, 77766, 68223 (MH); B.V. Shetty 23501, 23502, 68222 (MH). Shanavas Khan 23373, 37628, 11048 (CALI). MCT 92, 261.

Host plants: *Macrosolen capitellatus* (Wt. & Arn.) Dans. (Loranthaceae).

Note: Recorded always as an hyperparasite on *Macrosolen capitellatus* which in turn parasitic on *Mangifera indica* and *Semecarpus anacardium*. Several authors have considered this taxon as a variety of *V. capitellatum* except Talbot and Danser. No doubt that they are closely related but when we compared the Talbot's specimens in the herbarium with that of *V. capitellatum* it is clear that both are different in leaf character. The leaves of *V. trilobatum* distinctly broader with slightly triangular apex where as in *V. capitellatum*, the leaves are conduplicately folded or even sometimes rolled upwards. From the above observation it is evident that *V. trilobatum* is clearly distinct from *V. capitellatum*.

Many authors have commented that the name 'trilobate' was not appropriate to this species. But much importance has been given in the choice of specific name 'trilobate' for Talbot's specimen.

Distribution of Mistletoes

Impact of Mistletoes on their Hosts

The mistletoes are usually found in aggregates on their host branches. Some of the host plants with mistletoe infection show branch hypertrophy and tissue malformation due to the disruption of host tissues at the point of infection (Plate 18.22 A). Most mistletoe species are morphologically diverse which produce relatively large shoots with large, brightly coloured flowers. The host trees with mistletoe infection have stunted growth in terms of nutrient loss. Since mistletoes dependent on their host for resource acquisition, they can exert deleterious effects on the growth and development of host plants.

Branch dieback is another common symptom of mistletoe infection. Over a period of time, the host branch distal to mistletoe infection dries up and dies; where as the host branch segment proximal to the trunk remains alive and continues to supply water and nutrients to the parasite. This was very clearly observed on Neem trees which were infected with *Dendrophthoe falcata* throughout the State.

During the field collections in isolated and dense forest regions it was noticed that several host species are highly infested with the mistletoes to such an extent that

Plate 18.22: Impact of Mistletoes on their Hosts.

A: Branch hypertrophy at the point of infection; B: An entire tree of *Holoptelia integrifolia* is parasitized by *Scurrula parasitical*; C: *Viscum angulatum* on *Schrebera swietenioides,* D: *Dendrophthoe trigona* on *Ficus religiosa* (note the defoliated hosts); E. *Viscum nepalense* masking the host tree; F: Goats feeding on *Dendrophthoe falcata.*

the host tree could not be identified due to the clumps (Plate 18.22 E) of parasite masking the entire tree without leaving even a single host branch/leaf (Plates 18.22 B-D). In such instances, it was difficult to distinguish the host plant from the parasite. It is also observed that many trees without any parasitic infestation although the neighbouring trees are highly infested appear to be an amazing fact during the collections. Sometimes two host trees of the same species growing close to each other of which one with more parasitic infection and the other without any has also been recorded. This appears that the one without any parasitic infestation may indicate a kind of resistance.

Host Range

Mistletoes usually attack a large number of diverse plants. In the present investigation, it is observed that mistletoes can infect a large number of taxonomically diverse hosts. All the recorded host plants were dicotyledonous angiosperms, without even a single taxon from the monocotyledons. This may be because in dicotyledons, secondary growth ensures continuous supply of nutrients for mistletoes, while the monocotyledons do not have cambium and secondary growth. This is one of the important reasons for the absence of monocotyledons as hosts. No host species were herbaceous (as expected for woody aerial parasites), although many a times, the host genus include some herbaceous species which are not infected by the Parasitic Angiosperms. Morever, the herbaceous taxa are annuals or biennials and there is no guarantee for the continuous supply of nutrients and hence mistletoes are not found on herbaceous plants. Most of our fruit, horticultural and timber yielding trees are severely infested with Loranthaceous mistletoes. The adaptability of mistletoes has made them to victimize a large number of host plant species, all of which are dicotyledons. Because of their wide host range, mistletoes are considered as host generalists (Thriveni *et al.*, 2010).

Table 18.2: List of New Host Plants Recorded for *Dendrophthoe falcata*.

Sl.No.	*Name of the Host Plant*	*Family Name*
1.	*Acacia decurrens* Willd.	Mimosaceae
2.	*Albizia amara* (Roxb.) Boiv.	Mimosaceae
3.	*Aphanamixis polystachya* (Wall.) Parker	Meliaceae
4.	*Delonix regia* Raf.	Caesalpiniaceae
5.	*Muntingia calabura* L.	Tiliaceae
6.	*Persea macrantha* (Nees) Kosterm.	Lauraceae
7.	*Canthium dicoccum* (Gaertn.) Teism & Binn.	Rubiaceae
8.	*Radermachera xylocarpa* K. Schum.	Bignoniaceae
9.	*Tabebuia argentea* Don.	Bignoniaceae
10.	*Ziziphus jujuba* Lamk.	Rhamnaceae

Among the mistletoes one of the best examples for wide host range is *Dendrophthoe falcata* which has attained a very important position amongst phanerogamic parasites because of the damage and heavy loss caused to various medicinal, ornamental, horticultural, timber yielding and other forest trees. It attacks a great number of native as well as exotic trees. Many botanists have attempted to record the host range for *Dendrophthoe falcata*. We have recorded 10 new host species for *Dendrophthoe falcata* for the first time (Thriveni and Shivamurthy, 2010). It is observed that this parasite usually attacks the monoculture plantations like Teak, Rose wood, Nandi, Honne, Mathi, *Eucalyptus*, Rubber, Silver oak and horticultural trees like Mango, Orange, Sapota, Pomegranate, Guava and Custard apple. Avenue trees and some of the exotic trees are also affected by *Dendrophthoe falcata*.

Host Preference

Despite the wide host range, many mistletoes show high levels of host preference. Some hosts appear to be preferred simply due to their abundance which is more frequently encountered by the parasite. Our observation revealed that *Viscum orientale* (Viscaceae) has been recorded mainly on *Pongamia pinnata*. This parasite has been exclusively recorded in each and every locality visited. This doesn't indicate the host specificity since in some instances it has also been recorded on other hosts. Like wise *Helicanthes elastica* grows mainly on Mango trees in Uttara Kannada, Udupi, Dakshina Kannada and Belgaum Districts. We have also seen that different parasitic species grow on different hosts in different localities. The host preference also differs from place to place between the same species. In one locality, a species of mistletoe will show preference to certain hosts where as in another, the preference may change to an entirely new group.

It was also noticed that *Macrosolen parasiticus* grows only on *Mangifera indica* in Southern parts of Karnataka. However, in Dharwad, Belgaum and Shimoga Districts, *Mangifera indica* was infested by another species *i.e., M. capitellatus*. There are no instances of *Macrosolen parasiticus* on *Mangifera indica* in northern districts of Karnataka. In Biligiri Rangana Hills (B.R. hills) and also in Bandipur National Park almost all the *Phyllanthus emblica* trees which are locally known as Amla (bettada nellikayi mara) are severely infested with *Taxillus tomentosus* (Loranthaceae). The *Phyllanthus* trees have become the source of commercial income for the local people through their highly medicinal fruits. The infestation of this parasite has increased the host tree mortality. In Manasagangotri campus of Mysore University more than 200 Mango trees are infested with *Macrosolen parasiticus*. For the last several years the fruit yield of Mango has significantly reduced because of the mistletoe infestation.

Neem tree has enjoyed a special status in traditional system of Indian medicine which is rightly called as the wonder plant by scientists because of its healing versatility. It has been used in Ayurvedic medicine to treat various skin diseases. Neem wood is termite resistant, its leaves are insect repellent, oil is used in soap industries and seed cake for soil amendment to enrich organic matter. Hence, this tree was aptly regarded as "sarva roga nivarini" (in Sanskrit), meaning 'the curer of all ailments'. Even this plant was found to be severely infested with *Dendrophthoe falcata* in almost all the areas visited.

Host Specificity

There is no host specificity among the mistletoes and no special criterion is involved in the choice of hosts. Because the parasite has no role in selecting the host plants but it is the birds that play a vital role in the choice of host plants. Host specificity is related to both the abundance and diversity of host plants. It is long been known that birds are the agents in dispersing the mistletoe fruits/seeds which locate the host trees and are responsible for non host specificity.

Mimicry

In all botanical explorations in Karnataka State, we made a spectacular observation that the mistletoes mimic their hosts in their vegetative characters. Usually the infection of mistletoes can be easily identified by their obvious occurrence on the aerial branches and trunks of their host plants. Most mistletoes produce relatively large shoots, sometimes with large leaves which are clearly visible, but some produce small shoots that may be overlooked without careful observation. The leaves of a number of mistletoe species and their hosts look similar. In such instances it is difficult to distinguish the host and the parasite. In many instances the leaf size, shape and their presentation is almost similar to their hosts. But close inspection reveal the differences in the size, colour and texture. The most extraordinary example of leaf mimicry observed during our study is *Dendrophthoe trigona* (Loranthaceae) which is usually found growing on Moraceae members (especially on *Ficus benghalensis*) mimics exactly similar to its host. The leaves in both the cases are thick and broadly elliptic in nature. But *D. trigona* is usually robust on the tree trunks, young branches are three angled (trigonous) with three leaves in each whorl. Because of this nature it can be distinguished when carefully observed. Some times in the field, the parasites get escaped from our eyes because of their similarity in leaves with that of their host. The other host-mistletoe combinations which mimic include the *Dendrophthoe falcata* with linear, flattened and falcate leaves on *Eucalyptus*, *D. falcata* on *Mangifera indica* and *Macrosolen parasiticus* on *Mangifera indica*.

Hyperparasitism and Double Parasitism

The most amazing and noteworthy observation in the present study is the propensity of mistletoes to parasitize other mistletoes. If the host choice involved is mistletoe, such an association is called as hyperparasitism (Plates 18.23 A-D). Hyperparasitism is usually defined as an establishment of one parasite on another. This is also referred to as epiparasitism. Another fascinating aspect of mistletoes is the parasitization of two or three different species on the same host is known as double parasitism (Plates 18.24 A-D) and triple parasitism respectively. We have recorded more than 10 such instances in mistletoes. For example; *Viscum capitellatum* (Viscaceae) is invariably recorded as hyperparasitic on *Dendrophthoe falcata* (Loranthaceae) which is recorded in Haliyal forest range, Huliyar near Sira and also at B.R. Hills. The new taxon *Viscum bandipurense* Thriveni *et al*. (2013) is also recorded as hyperparasite on *D. trigona* which inturn parasitic on *Ficus benghalensis*, *Dalbergia latifolia* and on *Pterocarpus marsupium*. Usually the members of Viscaceae are recorded as hyperparasites on Loranthaceous mistletoes. Nevertheless as an exception to this we have recorded *V. capitellatum* on *Helixanthera intermedia* which inturn parasitic on *Chukrasia* sp. in B.R. Hills. We have recorded *Taxillus bracteatus* (Loranthaceae) hyperparasitic on *Viscum nepalense* (Viscaceae) in Chitradurga. Hitherto this is the only record wherein a Loranthaceae member found as a hyperparasite on Viscaceae.

Plate 18.23: Hyperparasitism

A: *Viscum bandipurense* (Vb) on *Dendrophthoe trigona* (Dt); B: *Taxillus bracteatus* (Tb) on *Viscum nepalense* (Vn); C: *Scurrula parasitica* (Sp) on *Dendrophthoe falcata* var. *falcata* (Df); D: *Viscum capitellatum* on *Dendrophthoe falcata.*

Plate 18.24: Double Parasitism.

A: ***Dendrophthoe falcata*** var. ***falcata*** (Df) and ***Viscum orientale*** (Vo) on ***Albizia amara;***
B: ***Dendrophthoe falcata*** (Df) and ***Viscum orientale*** (Vo) on ***Punica granatum;***
C: ***Dendrophthoe trigona*** (Dt) and ***Viscum ramosissimum*** (Vr) on ***Ficus benghalensis;***
D: ***Macrosolen capitellatus*** (Me) and ***Dendrophthoe falcata*** (Df) on ***Terminalia paniculata.***

Bird Assisted Self Pollination

As it has been observed for several years that, birds usually visit the flowers in search of food, often primarily to seek insects concealed in inflorescences and also take nectar if it is available. Floral nectar is considered as the most important reward offered to potential pollinators. The Loranthacean flower demonstrates the classical complement of adaptation to bird assisted self-pollination. The corolla tube is narrow at the base and gradually widens towards the upper part. Shortly above the middle region there is a constriction followed by a swollen portion. The swollen part again becomes narrow and its terminal part form somewhat stiff and gray coloured five ribbed structure which plays a major role in pollination. This ribbed upper part of the corolla tube tip consists of five sutures and ridges which enclose five anthers of epipetalous stamens and upper part of style and stigma. The elongated anther lobes are slightly approximated together and they inturn encircle the style and stigma so that they become circumaxial. It looks as if the anthers are apparently syngenesious and encircle the stigma.

The structure of the mature flowers is such that they will not open instead, unless some external pressure is applied to fling them open at their tip. When the flowers at anthesis were subjected to mechanical disturbance by applying a moderately forceful touch by the fingers and/or upon mechanical shaking, suddenly the corolla tube tip splits into five lobes and they recurve back which helps in the release of pollen from the anther lobes. It is only after pollination complete, the stigma project slightly above the anthers due to the elongation of style. Similar kind of flower opening and anther dehiscence was done by birds which visit the flowers chiefly with the purpose of sucking the nectar and pollen hidden inside the corolla tube. The birds peck the flowers at their base with their beak to suck the nectar from the corolla tube and while doing so, it triggers the corolla lobes tip to recurve back explosively (Plates 18.25 A-D). During this act the anthers dehisce longitudinally to release the pollen dust. This is almost comparable to the release of pollen from male flowers of *Pouzolzia* of Urticaceae at anthesis by mechanical touch or shaking and opening of fruits of Balsams. The clouds of pollen of all the five anthers dusted on to the surface of stigma of the same flower which is situated at the center of the anthers at close proximity which ensures self pollination (Plates 18.26 E and F). This mechanism slightly varies among the different genera of Loranthaceae.

In *Taxillus tomentosus* the flower structure is similar to that of *Dendrophthoe* sp. During anthesis the corolla lobes recurve back and while the anthers bend towards inside; the stigma pierces between any of the two anthers and rubs against them during which sufficient amount of pollen gets deposited on to the stigmatic surface (Plate 18.26G).

In case of *Helicanthes elastica* the corolla tube is somewhat swollen than the other members of Loranthaceae. At anthesis, when the birds peck the corolla tube to seek nectar, the corolla lobes split longitudinally up to the base into five slender lobes which elastically coil suddenly towards inner side like a spring. During this act the pollen is dusted on to the stigmatic surface. At this stage, the corolla lobes appear as five closely coiled tendrils (Plate 18.26 H).

Plate 18.25: Bird Assisted Self Pollination.

A–D: The flower pecker sucking the nectar from the flower of *Dendrophthoe falcata* and assisting in self-pollination.

***Note*: Flowers at various stages of development and at anthesis with reflexed petal tips.**

Plate 18.26: Bird Assisted Pollination.

E and F: The stigma piercing through the anthers of *Dendrophthoe falcata:* G: Flower of *Taxillus tomentosus* showing anthers bent towards inside and recurved petals towards outside; H: Flowers of *Helicanthes elastica* showing spirally coiled (spring like) petals (a-anther; s-stigma). Note the position of stigma at earlier and later stages of anthesis.

In all these instances, it is clearly evident that birds are intricately involved in the pollination mechanism of mistletoes. The Loranthacean flowers are well adapted themselves for the bird assisted self pollination.

If self-pollination fails, in such instances the corolla tube withers and is detached from the base of the flower circumscissily and distal part of it remains hanging round the tips of the style. After detaching, the corolla tube is pushed further towards stigma. While doing so, the anthers on the corolla lobes touch the stigmatic surface and there by ensures the self-pollination.

During the collections, in many instances it is also observed that birds inserting their beak into the corolla tube from above. In this process also the bird's beak and its head gets dusted with pollen which is transferred on to the stigma simultaneously which ensures self pollination.

Seed Dispersal Mechanism

In the present study, it is observed that birds play a vital role as agents in disseminating the mistletoe seeds. Mistletoe does not have any role in selecting its host plant and it is the birds which disperse and locates the seeds to suitable host plants. The mistletoe fruits are rich in sugar and conspicuous in colour when ripe. The seeds are enclosed by a mucilaginous layer called viscin. It is viscous and sticky in nature. The birds after eating the pulp of the fruit, the seeds stuck to its beak. In order to detach these seeds, the bird wipes off/rub its beak against branches. During this process the seed gets detached from the beak of the bird and adhere to the host tree branches with the help of sticky viscin. This viscin serves as both adhesive and also provides better moisture for germination.

During the present investigation, several species of birds which found eating the mistletoe berries were seen. These include Blue-winged leaf bird *(Chloropsis cochinchinensis)*, Thick-billed flower pecker *(Dicaeum agile)*, Tickell's flower pecker *(Dicaeum erythrorhynchos)*, Red-vented Bulbul *(Pycnonotus cafer)* and Red-whiskered Bulbul *(Pycnonotus jocosus)* which are involved in dispersal of mistletoe seeds.

Mistletoes as Food for Domestic Animals

A wide host range and a long flowering period in many taxa make mistletoes an attractive resource for wildlife. Since the mistletoes exhibit relatively high transpiration rates and possess fleshy leaves, many animals and birds are dependent on them for food and fodder. The mistletoe foliage, flowers and fruits are consumed by many wildlife including birds and insects. The leaves of *D. falcata* var. *falcata* are fed to the Goats in Siddara betta of Tumkur District and are considered as the highly favoured dietary food for many of our domesticated animals (Plate 18.22F). It is believed that the mistletoe diet increases the fertility of domesticated animals in many rural areas. The mistletoes are also used by birds for nesting. Although many members of both Viscaceae and Loranthaceae have reported to contain secondary metabolites, additional research is required to know that how these compounds influence herbivory by animals and insects.

Role of Mistletoes in Human Life

Apart form the wild life, mistletoes also thought to play an important role in human life. Sri Chandrashekar, a retired teacher and native health practitioner in Baghamandala has treated more than 300 couples who were suffering from inferitlity. *Taxillus cuneatus* extracts particularly growing on *Artocarpus hirsutus* (Hebbalasu) of Moraceae and *Flueggea leucopyrus* (Bili huli) of Euphorbiaceae was given to couples who are incapable of having children. Almost all the couples shown positive results of fertility and got conceived.

Discussion

Status of the Genus *Loranthus* L. (Loranthaceae)

Loranthus americanus L. is the only species first described by Linnaeus in 1753, which is thus the type by monotypy but has been included since 1830 within the American genus *Psittacanthus* Mart. In 1762, Linnaeus enlarged the genus *Loranthus* to include not only the American type species but four other species, among them the Asiatic *Loranthus scurrula* which he had treated in the Species Plantarum (1753) as generically distinct under the name *Scurrula parasitica*. In 1762 he added two more species *L. uniflorus* Jacq. and *L. europeaus* Jacq. which had come to his attention too late for inclusion with the others? Thus logical following of the type method would necessitate the use of the name *Loranthus* instead of *Psittacanthus*. To avoid the inconvenience and confusion of the usage of the genus *Loranthus*, International Botanical Congress conserved the name *Loranthus* from 1762, rejecting *Loranthus americanus* as the type species in order to conserve the name of the American genus *Psittacanthus*. Since then, a different taxonomic view point has prevailed, the genus *Loranthus sensu lato* has been divided into several genera and the selection of *L. scurrula* as type has led to unanticipated difficulties. In accepting this species the Congress did not consider the reestablishment of the genus *Scurrula* by G. Don in 1834. Later in 1929 Danser also adopted the genus *Scurrula*. But Danser rejected and ignored the Botanical Congress decision.

According to most taxonomists such as Endlicher (1840), Van Tieghem (1895), Bentham and Hooker (1880), Engler (1889) the name *Loranthus* was primarily associated by them with the small group of holarctic species that included *L. europaeus* Jacq. These authors regarded *L. europaeus* as a Linnaean species. Because the description of *L. europaeus* was first published by Jacquin in May 1762 before the first volume of Linnaeus's second edition of *Species plantarum* which was published in September 1762. So chronologically it takes precedence over *L. scurrula* as a name in the genus *Loranthus*. Moreover, Jacquin was a correspondent of Linnaeus himself and his description was accepted with the approval by Linnaeus, hence it should be regarded as of equal status with Linnaean species (Balle, 1955). Most of the literature except a few published so far including regional floras, monographs have followed the Danser's (1929) revised system of nomenclature for Loranthaceae. He has divided the large genus *Loranthus* into several genera which is currently followed by most of the taxonomists and we have also accepted and adopted the same in the present study.

Taxonomic Treatment

The taxonomic classification is the only method for reviewing the natural diversity and nomenclatural treatment is the method to the unique designation of biological unit. In the present investigation, seven genera and 16 species of Loranthaceae and Viscaceae with one genus and eight species are collected. In addition, three morphological varieties of *Dendrophthoe falcata* have been noticed. One new species, *Viscum bandipurense* Thriveni *et al.* (2013) is recorded and described from Bandipur National Park. It is envisaged that further studies involving the data from various fields of botany like cytology, anatomy, embryology, palynology, phytochemistry and also their molecular analysis would definitely throw more light on the possibility of occurrence of many new taxa of mistletoes and on their biology in Karnataka State.

An artificial key up to species level has been formulated based on the morphological characters observed in the field. This could be used both in field and laboratory for identification. A detailed description for each taxon is provided based on the morphological analysis. All the taxa are provided with an updated nomenclatural citation including the available literatures that are published so far. An illustration and a photographic plate including different developmental stages are given for each species. The illustrations are hand written by looking at the freshly collected plant specimen.

The members of Loranthaceae are more diverse than those of Viscaceae. Loranthaceous mistletoes are frequently recorded on most of the monoculture plantations. Viscaceae members have rather restricted distribution. A complete list of host plants is provided for each mistletoe species. All these data are the result of survey conducted throughout the Karnataka State. Most of the floras published so far do not include information on the host range, distribution and diversity of mistletoes. Hence, a serious attempt has been made to give a comprehensive account on the basic biology of mistletoes which is justified in the present investigation.

Distribution of Mistletoes

The survey conducted throughout Karnataka State revealed that the mistletoes are not uniformly distributed. The members of the Viscaceae are not frequently recorded compared to those of Loranthaceae. Perhaps, the former members have rather restricted distribution than Loranthacean taxa. The mistletoes have been reported from a wide range of host plants from forest trees to horticultural and monoculture plantation crops. The random distribution of mistletoes is significantly correlated with their unusual preference of host plants. The density of the mistletoes is more in dry deciduous forests when compared to tropical evergreen forests. Since mistletoes are photosynthetic, they acquire fixed carbon and nitrogen from their host plants. They compete with their host for the assimilation of carbon and nitrogen (Jakub *et al.*, 2010). Hence, they always prefer host plants with sparse canopy where they can receive more amount of sun light required for their photosynthesis. This theory was also supported by the work of Matthies (1995), who showed that parasites which compete with their host for light and resources restrict the parasites to low productivity. The distribution of mistletoes is also affected by the birds which disperse the seeds to the host branches. The seed dispersal is also affected by the neighbouring trees which attracts the birds through their fruits as a reward as it was observed in the present study.

Impact of Mistletoes on their Host Plants

The hypertrophy of host tissues at the site of infection is due to the disruption of normal tissue development (Knutson, 1983). Swelling of the host branches may be associated with either an increase in host cambial activity resulting in the formation of secondary xylem cells or displacement of host xylem and phloem tissue (Kuijt, 1969). It has been hypothesised that extensive hypertrophy of host tissue is indicative of a mistletoe-host incompatibility (Kuijt, 1964). Host tissue is stimulated to grow around the parasite absorptive tissue. Large swellings on a host branch are also associated with profusely branched extra xylary absorptive structures produced by the mistletoe (Thoday, 1957).

Though the branch dieback has been reported earlier, the pathological effects on the hosts have not been adequately investigated and assessed. Branch dieback is particularly prevalent during hot, dry seasons and droughts when the host is facing the water scarcity, but mistletoe continues to draw its nutrients and water from its host (Lei, 2001). This may be due to the higher transpiration rate of the parasite than its host.

Host Range

Parasitic angiosperms have the potency to infect a large number plants *i.e.*, they have a broad host range (Press and Phoenix, 2005). Nickrent (2002) has said that a good parasite does not kill its hosts. This is very much true because the parasite has to survive and perpetuate. The distribution of a parasite is restricted by its host range (Franco and Gray, 1996). Host range is generally defined as the total number of potential hosts that can be parasitized in a given population (Nickrent and Musselman, 2004). Host specificity is dependent on the preference to particular hosts, host resistance to parasitic attack, and performance of parasite when growing with different hosts (Marvier and Smith, 1997). Usually host specificity would help in managing the parasite, particularly in developing greater host resistance (Shamoun and Dewald, 2002). But during the present study mistletoes are recorded on a large number of highly diverse host plants. The adaptability of mistletoes has made them to victimize a large number of host plants. Because of their wide host range, mistletoes are considered as host generalists (Thriveni *et al.*, 2010). Hence, it is difficult to manage and control these mistletoes. Before attempting to management practice one should have the knowledge on the host plants which is parasitized by these mistletoes. Therefore in the present study, a complete list of host plants is also recorded to throw light on the variations in the performance of mistletoes. It is also observed that no tree is immune to mistletoe attack.

Wide host range is advantageous, especially in a heterogeneous community, as it allows the parasite to grow successfully on many of the potential hosts encountered. If host populations are unpredictable and ephemeral, generalists are more likely to occur (Thompson, 1994). The relative abundance of host species is therefore a key factor in determining the degree of host specificity in a parasite (Norton and Carpenter, 1998). Brooks and Mc Lennan (1993) have stated that the degree of host specialization may also be influenced by the length of time that the parasite and host are associated. The long association between a parasite and host will result in greater host specificity. However, Shaw (1994) suggests that parasites have a narrow host range when they first arise as distinct species and then the host range may expand subsequently. Press and Phoenix (2005) were of the opinion that the hosts which are available as a resource for longer period such hosts will be preferred by the parasite. But during the collection, it is evident that the duration of association between the host and parasite do not confirm the host specificity. The degree of infection may increase with the increase in association period of host and parasite. At the same time the host may also result in nutrient loss.

Dendrophthoe falcata (Loranthaceae) has attained a very important position among the mistletoes. One of the characteristic features of this parasite is its non host specificity in the selection of host plants. Although the selection of host plants is entirely random, experimental evidence suggest that this parasite has its own spectrum of hosts controlled by some factors, one of which seems to be osmotic pressure relationships between the host and parasite (Sampathkumar and Selvaraj, 1980). Fischer (1926) suggested that this parasite fail to attack a particular host merely due to lack of nutrition in an unsuitable position (Johri and Bhatnagar, 1972). Factors affecting the host selection are not known clearly. Further, Fischer (1926) has listed 153 different hosts of *D. falcata* from Southern India. Singh (1962) expanded the host list to 319. Rao and Ravindranath (1964) added 24 additional new hosts bringing the total to 343. Since then there have been several scattered reports of new hosts of *D. falcata* in the literature but some of the reports included already known hosts with their valid or invalid species names as high as 438 distributed in 241 genera. To solve this confusion Hawksworth *et al.* (1993) have compiled a new hosts list with correct binomials and synonyms used by earlier workers which suggests 401 hosts distributed in 227 genera of 77 different plant families of dicotyledons. Interestingly only four of them are Gymnosperms. During the course of study, 10 new hosts for *Dendrophthoe falcata* are recorded (Thriveni and Shivamurthy, 2010) which makes the total number of host plants to 411. The foregoing list clearly indicates that *D. falcata* has a wide host range and no host specificity was observed. Because of its wide host range, it is considered as a versatile and cosmopolitan parasite present in each and every ecosystem.

Dendrophthoe falcata is also frequently recorded on most of the Neem trees in all the localities. Neem is regarded as the wonder plant because of its healing properties. Fruit trees such as Mango, Guava, Pomegranate and Sapota are also severely infested with *D. falcata*. In addition to this, almost all the *Phyllanthus emblica* trees in B.R. Hills and Bandipur National Park are severely infested with *Taxillus tomentosus* (Loranthaceae) to such an extent that the fruit yield has been significantly reduced. The fruits of these trees are the source of commercial income for the local people. Mistletoes have become serious pest to most of the natural forests. Therefore, it is suggested that the management of these parasites have to be undertaken immediately.

Host Preference

In spite of the fact that mistletoes are recorded on a wide range of host plants, many host species appear to be preferred. It is not clear that what makes to some hosts to be preferred than others. Wang *et al.* (2007) have shown that the high nitrogen and carbon content of the host plants makes them to be preferred by mistletoes. Present study is also in agreement with this, where *Viscum orientale* (Viscaceae) is exclusively recorded on *Pongamia pinnata* which is a leguminous plant. Although mistletoes are apparently autotrophic, they are completely dependent on their host plants for carbon and nitrogen content. Kelly (1992), Matthies (1996) and Seel and Press (1993) also demonstrated that both root and shoot parasites often perform better on hosts with high nitrogen content such as the members of the family Fabaceae. This feature was also recorded in a root hemiparasite *Osyris arborea* of Santalaceae (Niranjana, 1994).

Host Specificity

The term low host specificity is more often considered to describe the mistletoes which have a highly diverse host plants. But host specificity implies when mistletoe species is restricted to one or a few related host species. Most of the studies such as Kuijt (1969), Hariri *et al.* (1987) and Yan (1993) state that host plants play an important role in host-mistletoe interaction. Downey (2004) also suggested that the mistletoe infection sustains, if the host plant is susceptible. In the present study, it is also noticed that several number of plant species other than host plants, are free from mistletoe infection. Hence, it is inferred that the host response is also important in the establishment of mistletoe on a particular host plant.

Mimicry

Several authors have analysed the mimicry in the mistletoes. Among them, Drummond (1840) and Hamsley (1896) commented that the similarity exist only in the vegetative features of mistletoes. Moore (1899) and Le Plastrier (1920) considered the mimicry as a protective adaptation from the herbivores. But Blakely (1922) denied the existence of host mimicry in mistletoes and concluded that no marked power of mimicry exists in mistletoes. Barlow and Wiens (1977) have critically studied the mimicry exhibited by these mistletoes. According to them the cryptic mimicry in mistletoes has evolved as a means of avoiding predation by herbivores since they have high rate of transpiration and high nitrogen content. They also stated that the mimicry established in mistletoes in the vegetative structure is such that the whole plant blends cryptically with its background. All these facts were not supported by the experimental evidence. Ehleringer *et al.* (1985) formulated two hypotheses; one is mistletoes mimicking host leaves have higher leaf nitrogen levels than their hosts and they gain advantage from the predation. Another is mistletoes which do not mimic their hosts also have an advantage from noncrypsis resulting form visual advertisement of their reduced nutritional status. Further, they state that mistletoes which mimic their hosts have high protein content than their hosts. However, this need not be a reason for mimicry. Because under some circumstances host may also possess higher concentrations of anti-herbivore compounds and mistletoes could derive some protection in these cases by mimicking their hosts, irrespective of their relative levels of leaf protein. Apart from this, those which do not mimic their hosts have lower protein content and thus they are less nutritious. This also serves as an advantage to the mistletoes through their leaves by having leaf morphology which contrasts with that of their host. These contrasting leaf morphologies may signal to potential herbivores of their lower nutritional status. Our personal observation also suggests that the growth habit of the mistletoes provide a best example of mimicry in plants. The wide host range of these mistletoes and also predation pressure from the herbivores has made the mimicry to be evolved among the mistletoe-host relationship. However, experimental evidences are yet to be initiated to throw more light on this aspect.

Hyperparasitism and Double Parasitism

When a parasite establishes itself on another and if two parasite species use the same host, there may be direct competition between the host and the parasite and also between the parasitic species. Both these phenomena are reported earlier by several authors, most of them are anecdotal in nature. Puustinen *et al.* (2001) studied the double parasitism of *Trifolium pratense* and predicted that simultaneous parasitism by both reduced the biomass of *Trifolium* more than the parasitism by either parasite alone. Even our personal observation in many localities has also revealed that the host growth has been severely stunted due to hyperparasitism and double parasitism.

Bird Assisted Self Pollination

The relationship between mistletoes and birds has been studied to emphasize the role of birds as agents assisting in self-pollination. Many authors have suggested that the pollination in mistletoes is cross pollination and not self pollination. Balle (1964) in his observation suggested that due to protandry, *Dendrophthoe pentandra* fails to achieve self pollination. In addition to this cleistogamy is also believed to occur as a mechanism to prevent cross pollination which is not true.

On the contrary, the present study revealed that mistletoes are entirely dependent on birds as the only effective means for flower opening, anther dehiscence and pollen release which ensures the self pollination. When the birds peck the flowers at base of the corolla tube which triggers the tip of corolla lobes to recurve back and open explosively. During this process the pollen gets dusted on to the stigma of the same flower. This clearly demonstrates that self pollination is the rule in Loranthaceae and not cross pollination. These observations suggest that the anther dehiscence and subsequent pollen release are arrested by an inbuilt mechanism until the flowers are triggered by the birds. This mechanism is comparable to the release of pollen from the anthers of *Pouzolzia* of Urticaceae by mechanical touch or shaking.

Mistletoe flowers are very important source of nectar for birds, particularly when nectar production by other plants is at its lowest (Reid, 1986). George Everhard Rumphius (1680) was the pioneer who worked on the pollination of mistletoes. He made some important observation that the bird *Dicaeum vulneratum* visiting the mistletoe flowers to suck their nectar.

The mistletoe pollinators belong to the family Dicaeidae, often called 'flower peckers' or 'mistletoe birds'. This family encompasses seven or eight genera and about fifty-five species. It is related to the Nectariniidae and other genus *Zosterops*, is sometimes included in the family (Mayr and Amadon, 1947). A more detailed study on the pollination mechanism in Loranthaceae members would definitely reveal many interesting results.

Seed Dispersal Mechanism

The dispersal of mistletoe fruits/seeds has unexpected facets of great biological interest (Kuijt, 1969). Gill and Hawksworth (1961) states that viscin layer serves as an adhesive to anchor the seed to host surface and also aids in germination by providing better moisture conditions. According to Kuijt (1969), it is the pulp of the berry which is attractive. Reid (1991) was of the opinion that the birds are responsible for the mistletoe spread. Usually when the birds feed on the fruits, the seed enclosed in the sticky viscin may be extruded suddenly from the pericarp and fall either mainly on to the branches of other plants or rarely on the ground. Alternatively the seeds are swallowed by the birds and excreted with faeces that would fall on to the branches of trees. The viscin around the seeds play an important role, not only in getting attachment, but also providing congenial environment for germination. This viscin is a major factor in the dispersal and survival of mistletoes. The detailed studies of *Viscum album* indicate that viscin consists of an inner pectic layer and an outer cellulose layer. As it passes through the bird's digestive system, the outer layer is digested and the pectic layer surrounding the seed remains essentially intact (Thomann, 1906). According to Kuijt (1969) association of mistletoe seeds with bird excrement may have produced the Germanic name 'mistletoe'.

There is another method of seed dispersal, which was observed by Ali (1931). According to him the flesh is teased out by revolving the fruit in the mandibles of the bird and consumed. After the flesh reused the epicarp discarded and the seed, which stuck onto to the bird's bill, was rubbed off onto a branch where the parasite seed starts germinating.

Kannan (1966) and Davidar (1985) stated that majority of the seeds are dispersed by a group of bird's, the flower peckers *(Diaceum* sp.) which feed mainly on the fruits of mistletoe plants. *Diaceum hirundinaceum* has been regarded as mistletoe bird which is involved in dispersal. The diet of these birds is believed to be mistletoe fruits exclusively. The seeds pass through the digestive tract in almost the same condition as when swallowed. The time required for this passage has not been recorded, but may be it is short. During the present study, we observed that several species of birds eating the mistletoe fruit which is rich in sugar content. After eating the pulp the seeds stuck to the bird's beak due to the presence of sticky viscin. In order to detach the seeds, the bird wipes off/rub its beak to the host branches and locates the seeds. The seeds readily germinate with the help of moisture provided by the sticky viscin.

Mistletoes as Food for Domestic Animals

Mistletoes have long been used as food and fodder for domestic animals and also for birds. The mistletoe fruit is a favourite diet for birds because of the sweet pulp. The foliage of mistletoes is fed to various cattles and sheeps. Gill and Hawksworth (1961) stated in their review that mistletoes provide nectar for birds and foliage for domestic animals during winter season through out USA. Similar kind of observation was also recorded in the present study, where the young twigs and leaves of *D. falcata* were given to cattle and goats.

Role of Mistletoes in Human Life

They have long been credited with the mystic powers. They have found their way into religious rites, mythology, healing arts and superstitions around the world. Paine and Harrison (1992) have reviewed the importance of mistletoes in human life. They stated that many cultures have used mistletoes for medicinal purposes. Like many medicinal plants used in the past, mistletoes contain compounds that in fact do have physiological effects on humans. Further research with regard to their uses and bioactive compounds is very much required.

Viscum nepalense has been reported to contain alkaloids, flavonoids, tannins, saponins, cardiac glycosides, sterols and reducing compounds (Murali *et al.*, 2011). Recent reports revealed that *Dendrophthoe trigona* (Loranthaceae) also found to contain secondary metabolites, total phenolic contents, antioxidant and reducing power activities (Puneetha *et al.*, 2013).

Despite the risks involved in its use, medicinal practitioners of many cultures used mistletoes as a curative for a number of ailments. They have been used as a sedative, to stop haemorrhaging and diarrhoea, as a treatment for pleurisy, gout, heart disease, syphilis and epilepsy. They were also used externally for sores, swellings, ulcers and abscesses, applied to forsbitten limbs.

As they (Paine and Harrison, 1992) mentioned in their review, the mistletoe was thought to influence human fertility as well and was prescribed to people who had fertility problems of all kinds. In Austria, twig of mistletoe was placed in a couple's bed to encourage conception. It was used for treatment of male impotence and was consumed by women along with wine to enhance fertility, and many cultures have used mistletoe to induce labor or the expulsion of afterbirth. It was also thought to be a powerful abortifacient and has been used to terminate unwanted pregnancies. In the present investigation also we have recorded that some of the Loranthaceous mistletoes were used to treat various ailments. The extract of *Taxillus cuneatus* specifically growing on *Artocarpus hirsutus* and *Flueggea leucopyrus* were given to couples who are incapable of having children. Most of the couples have got positive results. Further research is required in this regard which would throw more light on curative properties of mistletoes.

Sometimes, after the parasite decays, the swollen and ruptured tissues of the host assume a somewhat radiating structure and are termed as 'wood flowers' (wood roses) or 'Holzrosen'. These wood roses are sold as ornaments to the tourists and have become a source of income to the families in underdeveloped rural areas (Kuijt, 1969; Paine and Harrison, 1992).

Control of Mistletoes

The management of mistletoes involves the pruning of infected host branches beyond the point of infection if possible at the early stages and burning is the only effective management strategy which is in practice. Some times the manual removal of infected branches is likely to be temporary due to the high intensity of mistletoe infestation because, the haustoria remain within the host branch especially in the Viscaceae where the haustorium is endophytic in nature. In the present investigation, it is evident that most of the forest trees and horticultural trees including many valuable medicinal plants are severely infested with mistletoe. This has significantly reduced the growth rate and fruit yield of many trees. Hence, it is high time that we need to safe guard our valuable trees by controlling these mistletoes. Otherwise the day may not be far off when our future generations have to face several risks including scarcity of both food and forest products. An effective mistletoe control requires an integrated management practices from both local Government and public agencies.

Acknowledgement

The first author is thankful to University Grants Commission (UGC), New Delhi for the financial assistance in the form of research fellowship and University of Mysore, Mysore for providing neessary facilities. The second author thank JSS Mahavidyapeetha for support.

References

Ali, S.A. 1931. The role of the sunbirds and the flowerpeckers in the propagation and distribution of the tree-parasite *Loranthus longiflorus* Desr. in the Konkan. J. Bombay Nat. Hist. Soc. **35**: 144.

Aukema, J.E. 2003. Vectors, viscin and Viscaceae: mistletoes as parasites, mutualists and resources. Fron. Ecol. and Env. **1 (3)**: 212-219.

Balle, S. 1955. A propos de la morphologie des "*Loranthus*" d Afrique. *Webbia.* **11**: 541-585.

Balle, S. 1964. Les Loranthacees de Madagascar et des Archipels voisins. Adans. **4**: 105-141.

Barlow, B.A. 1983. Biogeography of Loranthaceae and Viscaceae. In: Calder, D.M. and Bernhardt, P. (eds.) The biology of mistletoes, Academic Press, Sydney, Australia, pp. 19–46.

Barlow, B.A. and Wiens, D. 1977. Host-parasite resemblance in Australian mistletoes: the case for cryptic mimicry. Evol. **31**: 69-84.

Barlow, B.A., Hawksworth, F.G., Kuijt, J., Polhill, R.M. and Wiens, D. 1989. Genera of mistletoes. The Golden Bough. **11**: 1-4.

Becker, H. and Schmoll, H. 1986. Mistel, Arzneipflanze, Brauchtum, Kuntsmotiv im Jugenstil. Wissenschaftliche Verlagsgesellschaft. MBH, Stuttgart, 132.

Bentham, G. and Hooker, J.D. 1862-1883. Genera Plantatrum, 3 volumes, Lovell Reeve, London, UK.

Bhat, K.G. 2003. Flora of Udupi. Indian Naturalist, Udupi.

Bishop, T.H. 1951. Mistletoe, myth and medicine. Med World, **75 (16)**: 439-442.

Blakley, W.F. 1922. The Loranthaceae of Australia part-I. Proc. Linn. Soc. N. S. W. **47**: 1-25.

Brooks, D.R. and Mc Lennan, D.A. 1993. Parascript: parasites and the language of evolution. Smithsonian Institution Press, Washington and London.

Calder, D.M. 1983. Mistletoes in focus: an introduction. In: Calder, D.M. and Bernhardt, P. (eds.) The biology of mistletoes, Academic Press, Sydney, Australia, pp. 1-18.

Calder, D.M. and Bernhardt, P. 1983. The biology of mistletoes, Academic Press, Sydney, Australia.

Cooke, T. 1901-1908. The flora of the presidency of Bombay, in 3 volumes, Taylor and Francis, London (Reprinted ed. 1958) BSI, Calcutta.

Cronquist, A. 1968. The evolution and classification of flowering plants, Houghton Mufflin Boston, London.

Danser, B.H. 1929. On the taxonomy and nomenclature of the Loranthaceae of Asia and Australia. Bull. du Jar. Bot. Buitenz. **10**: 291-373.

Danser, B.H. 1933. A new system for the genera of Loranthaceae-Loranthoideae with a nomenclator for the Old World species of this subfamily. Verh. der Kon. Akad. van Wet. Amst. Afd. Nat. Sect. 2, **29 (6)**: 1-128.

Danser, B.H. 1941. The British–Indian species of *Viscum* revised and compared with those of South-Eastern Asia, Malaysia and Australia. Blumea. **4 (2)**: 261-319.

Dassanayake, M.D. and Fosberg, F.R. 1983-2000. A Revised Handbook to the Flora of Ceylon, in 7 volumes, Amerind Publishing Co. Ltd., New Delhi.

Davidar, P. 1985. Ecological interactions between mistletoes and their avian pollinators in South India. J. Bom. Nat. Hist. Soc. **82 (1)**: 45-60.

De Laubenfels, D.J. 1959. Parasitic Conifer found in New Caledonia. Sci. **130**: 97.

Dengler, H.W. and Dengler, D.J. 1946. Mystery of mistletoes. American For. **52**: 572-573.

Devkota, M.P. 2005. Biology of mistletoes and their status in Nepal Himalaya. Him. J. Sci. **3**: 85-88.

Don, G. 1834. General history of dichlamydeous plants – III. Rivington, London.

Downey, P.O. 2004. A regional examination of the mistletoe host species inventory. Cunnighamia. **8 (3):** 354-361.

Drummond, J. 1840. Letters from swan river. Hook. J. Bot. **2**: 343-372.

Dzerefos, C.M., Shakleton, C.M. and Witkowski, E.T.F. 1999. Sustainable utilization of wood-rose producing mistletoes (Loranthaceae) in South Africa. Eco. Bot. **54 (3)**: 439-447.

Ehleringer, J.R., Schulze, E.D., Ziegler, H., Lange, O.L., Farquhar, G.D. and Cowan, I.R. 1985. Xylem tapping mistletoes: water or nutrient parasites. Sci. **227**: 1479-1481.

Endlicher, S.L. 1840. Genera Plantarum. Vienna.

Engler, A. 1889. Loranthaceae. In: Engler, A. and Prantl, K. (eds.) Die Natur. Pflanzen. **16**: 98-203.

Fischer, C.E.C. 1926. Loranthaceae of Southern India and their host plants. Rec. Bot. Sur. India. **11**: 159-195.

Fleming, G.L. 1952. The mistletoe family. J. California Gar. **43 (4)**: 1-3.

Franco, J.G. and Gray, V.R. 1996. Distribution and host specicity in the holoparasite *Bdallophytum bambusarum* (Refflesiaceae) in a tropical deciduous forest in Veracruz, Mexico. Biotrop. **28:** 759–762.

Gamble, J.S. and Fischer, C.E.C. 1915-1936. Flora of the presidency of Madras, in 3 volumes, Adlard and Sons Ltd., London. (Reprinted ed. 1957), BSI, Calcutta.

George Everhard Ramphius. 1680. The Ambonese Curiosity Cabinet. Yale University press, New Haven.

Gibson, C.C. and Watkinson, A.R. 1989. The host range and selectivity of a parasitic plant: *Rhinanthus minor* L. Oecologia. **78**: 401-406.

Gill, L.S. and Hawksworth, F.G. 1961. The mistletoes: a literature review. US Department of Agriculture Forest Service Technical Bulletin, 1424, 87.

Hamsley, W.B. 1896. Some remarkable phanerogamous parasites. J. Linn. Soc. **31**: 306-311.

Hariri, E.B., Andary, C. and Salle G. 1987. New insights on the resistance against mistletoe (*Viscum album* L.). In: Weber, H.Chr. and Forstreuter, W. (eds.) Parasitic Flowering Plants, pp. 293–294.

Hawksworth, F.G. 1983. Mistletoes as forest parasites. In: Calder, D.M. and Bernhardt, P. (eds.) The Biology of Mistletoes. Academic press, San Diego, California, pp. 317-334.

Hawksworth, F.G., Pundir, Y.P.S., Shaw III, C.G. and Gells, B.W. 1993. The host range of *Dendrophthoe falcata* (L.f.) Ettingsh (Loranthaceae). Ind. J. For. **16 (3)**: 263-281.

Hooker, J.D. 1872-1897. Flora of British India, in 7 volumes. Reeve and Co. Ltd., NR, Ashford, Kent, London.

Hutchinson, J. 1926-1934. The Families of Flowering Plants, in 2 volumes, Macmillan Co., London.

Jakub, T., Lenka, P. and Cameron, D. 2010. Interactions between hemiparasitic plants and their hosts: importance of organic carbon transfer. Pl. Sig. and Behav. **5 (9)**: 1072-1076.

Johri, B.M. and Bhatnagar, S.P. 1960. Embryology and taxonomy of the Santalales. I Proceedings of the National Institute of Sciences of India, **26B**: 199-220.

Johri, B.M. and Bhatnagar, S.P. 1972. Loranthaceae. Botanical monograph No.8, CSIR, New Delhi, India.

Kannan, P. 1966. Ornithophily. *M.Sc. thesis*, University of Bombay.

Kanner, L. 1939. Mistletoe, magic, and medicine. Bull. His. Med. **7**: 875-936.

Kelly, C.K. 1992. Resource choice in *Cuscuta europaea*. Proceedings of the National Academy of Sciences of the USA, **89**: 12194-12197.

Kelly, C.K., Venable, D.L. and Zimmerer, K. 1988. Host specialization in *Cuscuta costaricensis*: an assessment of host use relative to host availability. Oikos 53: 315-320.

Keshavamurthy, K.R. and Yoganarasimhan, S.N. 1990. Flora of Coorg (Kodagu), Karnataka, India, Vimsat Publishers, Bangalore.

Knutson, D.M. 1983. Physiology of mistletoe parasitism and disease responses in the host. In: Calder, D.M. and Bernhardt, P. (eds.) The biology of mistletoes. Academic press, San Diego, California, pp. 295-316.

Kuijt, J. 1964. Critical observations on the parasitism of New World mistletoes. Canadian J. Bot. **42**: 1242-1278.

Kuijt, J. 1969. The biology of parasitic flowering plants. University of California Press, Berkeley, CA.

Kurz, S. 1877. Forest flora of British Burma, Vol. 2. Missouri Botanical Garden, 321.

Lamarck, M. 1782-1792. Encyclopedia Methodique Botanique, Paris.

Le Plastrier, C.M. 1920. Notes on Loranthaceae from Wagga district. Aust. Natur. **4**: 139.

Lei, S. 2001. Survival and development of *Phoradendron californicum* and *Acacia greggii* during drought. W. N. Amer. Natur. **61 (1)**: 78-84.

Linnaeus, C. 1753. Species Plantarum, I Edition, Stockholm.

Linnaeus, C. 1762. Species Plantarum, II Edition, Stockholm.

Maheshwari, P. 1954. Embryology and systematic botany. Proceedings of 8th International Botanical Congress, Sections 7 and 8, 254-255, Paris.

Maheshwari, P. 1958. Embryology and taxonomy. J. Ind. Bot. Soc. **1**: 1-9.

Manjunatha, B.K., Krishna, V. and Pullaiah, T. 2004. Flora of Davangere District, Karnataka, India, Regency publications.

Marvier, M.A. and Smith, D.L. 1997. Conservation implications of host use for rare parasitic plants. Cons. Biol. **11 (4)**: 839-848.

Matthies, D. 1995. Parasitic and competitive interactions between the hemiparasites *Rhinanthus serotinus* and *Odontites rubra* and their host *Medicago sativa*. J. Ecol. **83**: 245–251.

Matthies, D. 1996. Interactions between the root hemiparasite *Melampyrum arvense* and mixtures of host plants: heterotrophic benefit and parasite mediated competition. Oikos. **75**: 118-124.

Mayr, E. and Amadon, D. 1947. A review of the Dicaeidae. Amer. Mus. Nov. **1360**: 1-32.

Metcalfe, C.R. and Chalk, L. 1950. Anatomy of the Dicotyledons. Loranthaceae **2**: 1188-1194, illus, Oxford.

Moore, S.L.M. 1899. The botanical results of a journey into the interior of Western Australia with some observations on the nature and relations of the desert flora. Bot. J. Linn. Soc. **34 (237)**: 171-261.

Muller, C.H. 1939. Mistletoes in legend and in Science. Sci. **48**: 73-74.

Murali, M., Puneetha, G.K., Thriveni, M.C., Niranjan, M.H., Shivamurthy G.R., Niranjana, S.R., Prakash, H.S. and Amruthesh, K.N. 2011. Phytochemical screening and antioxidant activity of hemi-parasitic Indian mistletoe *Viscum nepalense* Sprengel. J. Phar. Res. **4**: 3348-3350.

Musselman, L.J. and Press, M.C. 1995. Introduction to Parasitic Plants. In: Press, M.C. and Graves, J.D. (eds.). Parasitic Plants, Chapman and Hall, London.

Nickrent, D.L. 2002. Parasitic plants of the world. Chapter 2. In: Saez, L., Catalan, P. and Saez, L. (eds.) Parasitic plants of the Iberian Peninsula and Balearic Islands, pp. 7-27.

Nickrent, D.L. and Musselman, L.J. 2004. Introduction to parasitic flowering plants. The Plant Health Instructor. DOI: 10. 1094/PHI-I-2004-0330-01.

Nilsson, C.H. and Svensson, B.M. 1997. Host affiliation in two subarctic hemiparasitic plants: *Bartsia alpina* and *Pedicularis lapponica*. Ecosci. **4**: 80-85.

Niranjana, R. 1994. Morphogenetic studies in some root parasitic angiosperms. Ph.D thesis, University of Mysore, Mysore.

Norton, D.A. and Carpenter, M.A. 1998. Mistletoes as parasites: host specificity and speciation. Trends Ecol. and Evol. **13**: 101-105.

Paine, L.K. and Harrison, H.C. 1992. Mistletoe: its role in horticulture and human life. Hort. Technol. **2 (3)**: 324–330.

Parker, C. and Riches, C.R. 1993. Parasitic weeds of the world: biology and control. CAB International, Wallinford, UK.

Pennings, S.C. and Callaway, R.M. 2002. Parasitic plants: parallels and contrasts with herbivores. Oecologia **131**: 479-489.

Prakash, V.R. 2004. Mistletoe–a dung on twig. Kisan Wor. **31 (1)**: 59-60.

Press, M.C. 1998. Dracula of Robin Hood? a functional role of root hemi-parasite in nutrient poor ecosystems. Oikos. **882**: 609-611.

Press, M.C. and Graves, J.D. 1995. Parasitic plants. Chapman and Hall, London.

Press, M.C. and Phoenix, G.K. 2005. Impacts of parasitic plants on natural communities. New Phytol. **166**: 737-751.

Puneetha, G.K., Thriveni, M.C., Murali, M., Shivamurthy, G.R., Niranjana, S.R., Prakash, H.S., Sadashiva, M.P. and Amruthesh, K.N. 2013. Evaluation of a parasitic flowering plant *Dendrophthoe trigona* (Wt. & Arn.) Danser for its phytochemical and antioxidant activities. J. Phar. Res. **7**: 20-23.

Puustinen, S., Jarvinen, O. and Tiilikkala, K. 2001. Asymmetric competition between a hemiparasitic plant and a cyst nematode on a shared host plant. Ecosci. **8**: 51-57.

Ramaswamy, S.N., Radhakrishna Rao, M. and Govindappa, D.A. 2001. Flora of Shimoga District, Karnataka. Prasaranga, University of Mysore, Manasagangotri, Mysore.

Rao, R.R. and Razi, B.A. 1981. A synoptic flora of Mysore district. Today and Tomorrow's Printers and publishers, New Delhi.

Rao, V.L.N. and Ravindranath, V. 1964. A further contribution to the host range of *Dendrophthoe falcata* (L.f.) Ettingsh. Bull. Bot. Sur. India. **6**: 103.

Razi, B.A. 1957. An annonated list of phanerogamic parasites from India and Pakistan. Lloydia, **20**: 238-254.

Reid, N. 1986. Pollination and seed dispersal of mistletoes (Loranthaceae) by birds in Southern Australia. In: Ford, H.A. and Paton, D.C. (eds.) The dynamic partnership, birds and plants in Southern Australia. Government printer, South Australia.

Reid, N. 1991. Co-evolution of mistletoes and frugivorous birds. Aust. J. Ecol. **16**: 457-469.

Riopel, J.L. and Timko, M.P. 1995. Haustorial initiation and differentiation. In: Press, M.C. and Graves, J.D. (eds.) Parasitic Plants, Chapman and Hall, London, 39-79.

Saldanha, C.J. 1996. Flora of Karnataka. Oxford and IBH Publishing Co. Ltd., New Delhi.

Saldanha, C.J. and Nicolson, D.H. 1976. Flora of Hassan District, Karnataka, India. Amerind Publishing Co. Pvt. Ltd., New Delhi.

Sampathkumar, R. and Selvaraj, R. 1980. Some new hosts for *Dendrophthoe falcata* (L.f.) Ettingsh *(Loranthus longiflorus* Desr.). J. Bom. Nat. Hist. Soc. **78**: 200.

Sanajai, V.N. and Balakrishnan, N.P. 2006. Revision of Indian Viscaceae. Rheedea, **16 (2)**: 73-109.

Schulze, E.D. and Ehleringer, J.R. 1984. The effect of nitrogen supply on growth and water use efficiency of xylem-tapping mistletoes. Planta, **162**: 268-275.

Seel, W.E. and Press, M.C. 1993. Influence of the host on three sub-Arctic annual facultative root hemiparasites. New Phytol. **125**: 131-138.

Seshagiri Rao, R. 1957. Revision of the Indo-Malayan species of *Viscum* Linn. J. Indian Bot. Soc. **36 (2)**: 113-168.

Shamoun, S.F. and Dewald, L.E. 2002. Management strategies for dwarf mistletoes: biological, chemical and genetic approaches. In: Geils, B.W., Tovar, J.C. and Moody, B. (Tech. co-ords.) Mistletoes of North American Conifers, USDA Forest Service General Technical Report, Rocky Mountain Research Station.

Shaw, M.R. 1994. Parasitoid host ranges. In: Hawkins, B.A. and Sheehan, W. (eds.) Parasitoid Community Ecology, Oxford University Press, New York, pp. 111-144.

Singh, B. 1962. Studies in Angiospermic parasites: No.1. *Dendrophthoe falcata* (L. f.) Ettingsh, its life history, list of hosts and control measures. Bull. Nat. Bot. Gard. Lucknow, **69**: 1-75.

Singh, N.P. 1988. Flora of Eastern Karnataka, vol. 2. Mittal Publications, New Delhi.

Takhtajan, A. 1969. Flowering Plants: origin and dispersal (Translated by C. Jeffry). Smithsonian Institution Press, Washington.

Talbot, W.A. 1909-1911. Forest flora of the Bombay presidency and Sind, 2 volumes, Govt. of Bombay, Poona.

Thoday, D. 1957. Modes of union and interactions between parasite and host in the Loranthaceae. II. Phoradendreae. Proceedings of the Royal Society, London, **146**: 320-328.

Thomann, G. 1906. Vergliechende Untersuchunagen über die Beschaffenheit des Fruchtscheimes Von *Viscum album* L. und *Loranthus europaeus* und dessen biologische Bedeutung. Akad. der Wiss. Wien, Math. Natur. Klass. A bteilung I, **115**: 353-365.

Thompson, J.N. 1994. The Co-evolutionary process, University of Chicago Press, Chicago.

Thriveni, M.C. and Shivamurthy, G.R. 2010. Distribution, host range and mode of seed dispersal in *Dendrophthoe falcata* (L.f.) Ettingsh (Loranthaceae). Adv. Pl. Sci. **23 (2)**: 605-608.

Thriveni, M.C., Shivamurthy, G.R., Amruthesh, K.N. and Sadanand, K.B. 2013. Diversity, distribution and host range of mistletoes in Karnataka. JSSCWM Quest, **1 (1)**: 1-19.

Thriveni, M.C., Shivamurthy, G.R., Amruthesh, K.N., Vijay, C.R. and Kavitha, G.R. 2010. Mistletoes and their hosts in Karnataka. J. Amer. Sci. **6 (10)**: 827-835.

Thriveni, M.C., Shivamurthy, G.R., Amruthesh, K.N., Vijay, C.R. and Sadanand, K.B. 2013. A new species of *Viscum* (Viscaceae) from Bandipur, Karnataka, India. J. Sys. and Evol. **51 (2)**: 224. DOI: 10.1111/jse.12008_2.

Thwaites, G.H.K. and Hooker, J.D. 1858-1864. Enumeratio Plantarum Zeylaniae: an enumeration of Ceylon plants, London.

Tubeuf, C. 1923. Mongraphie der mistel. 832. illus. Munich and Berlin.

Tubeuf, C. 1936. Holzrosen als Reste des Kampfes Zwischen Parasiten und Wirten, Ztschr. f. Pflanzenkrank **46**: 586-608, illus.

Van Tieghem, P. 1895. Sur la structure de et les affinities des *Tupeia*, *Ginalloa*, *Phoradendron*, et *Dendrophthora*, de la famille des Loranthacées. Bull. de la Soc. Bot. de Fr. **30**: 643-652.

Wang, L., Kgope, B., D'Odorica, P. and Macko, S.A. 2007. Carbon and nitrogen parasitism by a xylem-tapping mistletoe *(Tapinanthus oleifolious)* along the Kalahari transect: a stable isotope study. Afri. J. Ecol. **46**: 540-546.

Westbury, D.B. 2004. Biological Flora of the British Isles. *Rhinanthus minor* L. J. Ecol. **92**: 906-927.

Wiens, D. 1973. Critical notes on the Viscaceae and Loranthaceae of Ceylon. Ceylon J. Sci. (Biol. Sci.) **9**: 43-49.

Yan, Z. 1993. Resistance to haustorial development of two mistletoes, *Amyema preisii* (Miq.) Tieghem and *Lysiana exocarpi* (Behr.) Tieghem ssp. *exocarpi* (Loranthaceae), on host and non host species. Inter. J. Pl. Sci. **154**: 386-394.

Index to Families, Genera and Species

Italicized Names Indicate Synonymy

www.ingramcontent.com/pod-product-compliance
Ingram Content Group UK Ltd.
Pitfield, Milton Keynes, MK11 3LW, UK
UKHW022216051026
15624UKWH00001B/190

9 789351 302506